An Integrated Approach to Agricultural Science

An Integrated Approach to Agricultural Science

Edited by Adriana Winkler

SYRAWOOD
PUBLISHING HOUSE

New York

Published by Syrawood Publishing House,
750 Third Avenue, 9th Floor,
New York, NY 10017, USA
www.syrawoodpublishinghouse.com

An Integrated Approach to Agricultural Science
Edited by Adriana Winkler

International Standard Book Number: 978-1-68286-853-9 (Hardback)

Cataloging-in-Publication Data

An integrated approach to agricultural science / edited by Adriana Winkler.
    p. cm.
Includes bibliographical references and index.
ISBN 978-1-68286-853-9
1. Agriculture. 2. Agriculture--Research. 3. Agricultural systems. 4. Agricultural resources. I. Winkler, Adriana.
S419 .I58 2020
630--dc23

TABLE OF CONTENTS

PREFACE

Agricultural science is the field of science concerned with food and fiber production and processing. It is a broad field of biology encompassing diverse social, natural and economic sciences, which are used in the practice and understanding of agriculture. Modern agriculture draws insights from plant breeding and genetics, horticulture, soil science, plant pathology, theoretical production ecology, entomology, etc. The techniques of crop cultivation and harvesting, soil cultivation, animal husbandry, etc. are vital to the practice of agriculture. Integrated pest management, agricultural chemistry, agricultural biotechnology, agricultural economics and agricultural engineering are some of the branches of agricultural science. This book studies, analyzes and upholds the pillars of agricultural science and its utmost significance in modern times. It brings forth some of the most innovative concepts and elucidates the unexplored aspects of this field. It is meant for students who are looking for an elaborate reference text on this discipline.

The researches compiled throughout the book are authentic and of high quality, combining several disciplines and from very diverse regions from around the world. Drawing on the contributions of many researchers from diverse countries, the book's objective is to provide the readers with the latest achievements in the area of research. This book will surely be a source of knowledge to all interested and researching the field.

In the end, I would like to express my deep sense of gratitude to all the authors for meeting the set deadlines in completing and submitting their research chapters. I would also like to thank the publisher for the support offered to us throughout the course of the book. Finally, I extend my sincere thanks to my family for being a constant source of inspiration and encouragement.

Editor

1

Re-orienting crop improvement for the changing climatic conditions of the 21st century

Chikelu Mba[1*], Elcio P Guimaraes[2] and Kakoli Ghosh[1]

Abstract

A 70% increase in food production is required over the next four decades to feed an ever-increasing population. The inherent difficulties in achieving this unprecedented increase are exacerbated by the yield-depressing consequences of climate change and variations and by the pressures on food supply by other competing demographic and socioeconomic demands. With the dwindling or stagnant agricultural land and water resources, the sought-after increases will therefore be attained mainly through the enhancement of crop productivity under eco-efficient crop production systems. 'Smart' crop varieties that yield more with fewer inputs will be pivotal to success. Plant breeding must be re-oriented in order to generate these 'smart' crop varieties. This paper highlights some of the scientific and technological tools that ought to be the staple of all breeding programs. We also make the case that plant breeding must be enabled by adequate policies, including those that spur innovation and investments. To arrest and reverse the worrisome trend of declining capacities for crop improvement, a new generation of plant breeders must also be trained. Equally important, winning partnerships, including public-private sector synergies, are needed for 21st century plant breeding to bear fruits. We also urge the adoption of the continuum approach to the management of plant genetic resources for food and agriculture as means to improved cohesion of the components of its value chain. Compellingly also, the National Agricultural Research and Extension System of developing countries require comprehensive overhauling and strengthening as crop improvement and other interventions require a sustained platform to be effective. The development of a suite of actionable policy interventions to be packaged for assisting countries in developing result-oriented breeding programs is also called for.

Keywords: Plant genetic resources for food and agriculture, PGRFA, Plant breeding, Crop improvement, Climate change, Biotechnology, Marker-aided selection, Genetic transformation, Induced mutations, Phenomics

Introduction

Population growth rates globally have so outstripped the linear rate of increases in food production that the Food and Agriculture Organization of the United Nations (FAO) estimated that 70% more food [1] must be produced over the next four decades in order to nourish adequately a human population projected to exceed 9 billion by the year 2050. The odds for attaining such an unprecedented increase, which would require the raising of the historically linear increases in annual food production by 37% [2], is substantially lessened by the consequences of climate change and variations on crop production systems [3,4].

The scope of the problem

The frequent occurrences of drought and floods, that invariably result in acute food shortages such as the very recent ones in the Horn of Africa [5], are symptomatic of the grave implications of extreme weather conditions for crop production and, hence, food security. Chatham House [6] had, relying on data provided by the United Nation's Intergovernmental Panel on Climate Change (IPCC), concluded that an additional 40 to 170 million more people will be undernourished as a direct consequence of climate change. Indeed, the overwhelming prognosis is that extreme weather events such as heavy precipitation, heat waves, and rising sea levels will occur in many parts of the world during the 21st century [7] with resulting floods, drought, and salinity as the most critical consequences. The strategies for devising solutions

* Correspondence: Chikelu.Mba@fao.org
[1]Plant Genetic Resources and Seeds Team, Plant Production and Protection Division, Food and Agriculture Organization of the United Nations (FAO), Rome, Italy

to these constraints will vary across geographical regions as the types and magnitudes of the problems will vary. For instance, though there is the consensus that rainfall is expected to increase globally overall, some places will actually be receiving less annual rainfalls while the seasonality of rains and hence the timing of the cultivation of crops will also change. More worrisome yet, the frequencies of occurrence and durations of the extreme weather events are also expected to increase. Table 1 summarizes some of the expected negative impacts on crop production by regions of the world.

This generational challenge of producing enough food for a rapidly growing population under extreme and changing weather conditions is further exacerbated by dwindling agricultural land and water resources. There are no more redundant water resources and arable lands to deploy in augmenting the already over-stretched ones in many parts of the world. Other noteworthy drivers

Table 1 Some expected negative impacts of climate change on crop production by regions[a]

Asia

· Crop yields could decrease by up to 30% in Central and South Asia

· More than 28million hectares (ha) in arid and semi-arid regions of South and East Asia will require substantial (at least 10%) increases in irrigation for a 1 °C increase in temperature.

Africa

· One of the most vulnerable continents to climate change and climate variability

· With many semi-arid regions and projected increase of 5% to 8% by the 2080s, likely reduction in the length of growing seasons will render further large regions of marginal agriculture out of production

· Projected reductions in crop yields of up to 50% by 2020

· Fall in crop net revenues by up to 90% by 2100

· Population of 75 to 250 million people at risk of increased water stress by the 2020s and 350 to 600 million people by the 2050s

Australia and New Zealand

· Agricultural production may decline by 2030 over much of southern and eastern Australia, and over parts of eastern New Zealand, due to increased drought and fire

· Change land use in southern Australia, with cropping becoming non-viable at the dry margins

· Production of Australian temperate fruits and nuts will drop on account of reduced winter chill

· Geographical spread of a major horticultural pest, the Queensland fruit fly (*Bactrocera tryoni*), may spread to other areas including the currently quarantined fruit fly-free zone

Europe

· Crop productivity is likely to decrease along the Mediterranean and in south-eastern Europe

· Differences in water availability between regions are anticipated to increase

· Much of European flora is likely to become vulnerable, endangered or committed to extinction by the end of this century

North America

· Increased climate sensitivity is anticipated in the south-eastern USA and in the USA corn belt making yield unpredictable

· Yields and/or quality of crops currently near climate thresholds (for example, wine grapes in California) are likely to decrease

· Yields of cotton, soybeans, and barley are likely to change

Latin America

· Risk of extinctions of important species

· By the 2050s, 50% of agricultural lands in drier areas may be affected by desertification and salinization

· Generalized reductions in rice yields by the 2020s

· Reductions in land suitable for growing coffee in Brazil, and reductions in coffee production in Mexico

· The incidence of the coffee leaf miner (*Perileucoptera coffeella*) and the nematode *Meloidogyne incognita* are likely to increase in Brazil's coffee production area

· Risk of *Fusarium* head blight in wheat is very likely to increase in southern Brazil and in Uruguay

Small islands

· Subsistence and commercial agriculture on small islands will be adversely affected by climate change

· In mid- and high-latitude islands, higher temperatures and the retreat and loss of snow cover could enhance the spread of invasive species including alien microbes, fungi, plants, and animals

[a]Adapted from the Second Report on the State of the World's Plant Genetic Resources for Food and Agriculture [38].

for food insecurity include the competing demands on scarce, depleted, and over-used arable lands and scarce food stuff for productions of bioenergy and livestock feeds. Equally confounding current conventional efforts to increase crop production sustainably is the prohibitive economic and environmental costs of the deployment of further agricultural chemicals as means for boosting yields.

The most vulnerable segments of society will be in poor developing countries, particularly in South Asia and sub-Saharan Africa, as they will suffer the most consequences of these changes to their food production systems [8-10]. In fact, Ejeta [11] estimated yield decreases of 10% to 20% for Africa's most important food crops in the coming decades. Similarly, Tester and Langridge [2] inferred that the greatest demand for yield increases as population continues to increase will be in the developing countries of the world though interestingly, Foresight [12] averred that the applications of already existing knowledge and technology could increase yields two- to three-fold in the medium and low income countries of the world.

Success in attaining the imperative of producing more food under worsening climatic conditions and with a severely constrained natural resources base hinges on enhanced efficiencies, that is achieving more yield per unit of input. This consideration informed the advocacy by Chatham House [6] for the eco-friendly 'knowledge-intensive' 21st century Green Revolution that will replicate the dramatic yield increases of its 20th century 'input intensive' precursor [11,13]. The growing of diverse 'smart' crop varieties that are capable of producing 'more with less' is in accord with this 'greener' perspective and will be critically important to achieving the *sine qua non* of enhanced efficiencies. This will of course require the re-orientation of many aspects of crop production systems with plant breeding and the cultivation of the resulting high yielding, well-adapted, input use-efficient, and resilient crop varieties constituting a major component of the interventions. In line with this perspective, Beddington *et al.* [4] aptly surmised that the concomitant attainment of food security and environmental sustainability would require innovative interventions as main driver for change.

Genetic gains translate to 'smart' crop varieties

Crop yields represent the net result of the intricate interactions between two main critical determinants, of approximately equal contributory effects, namely, the inherent genetic constitution of the crops and agronomic management practices [14]. Indeed, over the past seven decades in the United States, the percentage contribution of genetic gains to total on-farm yield increases in maize ranged between 33% and 94% with an average of about 50% to 60%

[15-17]. Genetic gains, accruable from harnessing the potentials coded into the genetic blueprints of plant genetic resources for food and agriculture (PGRFA), could therefore make significant contributions to attaining this required 70% increases in food production.

Instances of the dramatic effects of genetic gains on crop yields include the development and massive dissemination of high yielding and resilient cereal crop varieties around the world in the course of the aforementioned Green Revolution starting in the late 1960s. The consequent marked increases in food production in many food deficit countries was credited with saving billions of people from starvation especially in Asia [11,13]. More recently, the introduction of high yielding rice varieties, the New Rice for Africa (NERICA), in sub-Saharan Africa has also been credited with substantial increases in the production of the crop in the region [18-21].

Improved crop varieties, that possess superior agronomic and quality traits, are the direct outputs of plant breeding, described by the Columbia Encyclopedia as the science of altering the heritable patterns of plants to increase their value [22]. Foresight [12] had, in recommending the use of new scientific and technological tools to address the significant challenges of producing substantially more food with minimal environmental footprints, specifically identified 'plant breeding using conventional and new techniques to improve yields ... increase water, nutrient and other input efficiencies' as means to attaining this goal. The World Economic Forum [23] also situated the breeding of new crop varieties at the top of the agenda of its industry partners' coalition of global companies to address food insecurity. This paper contributes to the ongoing discussions on how plant breeding could be rendered more responsive to these challenges. We highlight some of the strategic policy, scientific, technological, and partnership interventions that can aid national programs, especially of developing countries, to have responsive result-oriented crop improvement activities.

Profile of the desired 'smart' crop varieties

FAO [24] posited that 'a genetically diverse portfolio of improved crop varieties, suited to a range of agroecosystems and farming practices, and resilient to climate change' is key to sustainable production intensification. In addition to high yields, the new elite varieties envisioned to address the bourgeoning drivers for food insecurity must be adapted to extreme weather conditions and the attendant continually evolving new strains and biotypes of pests and diseases. Extreme and changing patterns of drought and salinity are probably the most critical consequences of climate change and variations for which plant breeding must develop well-adapted varieties. Additionally, the 21st century plant breeding must cater to different prevailing farming systems and conditions - including rain-fed agriculture that accounts for a significant proportion of global food production in places where erratic rainfall patterns are expected. The new elite varieties must make more efficient use of inputs, and have improved nutritional qualities that meet the myriad dietary preferences of an increasingly more affluent, health-conscious, and generally more discerning consumer. Breeding objectives and strategies must also lead to those crop varieties that fit into ecosystem-based approaches such as conservation agriculture that emphasizes zero tillage. The breeding of multipurpose crop varieties which biomass are severally suited for use as food, bioenergy substrates, livestock feeds, and fiber will contribute to assuaging the effects of the ever increasing competing demands from these industries on arable lands, water resources, and even foodstuff.

Unlocking the inherent potentials of PGRFA

Deliberate human interventions, including hybridizations and selection pressures, in the last 10,000 years have resulted in the domestication of wild ancestors into the hundreds of thousands of breeds of both plants and animals that now form the basis for food and agriculture [25,26]. An unintended consequence of this human intervention in the otherwise natural process of evolution and speciation has been the narrowing of the genetic base of the plants cultivated for food [2]. The extremely narrow genetic base of crops, as evidenced in the similarities and shared close ancestries of cultivars, imperil food security grievously as a majority of the cultivars of the world's most important food crops would be vulnerable to the same stresses. In Russia, for instance, 96% of all winter wheat varieties are descendants of either one or both of two cultivars, Bezostaya 1 and Mironovskaya 808 [27]. This scenario evokes the specter of the potato blight and ensuing famine in Ireland in the mid-19th century and more recently in the summer of 1970, the major devastation of corn fields by a strain of *Helminthosporium maydis* in the middle and south central part of the United States. With climate change and variations, the threat of wide ranging major crop failures as a result of biotic and abiotic stresses is all too real. This threat can be mitigated by sourcing and/or inducing and deploying new allelic variations in plant breeding.

Widening the sources of heritable variations

Scientists are mindful of the shortcomings in the genetic diversity - and hence, increased vulnerabilities - of crops. Wild relatives of crops, land races, and other non-adapted genetic materials, even if usually low yielding and harboring undesirable traits, should be used more routinely in genetic improvement as means to addressing this shortcoming [2,25]. The investments of efforts in the use of such non-adapted materials in plant breeding have been quite rewarding. Instances include the use of genes located on a translocated chromosome arm of rye in the genetic improvement

of wheat [28]. Gur and Zamir [29] also demonstrated that the introduction of genes from the wild relative of tomato, the drought-tolerant green-fruited *Solanum pennelli*, increased yields by up to 50%. Two centers of the Consultative Group on International Agricultural Research (CGIAR), the International Institute of Tropical Agriculture, Ibadan, Nigeria and the International Center for Tropical Agriculture, Cali, Colombia, have severally used wild relatives of cassava to enhance disease resistance, improve nutritional qualities and extend shelf life of the fresh roots of the crop [30-34]. The legendary contribution of the reduced height gene from the Japanese wheat variety, Norin 10, to the Green Revolution is widely chronicled and certainly, other efforts have yielded significant results as well.

In general, crop wild relatives (CWRs), underutilized crops, and neglected species, that are conserved *ex situ*, on-farm, and *in situ*, are veritable repositories of the beneficial heritable traits lost in the course of domestication [29], including those for adapting to climate change [35]; these can be assembled into the envisaged 'smart' crop varieties. McCouch [25] had aptly surmised that in crop improvement, 'the surest way to succeed in a reasonable amount of time is to have access to a large and diverse pool of genetic variation'. This imperative is at the core of the work of the International Treaty on Plant Genetic Resources for Food and Agriculture (the International Treaty) which aims at the conservation, access, and sustainable use of PGRFA [36,37].

It is indeed paradoxical that PGRFA is the least tapped resource [38] in the quest for increased food production under worsening climate change and variations scenarios even though there is ample compelling evidence to the contrary. We recommend the harnessing of the widest possible spectrum of the inherent potentials of crops and their relatives as reversal to this trend of sub-optimal use of PGRFA in crop improvement. The accruable benefits from using these non-adapted materials certainly outweigh the additional efforts and costs in time and resources for breaking linkage drags and eliminating unwanted deleterious alleles - the main reason why breeders repeatedly and largely invariably always use the same set of 'safe bet' parents. A large scale global project aimed at collecting and using wild relatives of crops in plant breeding being implemented by the Global Crop Diversity Trust, for instance, is an example of internationally driven multi-stakeholder efforts to redress this shortcoming [39]. Pre-breeding, whereby germplasm curators and plant breeders work together to use heritable variations from non-traditional gene donors to produce populations of intermediate materials that can then be used in breeding, should be adopted universally in achieving this diversification of the genetic base of improved crop varieties. The e-learning course on pre-breeding [40,41] developed by FAO and partners under the auspices of the Global Partnership Initiative for Plant

Breeding Capacity Building (GIPB; [42]), is contributing to capacity development in this novel aspect of crop improvement. Pre-breeding facilitates the broadening of the genetic base of crops through the integration of new alleles of genes into elite novel crop varieties.

Through its Global System on PGRFA [43], FAO makes available relevant policy instruments, information systems, and other mechanisms that facilitate the conservation and sustainable use of PGRFA for food security. These include the World Information and Early Warning System (WIEWS; [44]) which provides online access to 19 databases and 13 organizations, instruments, and entities relevant to PGRFA and the World Information Sharing Mechanism on the implementation of the GPA [45] which provides access to PGRFA information of 71 countries, most of which also have their own portals. FAO's Global System for PGRFA also includes landmark publications such as the Second Report on the State of the World's PGRFA [38] which provides a periodic comprehensive report on not only the status of conservation and use of PGRFA worldwide but also the relevant emerging trends. Most recently in 2011, the Second Global Plan of Action for Plant Genetic Resources for Food and Agriculture (the Second GPA; [46]) was adopted by countries as a global framework to strengthen the capacities of

countries in the conservation of crop diversity and the development and deployment of a genetically diverse portfolio of improved varieties with new traits that meet food and nutritional security needs (Table 2).

All these information repositories are aiding the access to, and use of, genetic variability even across national boundaries. They facilitate access to the 1,750 national, regional, and international genebanks around the world which collectively hold about 7.4 million accessions [38]. These genebanks have been particularly successful with the collection, characterization, evaluation, and conservation of crop germplasm. Complementing the roles of these *ex-situ* gene repositories are about 2,500 botanical gardens which provide refuge for innumerable CWRs *in-situ* and the Svalbard Global Seed Vault, Norway which holds over 400,000 duplicate copies of crop germplasm from around the world [38]. Continued support, through sustained funding and enabling policies, is important for these repositories to be able to avail access to the widest possible genetic variation for improving crops. A major critical weakness in the conservation of PGRFA is the absence of a concerted, possibly global mechanism that mirrors the management of *ex-situ* collections, for *in-situ* conservation. CWRs continue to be lost as their refuges are appropriated

Table 2 Priority Activities of the Second Global Plan of Action on PGRFA[a]

Theme	Priority activity
In situ conservation and management	1. Surveying and inventorying plant genetic resources for food and agriculture
	2. Supporting on-farm management and improvement of plant genetic resources for food and agriculture
	3. Assisting farmers in disaster situations to restore crop systems
	4. Promoting *in situ* conservation and management of crop wild relatives and wild food plants
Ex situ conservation	5. Supporting targeted collecting of plant genetic resources for food and agriculture
	6. Sustaining and expanding *ex situ* conservation of germplasm
	7. Regenerating and multiplying *ex situ* accessions
Sustainable use	8. Expanding characterization, evaluation, and further development of specific subsets of collections to facilitate use
	9. Supporting plant breeding, genetic enhancement, and base-broadening efforts
	10. Promoting diversification of crop production and broadening crop diversity for sustainable agriculture
	11. Promoting development and commercialization of all varieties, primarily farmers' varieties/landraces and underutilized species
	12. Supporting seed production and distribution
Building sustainable institutional and human capacities	13. Building and strengthening national programmes
	14. Promoting and strengthening networks for plant genetic resources for food and agriculture
	15. Constructing and strengthening comprehensive information systems for plant genetic resources for food and agriculture
	16. Developing and strengthening systems for monitoring and safeguarding genetic diversity and minimizing genetic erosion of plant genetic resources for food and agriculture
	17. Building and strengthening human capacity
	18. Promoting and strengthening public awareness on the importance of plant genetic resources for food and agriculture

[a]Adapted from the Second Global Plan of Action for PGRFA, Food and Agriculture Organization of the United Nations, Rome, Italy [46].

for agricultural production or development projects so time is of essence in this regard.

Induced mutations

In situations where it is either impossible or impractical to source heritable variations from existing germplasm, the induction of allelic variations becomes an appealing option. Mutation, the heritable alteration to the genetic blueprint, has been the main driver for evolution and hence speciation and domestication of both crops and animals. Following the sublime discovery of X-rays and other forms of radiation in the early 20th century and the subsequent demonstration that these could alter the genetic material permanently, scientists have induced mutations in plants using both physical and chemical agents [47-49]. Induced mutation is hence an established crop improvement strategy and is credited with the development of over 3,200 officially released elite crop varieties and ornamental plants being cultivated all over the world [50].

The induction of mutation is a chance event so scientists traditionally enhance their chances of success at inducing useful mutation events by generating massive numbers of putative mutants that are then subsequently screened. This is expensive and time-consuming with the associated sheer drudgery cited as main reason for seeking other means for exploiting heritable variations in crops. Biotechnology applications are now being used to enhance the efficiency levels for producing and evaluating large populations. For instance, the high throughput reverse genetics technique, TILLING, short for Targeted Induced Local Lesions IN Genomes [51-53] permits the efficient screening of large populations of plants for specific mutation events [54-64]. The specificity, and hence efficiency, of TILLING - it identifies mutation events in predetermined genome regions - holds great promise for the use of induced mutations to broaden the genetic base of crops.

Cell and tissue biology techniques are also used to enhance the efficiency of mutation induction. For instance, with doubled haploidy [65,66], homozygosity of the mutated segments of the genome is achieved rapidly while *in vitro* propagation techniques are used to dissociate chimeras quickly (to generate solid homohistonts) and to produce and manage large mutant populations in cost-, time-, and space-efficient manners [67]. The critical importance of other uses of cell biology techniques, for instance, in germplasm conservation, in overcoming hybridization barriers and in the rapid multiplication of disease-free planting materials makes it an indispensable tool in crop improvement in general.

A re-invigorated plant breeding for a changing world

Translating the combinations of the widest possible sources of heritable variations efficiently into crop varieties whose increased yields, improved nutritional quality attributes and enhanced adaptations to abiotic and biotic stresses exceed those of the prior gains of the 20th century Green Revolution cannot be attained with a business-as-usual mindset. The current yield-centric breeding practices, of oftentimes weak breeding programs, whose objectives are largely conceived solely by the plant breeders, must evolve into participatory, multidisciplinary, and demand-driven programs that, underpinned by nurturing policy environments, make use of the most suitable scientific and technological tools to harness the potentials of PGRFA. Plant-breeding activities must perforce be re-oriented in order to have a reasonable chance of succeeding in the development of the envisaged portfolio of 'smart' crop varieties. We discuss some of the specific attributes that must characterize the result-oriented crop improvement programs of the 21st century.

Participatory plant breeding

Factoring in the perspectives of the growers and other stakeholders such as consumers, extensionists, vendors, industry, and rural cooperatives in the crop improvement endeavor of developing new varieties is known as Participatory Plant Breeding (PPB; [68]). The need for this paradigm in plant breeding is probably greatest in developing countries relative to the industrialized countries where market forces determine agricultural research and development (R&D) themes including plant-breeding objectives. By having farmers and other end-users involved in the development of varieties, feedback mechanisms are enhanced hence improving the relevance of the breeding activities to the needs of the growers. Farmers' participation in plant breeding can be categorized under the three stages of design, testing, and diffusion [69]. During the design stage, breeding goals are set and variability to be used created while at the testing stage, the breeding materials are evaluated and narrowed down to the few promising ones. The diffusion stage encompasses activities spanning varietal release, on-farm trials under farmer management and the identification of the mechanisms for the dissemination of the seeds and planting materials of the improved varieties.

Farmers, as the custodians of PGRFA, have over the several millennia of selecting from, improving, and exchanging local genetic diversity contributed immensely to the diversity of plants we grow. With the upsurge in the ready availability of modern crop varieties bred in research institutes, the roles of farmers in ensuring diversity and adding value to PGRFA have waned significantly. One effect of this shift is the precariously narrow genetic base of the modern crop varieties. The obvious threat that this poses to food security calls for the systematic re-integration of farmers' knowledge and perspectives in the developing of modern crop varieties. PPB is a veritable and validated means for ensuring this. The International Treaty, through its Article 9, also requires of contracting parties the safeguarding of farmer's rights to access and benefit from PGRFA. Those rights are not safeguarded when crop varieties that do not meet their food security and nutritional needs and/or do not enhance the resilience of their farming systems are all that are available to them.

In general, PPB facilitates the rapid and enthusiastic adoption of crop varieties [70]. The related Participatory Varietal Selection (PVS) is a means for involving these stakeholders in breeding when elite materials are already available to select from and is relatively more rapid and cost-effective than the more resource-intensive PPB [71]. Ashby [69] identified the impact pathways for PPB and PVS and concluded that their characteristic of producing more acceptable varieties and hence increasing adoption was the most compelling incentive for plant breeders to adopt this paradigm. Indeed, a CGIAR-wide review of plant breeding had recommended that PPB constitute 'an organic part of each center's breeding program' [72].

Novel plant-breeding techniques

The incredible advances in biotechnology demonstrably hold great promise for crop improvement [73]. For instance, molecular breeding, the integration of molecular biology techniques in plant breeding [74], through enhanced efficiencies, has great potentials for changing permanently the science and art of plant breeding. Molecular breeding encompasses both the use of distinguishing molecular profiles to select breeding materials and the applications of recombinant deoxyribonucleic acid (DNA) methods, that is genetic transformation, to add value to PGRFA. There are also a number of other emerging molecular biology-based techniques that hold promise for enhancing the efficiency levels of plant breeding activities. We provide some overview of the use of these technologies and techniques in developing novel crop varieties.

Marker-assisted selection

The increasingly available rapid, efficient, high throughput, and cost-effective molecular biology tools for identifying the sources, and tracing the inheritance, of desired traits are revolutionizing the management of PGRFA in general and plant breeding in particular. Advances in molecular biology, including the ever cheaper sequencing of whole genomes, have resulted in the availability of significant amounts of information on, and hence tools for assaying, the totality of an individual's genetic make-up, that is the genome; this is known as genomics. The related proteomics (the study of proteins) and metabolomics (the study of metabolites), made possible by an ever growing volume of publicly accessible DNA, gene, and protein sequence information, are also novel ways for investigating the heredity of traits. Equally significant, advances in bioinformatics and computational molecular biology which are facilitated greatly by the novel

sophisticated and powerful information technology platforms for storing and analyzing the huge volumes of data generated through these molecular biology strategies, permit the making of valid inferences in the molecular characterization of germplasm, assessments of genetic diversity and for the selections of breeding materials.

The ability to use appropriate molecular approaches in identifying genome segments that discriminate between individuals (that is molecular markers) and to apply statistical algorithms in identifying precisely where these 'landmarks' are located on the genome has changed plant breeding permanently and will be key in developing the 'smart' crops of the 21st century. Molecular markers are now demonstrably the tools of choice for tracing the inheritance of target regions of genomes in breeding materials, a plant breeding methodology known as marker-assisted (or -aided) selection (MAS).

MAS entails the use of environment-neutral molecular markers to trace the inheritance of genes, and hence the trait(s) they control, in a breeding program with or without phenotypic selection [75]. The utility of MAS is greatest for genes whose effects are difficult, time-consuming, or otherwise expensive to evaluate in a population. This may be on account of the phenotypic effects being evident only at maturity, low heritabilities, the absence of the particular stress factor being bred for or as a result of confounding environmental influences on the trait.

The use of MAS is relatively straightforward in breeding for qualitative monogenic traits with clear-cut differences between phenotypes, such as disease resistance in plants, as the genetic mapping of the associated marker results in the mapping of the trait also and vice versa. For quantitative traits, the validation of the trait-marker association through large-scale field experimentations and statistical methods in order to more precisely identify the target genome segments, that is quantitative trait loci (QTL), is additionally required [76,77]. In general, once the marker-trait association has been verifiably established, the transmission of trait genes from parent to offspring is monitored by querying segregating materials for closely linked markers using suitably designed marker-assisted backcrossing, for instance. The utility of MAS in breeding for polygenic traits can also be derived in gene pyramiding, that is the accumulation of two or more genes, say for disease and pest resistance, which seems feasible only with this method [2].

It has been demonstrated that consistently, MAS, either as a standalone strategy or in combination with phenotyping, significantly reduces the number of generations for evaluating segregating breeding materials and generally increases efficiency levels [2,74,75,78-93]. Indeed, it has been demonstrated that MAS permits a seven-fold increase in data handling and ultimately halves the time required for breeding a new crop variety [94]. Nonetheless, the cost-benefit analysis for adopting MAS relative to phenotypic selection is always a critical consideration that must be borne in mind in devising breeding strategies especially for developing countries.

Already routinely applied in the private sector breeding companies, such as the multinational companies, Monsanto [94]; Pioneer Hi-Bred [95] and Syngenta [96], MAS is yet to take hold in public crop improvement programs mostly on account of high set-up costs and intellectual property rights (IPR) restrictions. This implies that public sector plant breeding is clearly missing out on this singularly promising opportunity to innovate. Thro et al. [97] captured the immense expectations riding on the investments in plant genomics in relation to crop improvement in characterizing plant breeding as the 'translator' of knowledge into improved crop varieties. Public sector plant breeding is yet to assume this 'translator' role in the new dispensation of crop improvement that must be 'knowledge-intensive'.

An encouraging trend, though, is the progressive decline in the cost and the concomitant improvement in the high throughput applicability of molecular biology assays and equipment. It is logical to assume that at some point in the near future, set-up costs would be generally affordable and routine assays sufficiently efficient [98] as to permit wide adoption of MAS in the public sector. The continued successful use of MAS in the private sector is providing the much needed validation and proof of concept for this paradigm. This is critically important as capacity for this breeding methodology will be critical in handling the large populations of new breeding materials to be produced from pre-breeding activities using non-adapted genetic resources, for instance. The Integrated Breeding Platform (IBP) of the Generation Challenge Program of the CGIAR [99] is an example of multi-stakeholder efforts to extend the use of MAS to developing elite varieties of food security crops in developing countries.

Genetic transformation

Recombinant DNA technology, involving the use of molecules containing DNA sequences derived from more than one source to create novel genetic variation, has become an important crop improvement option. This is known as genetic modification (or transformation) with the new variants referred to as transgenics or simply genetically modified organisms (GMOs). The procedures involve the incorporation of exogenous DNA or ribonucleic acid (RNA) sequences, using either biolistics or vectors, into the genome of the recipient organism which, as a result, expresses novel and agronomically useful traits. Though transgenic varieties of only four crops, maize, soybean, canola, and cotton, harboring two transformation events, that is herbicide tolerance and insect resistance or their combinations,

have been grown commercially since the first approvals in 1996, James [100] estimated that there had been a 94-fold increase in hectarage in the 16 years of the commercialization of genetically modified (GM) crops (from 1.7 million hectares in 1996 to 160 million hectares in 2011). Grown in 29 countries (19 developing and 10 industrial), the author estimated the value of the GMO seed market at US$13.2 billion in 2011 while the produce for GM maize, soybean, and cotton were valued in excess of US$160 billion for the same year.

In spite of the low numbers of commercial GM crops and the transformation events that confer the modified agronomic traits, four and two, respectively, the development and deployment of GM crops signal a trend in crop improvement that can no longer be ignored. This is more so as approvals for the importation of GM crops and release to the environment had been approved in 31 other countries [100]. Tester and Langridge [2] pointed out that, though the major contributions to crop improvement for this decade will be non-GM, the production and evaluation of GM crops remained an actively researched theme with only political and bioethical considerations (both driven mostly by public negative perceptions for the technology) constituting the main hindrances to wider access to the technology by growers in more countries.

Technically, the drawbacks to more widespread development of GM varieties include the lack of efficient genotype-independent regeneration systems for most crops. Also, the lingering technical difficulties with the stacking of transformation events severely limits the utility of genetic transformation in breeding for polygenic straits such as resistance to the abiotic stresses, for example salinity and drought, being caused by climate change and variations. However, the successful stacking of genes conferring insect resistance and herbicide tolerance [100] is indicative of progress in addressing this constraint. Also, research efforts must target the increasing of the range of agronomic traits being improved through this method; the two transformation events in commercial varieties are simply inadequate for GM technology to become a dominant crop improvement method.

Probably the most limiting of all factors, however, is the associated intellectual property rights (IPR) protections that restrict access to the technology. Such IPR regimes have made GMOs remain the exclusive preserve of multinational plant breeding and seed companies in developed countries that effectively use patents to restrict access to several technologies relevant to the R&D efforts for the production of the transgenic crops. These constraints must be addressed in order that this technology be used fully in realizing its possible contributions to the development of the 'smart' crop varieties of this century. With GMO crops currently grown in developing countries, for example about 60 million hectares in South America in 2011 and with millions of small holder farmers cultivating transgenic cotton in both India and China [100-102], it is plausible to expect that the IPR regimes will be changing in the future. Another hindrance to wider adoption of the GM technology is the absence of biosafety regulatory frameworks as specified by the Cartagena Protocol on Biosafety to the Convention on Biological Diversity [103] in many countries.

Efforts to address the constraints that impede both the use of the GM technology in R&D and the cultivation of GMOs have been significant as well. For instance, the African Agricultural Technology Foundation (AATF; [104]), based in Nairobi, Kenya, is acquiring and deploying proprietary agricultural technologies in sub-Saharan Africa. In one instance, AATF obtained 'a royalty-free, nonexclusive license to Monsanto technology, a *Bacillus thuringiensis* (Bt) gene (cry-1Ab)' which is being used in the development of cowpea varieties with resistance to the cowpea pod borer [105]. Similarly, the US-based Public Sector Intellectual Property Resource for Agriculture (PIPRA; [106]), assists 'foundations, not-for-profit organizations, universities, international aid agencies, and governments' in dealing with IPR issues in order to enable access to proprietary technologies. Also, Cambia, an Australian private, non-profit research institute, publishes relevant patents, white papers, and provides tutorials as means 'to provide technical solutions that empower local innovators to develop new agricultural solutions' [107]. The activities of these organizations underscore the seriousness of the impediments that IPR protections pose for innovations in agriculture and the countervailing efforts to extend the reach of the technologies and applications especially into the public goods and commons R&D domains.

Emerging biotechnology techniques of relevance to plant breeding

The integration of biotechnologies into crop improvement is a very dynamic field of endeavor that is changing continually. A snapshot of the status of emerging technologies is provided by Lusser et al. [108] in response to a request by the European Commission 'to provide information on the state of adoption and possible economic impact of new plant breeding techniques'. The authors identified eight new such techniques and concluded that the new varieties ensuing from these techniques might be released within 3 years. These new techniques and their features are:

- Zinc finger nuclease (ZFN): Single mutations or short indels are generated or new genes are introduced into pre-determined target sites of the genome

- Oligonucleotide directed mutagenesis (ODM): Targeted mutations of one or a few nucleotides are induced
- Cisgenesis and intragenesis: GMOs are produced by the insertion of hereditary materials derived from the species itself or from a cross-compatible species and are contiguous and unchanged (cisgenesis) or the inserted DNA may be a new combination of DNA fragments but must still be from the species itself or from a cross-compatible species
- RNA-dependent DNA methylation (RdDM): Still being refined, modified gene expressions are epigenetic with the new phenotypes inherited only over a few generations
- Grafting (on GM rootstock): Desired improvements are achieved by the grafting of non-transgenic scions onto GM rootstock
- Reverse breeding: A combination of recombinant DNA techniques and cell biology procedures is used to generate suitable transgene-free homozygous parental lines rapidly for reconstituting elite heterozygous genotypes
- Agro-infiltration: Used mostly in research settings, for example to study plant-pathogen interaction in living tissues, to select parental lines or to evaluate the efficacy of transgenes, a liquid suspension of *Agrobacterium* sp. containing the desired gene(s) is used to infiltrate plant tissues, mostly leaves, so that the genes are locally and transiently expressed at high levels
- Synthetic genomics: Large functional DNA molecules that are synthesized without any natural templates are used for constructing viable minimal genomes which can serve as platforms for the biochemical production of chemicals such as biofuels and pharmaceuticals

Lusser et al. [108] concluded that ODM, cisgenesis/intragenesis, and agro-infiltration were the most commonly used techniques with the crops developed using them having reached the commercial development phase. On the other hand, the ZFN technology, RdDM, grafting on GM rootstocks, and reverse breeding were the less used techniques in breeding. The authors further projected that the first commercial products derived from these technologies that will be released for production would be herbicide resistant oilseed rape and maize using ODM and fungal resistant potatoes, drought tolerant maize, scab resistant apples, and potatoes with reduced amylose content developed using cisgenesis and/or intragenesis.

The clearly identified needs for the further fine-tuning of technical impediments to the routine adoptions and use of these new techniques notwithstanding, it would appear that policy regulations that are expensive to comply with and public perceptions, rather than the ability to innovate, are holding back the unleashing of the incredible advances of science and technology in crop improvement. Considering that Blakeney [109] opined that 'the right to patent agricultural innovations is increasingly located within a political context', it is plausible that the magnitude of the worsening threats to global food security may ultimately serve as the critical inducement for policy-makers, interest groups, and leaders of thought and industries to unravel the thorny issues that constrain the scope of the integration of biotechnology into crop improvement.

High throughput phenotypic evaluations

The selections of few promising individuals out of large populations of segregating materials can be a very daunting task. With MAS, the volume of assays that can be carried out and data points generated per unit time has increased substantially. For the workflow to be wholly efficient, the assessments of the phenotypes must also keep pace with high throughput molecular assays. Indeed, for molecular data used in breeding to be reliable, the corresponding phenotypic data for which inferences are made, must also be accurate [110]. Phenomics, the study of phenomes - the sum total of an individual's phenotype is the term that describes the novel high throughput measurements of the physical and chemical attributes of an organism. Somewhat imprecisely named in this seeming analogy to genomics, it is defined by Houle et al. [111] as 'the acquisition of high-dimensional phenotypic data on an organism-wide scale'. High throughput imaging of parts of a living plant, for example roots and leaves, using thermal infra-red, near infra-red, fluorescence, and even magnetic resonance imaging permit non-destructive physiological, morphological, and biochemical assays as means for dissecting complex traits such as drought and salinity tolerances into their component traits [112,113]. Though significant technical challenges, such as data management, still require addressing, phenomics facilities are increasingly being set up with a number of them providing high throughput phenotyping services to requestors. These new facilities include the High Resolution Plant Phenomics Centre in Canberra and the Plant Accelerator in Adelaide, both in Australia [114]; LemnaTec in Wuerselen [115] and Jülich Plant Phenotyping Centre in Jülich [116] both in Germany; and Ecotron [117] and Ecophysiology Laboratory of Plant Under Environmental Stress (LEPSE; [118]) both in in Montpellier, France. In Canada, there is the The Biotron Experimental Climate Change Research Centre in London, Ontario [119]. The high set-up costs and technical know-how may impede the access of developing countries to such platforms for some considerable time.

Overarching policy environment for the PGRFA management continuum

The benefits of value addition to PGRFA, that is improved crop varieties that meet the needs of the growers, can be derived sustainably, especially for the most at-risk food insecure countries in the developing world, only with the comprehensive strengthening of, and forging of linkages between, the three components of the PGRFA value chain: (1) conservation; (2) plant breeding; and (3) the delivery of high quality seeds and planting materials to growers. This is the 'PGRFA continuum' [120], the seamless dovetailing of the three components, as distinct from targeting the strengthening of any of the three in isolation. Based on the cohesion in this value chain - that characterizes the activities of private sector commercial breeding companies and the PGRFA management of some emerging countries such as Brazil, China, and India [94] - it is logical to conclude that the real value of crop germplasm lies in its use in plant breeding. Pragmatically also, the efforts invested in breeding come to naught if there is no effective delivery system for the seeds and planting materials underscoring therefore the need to interlock all three components.

The successful implementation of the Second GPA [46] also envisages the adoption of this continuum approach. The 18 priority activities (Box 1) of the GPA provide a most practical template for countries for concerted interventions at the three components of the PGRFA value chain. These PAs are subdivided into four main themes: in-situ conservation and management; ex-situ conservation; sustainable use; and building sustainable institutional and human capacities.

The sustainable use of PGRFA encompasses activities relating to direct utilization of PGRFA by farmers and to their uses in crop improvement. The International Treaty, especially in its Article 6, equally requires of contracting parties not only to conserve their genetic resources but to use them (for value addition) and to deliver the improved varieties efficiently. FAO [121] opined that 'any weakness in this continuum truncates the value chain and effectively scuttles all the efforts to grow the most suitable crop varieties'. It is in this vein that FAO and partners are working with developing countries to articulate National PGRFA Strategies for institutionalizing the continuum approach to managing PGRFA [120]. The strategy identifies priority crops and relevant stakeholders; prescribes time-bound action plans across the continuum and enunciates governance mechanisms and means for monitoring implementation. Nurturing policy environments, especially those that enable countries adopt the continuum approach to the management of PGRFA, are critically important for reaping the most sustainable benefits from PGRFA, namely, the improved crop varieties.

FAO's normative activities provide support for the implementations of the International Treaty and the Second GPA and for developing the necessary policies, and legislations as means for attaining this goal.

Winning partnerships

The reorientation of crop improvement in order to be responsive to the drivers of food insecurity, especially in developing and emerging economies, will require a wider range of partnerships beyond the traditional National Agricultural Research and Extension Systems (NARES). FAO [38] reported the prevailing trend whereby the private sector (multinational and local commercial plant breeding and seed companies) is increasingly developing and deploying elite crop varieties especially in instances where markets, favorable policy regimes, and legal frameworks that spur investments are in place. In tandem, public investment in crop breeding programs is contracting implying therefore that the breeding and dissemination of elite varieties of crops that fall outside of the business remit of the private sector could, as is increasingly the case, be neglected to the detriment of food security. Equally important is the role of non-governmental organizations and myriad civil society actors in the provision of agricultural extension services in developing countries. These bourgeoning dynamics must influence the articulation of policies and the building of collaborations and wide-ranging partnerships. For such partnerships to succeed, local knowledge must be integrated just as relevant private and public sector entities including the NARES, centers of the CGIAR, and regional R&D networks are assembled. The safeguarding of intellectual property rights, including plant variety protection, and the respect of patents are means for attracting private sector investments. Public-private partnerships, for example the ongoing joint activities between Syngenta and public African NARES [122,123], are particularly important for technology transfer, a critical vehicle for increasing the access of developing countries to novel biotechnologies that impact on crop improvement, for instance. On the other hand, public sector investments in food security must be ensured as the private sector, especially in developing countries, do not cater for all crops that are important for food security. Partnerships must also be cross-sectoral, for instance between ministries responsible for the environment, science and technology, commerce, education, and the ministry of agriculture. This ensures access to the full spectrum of PGRFA that may be needed for value addition while also ensuring a means for delivering the planting materials efficiently to the growers in gainful manners.

National capacities for crop improvement

The GIPB surveyed 81 countries for capacities in plant breeding and related biotechnologies [124] and subsequently conducted in-depth analysis of the plant-breeding and seed systems sectors of six of the countries: Ghana, Kenya, Malawi, Bangladesh, Thailand, and Uruguay [125]. The findings reflected the deduction by FAO [38] that, in general, the scope of funding, staffing and hence, activities per capita, of publicly-funded plant-breeding programs were either dwindling progressively or had stagnated over time. In Africa, instances of up to a 10-fold decrease in funding of plant breeding activities have occurred between 1985 and 2001 [126,127]. The worrisome global trend of ageing and retiring plant breeders that were not being replaced by younger ones was captured in these surveys also; over 40% of plant breeders in the countries surveyed were aged 50 years and above. Indeed, to compound the problem, too few new plant breeders are being trained in universities in both developed and developing countries [127-129]. It would appear though that there was no perceptible downward trend in the award of plant breeding degrees in the USA between 1995 and 2000 [130] implying that this problem might either have been more acute in developing countries [128] or had assumed a global dimension only in the last decade. Currently, there is a general consensus however that the current capacity for plant breeding is inadequate to deal with the generational challenges of food insecurity with Knight [131] encapsulating the sense of despair in the somberly titled article, 'A Dying Breed'.

The training of future plant breeders is generally considered a major component of the preparedness for sustained food security and has been the subject of copious analyses and studies. For instance, the symposium 'Plant Breeding and the Public Sector: Who Will Train Plant Breeders in the U.S. and around the World?' held at Michigan State University in the US was aimed at charting a course for addressing this critical constraint through the devising of curricula, raising awareness, and fostering partnerships [126,132-139]. The symposium concluded that future plant breeders, at PhD level, must in addition to possessing skills in the traditional disciplines of experimental design, applied statistics, Mendelian (transmission) genetics, population and quantitative genetics, and principles and practice of plant breeding also be trained in myriad areas ranging from subjects in the biological sciences including plant physiology, ecology, pathology, entomology, molecular biology, and genomics through business management to law, especially IPR [137]. More recently, Repinski *et al.* [129] in analyzing a very wide ranging Delphi study for articulating the curriculum of the future plant breeder came to the same conclusions regarding the need for broadening the scope of the curriculum to reflect the realities of modern breeding techniques and the fact that a significant number of plant breeders work in the private sector where legal and policy issues are critically important. Multidisciplinary teams, staffed by personnel with specialized skills in these areas, will compensate for the reality that no one plant breeder will be adept at sufficient levels of skill in all these disciplines.

Granted, most private sector plant breeders graduated from publicly-funded institutions ([132] estimated that most private sector breeders in the US attended publicly-funded land grant universities, for instance) but the public sector's role in the training of plant breeders is very critical and must be considered a contribution to public good [133] that cannot be ceded wholly to the private sector without compromising the future of plant breeding and hence food security. While the role of the private sector is also critical in this regard, in the provision of fellowships, for instance [139], it should not be expected to play the leading role as funding could not be guaranteed this way.

The centers of the CGIAR are also considered valuable partners in the training of plant breeders [133]. With improved funding, these centers, appropriately located in developing countries and working on food security crops, could provide the much-needed training facilities that many developing country governments cannot provide. The IBP, for instance, is spearheading the training of plant breeders from developing countries in molecular breeding techniques. The African Centre for Crop Improvement (ACCI; [139]) at the University of Kwazulu-Natal, South Africa and the West Africa Centre for Crop Improvement (WACCI;) at the University of Ghana, Legon, Ghana, both funded under the auspices for the Alliance for a Green Revolution in Africa, are producing highly skilled plant breeders that are trained in Africa to work on African food security crops. Both universities partner with Cornell University, Ithaca, New York in the US in this endeavor. This is a very laudable model that is bridging the gap created by the continued inability of countries to establish and fund training facilities adequately.

Conclusions and future perspectives

There is a compelling urgency to institute measures that ensure that farmers worldwide, but especially the small-scale farmers that produce the majority of the food in food insecure countries, can grow the portfolio of suitable crop varieties that are amenable to the eco-efficient production systems of the sustainable crop production intensification (SCPI) paradigm needed to feed the world in the 21st century. The major hindrances to the attainment of SCPI include: inadequate investment; sub-optimal human resources; inability to innovate as evidenced in prevailing inadequate deployment of appropriate science and technology; weak institutions; sub-optimal R&D infrastructure; and poor policy regimes. Crop improvement, by fostering genetic gains that aid food production through enhanced

productivities, is a very critical component of SCPI. We make the case therefore that plant breeding, by translating the potentials inherent in PGRFA into 'smart' crop varieties, can engender a most significant impetus for sustained food security even as human population increases and extremely inclement weather conditions constrain crop production. To achieve this, plant breeding must be re-oriented in a number of very critical ways.

Broadened genetic diversity of crops

Firstly, the extremely narrow genetic base of crops, which puts food security at risk, must be broadened at both the intra- and inter-specific levels. Conserved PGRFA, *ex-situ* and *in-situ*, and the heritable diversity available on-farm, including in landraces, must be explored to source the novel alleles that confer enhanced productivities. FAO through its Global PGRFA System, the International Treaty and the Global Crop Diversity Trust; the CGIAR centers, regional networks, and the NARES around the world must continue to invest considerable efforts to ensure that breeders have access to the genetic variations they require for their work. Some harmonization of the information dissemination mechanisms is called for to ensure enhanced efficiencies. International norms are now being leveraged to facilitate the sourcing of these much needed genetic variations even across national boundaries. Induced mutations, an established scientific method that has been used for almost one century to mimic nature, is increasingly important for inducing the unmasking of novel alleles of genes to which plant breeders do not otherwise have access. The current constraints to crop productivities deny humanity the limitless space and time for the natural process of spontaneous mutations to make these novel heritable variations available. Pre-breeding is critical in achieving this broadened genetic base of crops. The introduction of new genes and their variants into crops from novel sources will be critical to replicating the impacts of the Green Revolution as the current generational challenges demand.

Defining the breeding objectives

A second area for re-orienting plant breeding is in the 'what'. What should be the breeding objectives? Without de-emphasizing yield, resistances to biotic and abiotic stresses of import in climate change adaptation, enhanced nutritional quality traits, and the multipurpose use of crop biomass (including for bioenergy, livestock feed, and fiber) are key objectives. Also, the amenability to low-input eco-efficient farming systems will increasingly constitute standard breeding objectives. The enthusiastic adoption of NERICA in sub-Saharan Africa is an example of the efficacy of the alignment of breeding objectives to addressing the constraints posed by empirically determined drivers. In general, market forces which reflect end-user preferences will be the main driver in the definition of breeding objectives.

Innovating for result-oriented plant breeding

Thirdly, the 'how' of plant breeding will probably attract the most innovative interventions. How should crops be bred? Increased use of the immensely powerful biotechnologies that have revolutionized the biological sciences is imperative. Demonstrably, MAS, supported by the tools of genomics and the other -omics and information technology platforms, permits high throughput evaluations of breeding materials. Genetic transformation and the resulting GM crops are increasingly cultivated around the world; the technology holds promise and countries need capacity building in order to, at the minimum, make evidence-based decisions as to its adoption. Equally, the other emerging biotechnologies such as ZFN, ODM, transgenesis and cis-genesis, RdDM, grafting on GM stock, reverse breeding, agro-infiltration, and synthetic genomics, though requiring further refinements to varying degrees, will also become quite important in the very near future. Countries will increasingly require support in navigating the IPR regimes that govern access to these technologies and the regulatory issues pertaining to their adoptions. As massive numbers of new breeding materials are generated through pre-breeding , MAS must be complemented by phenomics in order that reliable predictions of the breeding values can be made. Private sector plant breeding and seed companies have taken the lead in leveraging these innovations in producing highly successful crop varieties and provide models for retooling the public sector crop improvement programs.

Policy and strategic interventions

A fourth consideration is the 'where' in the agricultural R&D environment for situating plant breeding. Certainly, an enabling environment is required for breeding to be relevant and, hence, thrive. The erstwhile piecemeal interventions at the three components of the PGRFA value chain, namely, conservation, breeding, and dissemination of seeds and planting materials is, simply, inadequate. A result-oriented plant breeding must have access to the widest possible source of heritable variations just as it needs an effective mechanism to deliver high quality seeds and planting materials to the growers. This is the PGRFA continuum that significantly enhances the ability of plant breeding to deliver need-based outputs. We posit that not only all three individual components but their intervening linkages must be strengthened in tandem. A National PGRFA Strategy helps to institutionalize this paradigm that demonstrably mirrors the operations of the highly successful private sector crop improvement multinationals.

Winning partnerships for the reinvigorated crop improvement

The 'who' of the 21st century plant breeding is the fifth critical consideration. Who are the main stakeholders in the crop improvement component of the PGRFA management continuum? The increasingly pivotal roles of the private sector must be factored into policy-making and in the development of strategies. The private sector is not only marketing seeds and planting materials but also breeding the new varieties; its continued participation in these activities must be encouraged especially where comparative advantages are demonstrated. Enabling policy, legal, and market environments that spur innovation and investments of capital are key to fostering the much needed public-private partnerships required for operating at scale. A healthy balance must be struck between IPR (and the innovations and investments that they encourage) and the imperative of contributing to public good. The roles of the International Convention for the Protection of New Varieties of Plants, that is UPOV, and various national, regional, and global industry interest groups will be critically important in this regard.

Capacity enhancements for the 21st century plant breeder

A sixth consideration is the 'by whom'. What is the profile of the 21st century plant breeder? In fact, the 'plant breeder' is the multidisciplinary team that makes use of the most appropriate scientific and technological tools in generating new crop varieties and the germplasm curators, farmers, and seed marketers that they work with. Technically, the multidisciplinary team driving a breeding program will include persons skilled in the traditional disciplines of plant breeding as well as those with in-depth knowledge of various ancillary biotechnological techniques. Skills in information technology, business management, law, and so on will also be required in such teams. Aside from private sector plant breeding and seed companies, such a suite of expertise does not exist in most public sector breeding concerns. The training of the future plant breeder, though mentioned often now, is still not receiving as much attention, in terms of funding, facilities, skilled trainers, and the number of available opportunities, which it deserves. Capacity building will require wide-ranging public-private partnerships in order that the curriculum being developed can be effective. The role of the CGIAR centers will remain critical. The regional training hubs, ACCI and WACCI, provide models worth emulating and scaling up. The highly successful land grant universities scheme of the United States demonstrates the lasting impacts that concerted investment of resources in training can have.

Strengthening the NARES

Finally, the re-oriented crop improvement programs require a sustaining platform, in this case, the NARES. As we have indicated, the continued decline in funding for agricultural R&D has led to weakened NARES; breeding programs are ill-staffed and poorly equipped while extension systems have become moribund in many developing countries. Equally disturbing is the dearth of reliable mechanisms for the dissemination of high quality seeds and planting materials of improved varieties. Indeed, while the work of the CGIAR centers in filling this gap cannot but be commended, the manifest over dependence of many NARES on these centers can only be injurious in the long run. For one thing, the mandates of these centers preclude work on many important food security crops. United in the recognition of the imperative for re-orienting agriculture, development organizations including FAO, the World Bank, the International Fund for Agricultural Development (IFAD), the CGIAR, and so on have severely recommitted their resolves to stamp out hunger. The strengthening of the NARES, the ultimate bulwark between hunger and the populace in many developing countries, must be at the top of the agenda. Bold initiatives underpinned by political will have strengthened and re-oriented agriculture in the past. For instance, the contributions of the land grant universities, including the extension services, to the food security of the US are legendary. Many national governments sadly lack the political will to strengthen their NARES as means for ending hunger. Support to national governments must therefore include mechanisms that contribute to fostering the nurturing policy environments for investments to bear fruit. In the final analysis, the ultimate responsibilities for crop improvement, just as in safeguarding food security in general, lies with national governments and by extension, their NARES. These responsibilities may be abdicated only at the peril of food security and at the certain risk of consequent instability and retarded development. The well-funded and adequately staffed Embrapa, Brazil's Agricultural Research Corporation, for instance, demonstrates very clearly the recent significant impacts that government policies can have on the viability of a country's agricultural R&D sector.

The coalescence of the consequences of climate change and variations with other critical demographic, economic, social, and industrial pressures pose unprecedented monumental risks to food security and people's general well-being. Unarguably, crop improvement and its outputs of 'smart' crop varieties can contribute to mitigating these threats. Multilateral organizations, civil society, and national governments must ride the momentum of the current reinvigorated attention to food security and strengthen capacities for crop improvement in innovative manners. Countries need assistance with suites of actionable policy interventions that leverage validated technologies and strategies in aid of result-oriented crop improvement. Such policy items or measures that countries can adopt in strengthening the three components of, and the linkages between, the PGRFA continuum in tandem are not readily available in forms amenable to ease of dissemination. The re-orienting of crop improvement would require the packaging of validated measures into a 'toolbox' to act as a one-stop shop for actionable intervention instruments. The work of the GIPB and similar multi-stakeholder platforms in articulating and assembling such tools serve as examples of multi-stakeholder efforts that deserve continued support especially in order to operate successfully at scale.

Abbreviations
AATF, African Agricultural Technology Foundation; ACCI, African Centre for Crop Improvement; Bt, *Bacillus thuringiensis*; CBD, Convention on Biological Diversity; CGIAR, Consultative Group on International Agricultural Research; CWRs, crop wild relatives; DNA, deoxyribonucleic acid; FAO, Food and Agriculture Organization of the United Nations; GIPB, Global Partnership Initiative for Plant Breeding Capacity Building; GM, genetically modified; GMO, genetically modified organism; GPA, Global Plan of Action for Plant Genetic Resources for Food and Agriculture; IBP, Integrated Breeding Platform of the Generation Challenge Program of the CGIAR; IFAD, International Fund for Agricultural Development; IFPRI, International Food Policy Research Institute; IPCC, United Nation's Intergovernmental Panel on Climate Change; MAS, Marker-Assisted (or, Aided) Selection; NARES, National Agricultural Research and Extension Systems; NERICA, New Rice for Africa; ODM, Oligonucleotide directed mutagenesis; PGRFA, Plant Genetic Resources for Food and Agriculture; PIPRA, Public Sector Intellectual Property Resource for Agriculture; PPB, Participatory Plant Breeding; PVS, Participatory Varietal Selection; R&D, research and development; QTL, quantitative trait loci; RdDM, RNA-dependent DNA methylation; RNA, ribonucleic acid; SCPI, sustainable crop production intensification; TAC, Technical Advisory Committee of the Consultative Group on International Agricultural Research; TILLING, Targeted Induced Local Lesions IN Genomes; UPOV, International Convention for the Protection of New Varieties of Plants; WACCI, West Africa Centre for Crop Improvement; WIEWS, World Information and Early Warning System; ZFN, Zinc finger nuclease.

Competing interests
The authors declare that they have no competing interests.

Acknowledgements
Generous funding provided by the Bill and Melinda Gates Foundation - through the Global System Project implemented by the Global Crop Diversity Trust - in support of the activities of the Global Partnership Initiative on Plant Breeding Capacity Building is gratefully acknowledged.

Author details
[1]Plant Genetic Resources and Seeds Team, Plant Production and Protection Division, Food and Agriculture Organization of the United Nations (FAO), Rome, Italy. [2]International Centre for Tropical Agriculture (CIAT), Cali, Colombia.

Authors' contributions
CM conceived of the study, developed the outline for a publication, conducted literature reviews, coordinated inputs and drafted the manuscript. EPG conducted literature reviews and contributed to the design and draft of the manuscript. KG conducted literature reviews and contributed to the design and draft of the manuscript. All authors read and approved the manuscript.

References
1.　Food and Agriculture Organization of the United Nations: *How to Feed the World in 2050*. Rome: FAO; 2009.

2. Tester M, Langridge P: **Breeding technologies to increase crop production in a changing world.** *Science* 2010, **327:**818–822.

3. Food and Agriculture Organization of the United Nations: *Mitigation of Climate Change in Agriculture Series No. 1.* Rome: Global Survey of Agricultural Mitigation Projects; FAO.

4. Beddington J, Asaduzzaman M, Fernandez A, Clark M, Guillou M, Jahn M, Erda L, Mamo T, Van Bo N, Nobre CA, Scholes R, Sharma R, Wakhungu J: *Achieving food security in the face of climate change: Summary for policy makers from the Commission on Sustainable Agriculture and Climate Change. Copenhagen, Denmark: CGIAR Research Program on Climate Change.* Copenhagen: Agriculture and Food Security (CCAFS); 2011.

5. International Food Policy Research Institute (IFPRI): **Global Hunger Index: The Challenge of Hunger.** In *Taming Price Spikes and Excessive Food Price Volatility.* Edited by. Washington, DC:; 2011.

6. Evans A: *The Feeding of the Nine Billion: Global Food Security for the 21st Century.* London: Chatham House; 2009.

7. Intergovernmental Panel on Climate Change (IPCC): **Summary for Policymakers.** In *Managing the Risks of Extreme Events and Disasters to Advance Climate Change Adaptation.* Edited by Field CB, Barros V, Stocker TF, Qin D, Dokken DJ, Ebi KL, Mastrandrea MD, Mach KJ, Plattner G-K, Allen SK, Tignor M, Midgley PM. Cambridge: Cambridge University Press; 2012.

8. Nelson GC, Rosegrant MW, Koo J, Robertson R, Sulser T, Zhu T, Ringler C, Msangi S, Palazzo A, Batka M, Magalhaes M, Valmonte-Santos R, Ewing M, Lee D: *Food Policy Report No. 21. Climate Change: Impact on Agriculture and Costs of Adaptation.* Washington DC, USA: International Food Policy Research Institute; 2009:19.

9. Hertel TW, Burke MB, Lobell DB: **The poverty implications of climate-induced crop yield changes by 2030.** *Global Environmental Change* 2010, **20:**577–585.

10. Rosegrant MW: **Impacts of climate change on food security and livelihoods.** In *Food security and climate change in dry areas: proceedings of an International Conference, 1-4 February 2010, Amman, Jordan.* Edited by Solh M, Saxena MC. Aleppo: International Center for Agricultural Research in the Dry Areas (ICARDA); 2011:24–26.

11. Ejeta G: **Revitalizing agricultural research for global food security.** *Food Sec* 2009, **1:**391–401.

12. Foresight: *The Future of Food and Farming: Final Project Report.* London: The Government Office for Science; 2011.

13. Ejeta G: **African Green Revolution Needn't Be a Mirage.** *Science* 2010, **327:**831–832.

14. Fernandez-Cornejo J: *Agriculture Information Bulletin No. 786. The Seed Industry in US Agriculture: An exploration of Data and Information on Crop Seed Markets, Regulation, Industry Structure, and Research and Development.* Washington DC, USA: Economic Research Service, US Department of Agriculture; 2004:213.

15. Duvick DN: **Genetic contributions to advances in yield of US maize.** *Maydica* 1992, **37:**69–79.

16. Duvick DN: **Heterosis: Feeding People and Protecting Natural Resources.** In *The Genetics and Exploitation of Heterosis in Crops.* Edited by Coors JG, Pandey S, Madison WI. MAdison, USA: American Society of Agronomy, Inc., Crop Science Society of America, Inc., Soil Science Society of America, Inc; 1999.

17. Duvick DN: **Genetic Progress in Yield of United States Maize (Zea mays L.).** *Maydica* 2005, **50:**193–202.

18. Dalton TJ, Guei RG: **Productivity gains from rice genetic enhancements in West Africa: countries and ecologies.** *World Development* 2003, **31:**359–374.

19. Diagne A: **Diffusion and adoption of NERICA rice varieties in Côte d'Ivoire.** *The Developing Economies* 2006, **XLIV-2:**208–231.

20. Oikeh SO, Nwilene F, Diatta S, Osiname O, Touré A, Okeleye KA: **Responses of upland NERICA rice to nitrogen and phosphorus in forest agroecosystems.** *Agron J* 2008, **100:**735–741.

21. Wopereis MCS, Diagne A, Rodenburg J, Sié M, Somado EA: **Why NERICA is a successful innovation for African farmers.** *Outlook on Agriculture* 2008, **37:**169–176.

22. The Columbia Encyclopedia:. Sixth ednth editionNew York, NY: Columbia University Press; 2008 [http://www.encyclopedia.com/topic/plant_breeding. aspx].

23. World Economic Forum: *Realizing a New Vision for Agriculture: A roadmap for stakeholders.* Geneva: World Economic Forum; 2010.

24. Food and Agriculture Organization of the United Nations: *Save and Grow - A Policy Maker's Guide to the Sustainable Intensification of Smallholder Crop Production.* Rome: FAO; 2011.

25. McCouch S: **Diversifying selection in plant breeding.** *PLoS Biol* 2004, **2:**1507–1512.

26. Waines GJ, Ehdaie B: **Domestication and crop physiology: roots of green-revolution wheat.** *Annals of Botany* 2007, **100:**991–998.

27. Martynov SP, Dobrotvorskaya TV: **Genealogical analysis of diversity of Russian winter wheat cultivars (Triticum aestivum L.).** *Genetic Resources and Crop Evolution* 2006, **53:**386–386.

28. Rabinovich SV: **Importance of wheat-rye translocations for breeding modern cultivars of Triticum aestivum L.** *Euphytica* 1998, **100:**323–340.

29. Gur A, Zamir D: **Unused natural variation can lift yield barriers in plant breeding.** *PLoS Biol* 2004, **2:**1610–1615.

30. Hahn SK, Terry ER, Leuschner K: **Breeding cassava for resistance to cassava mosaic disease.** *Euphytica* 1980, **29:**673–683.

31. Akano AO, Dixon AGO, Mba C, Barrera E, Fregene M: **Genetic mapping of a dominant gene conferring resistance to the cassava mosaic disease (CMD).** *Theor Appl Genet* 2002, **105:**521–525.

32. Fregene M, Matsumura H, Akano A, Dixon A, Terauchi R: **Serial analysis of gene expression (SAGE) of host-plant resistance to the cassava mosaic disease (CMD).** *Plant Molecular Biology* 2004, **56:**563–571.

33. Chavez AL, Sanchez T, Jaramillo G, Bedoya JM, Echeverry J, Bolanos EA, Ceballos H, Iglesias CA: **Variation of quality traits in cassava roots evaluated in landraces and improved clones.** *Euphytica* 2005, **143:**125–133.

34. Blair MW, Fregene MA, Beebe SE, Ceballos H: **Marker-assisted selection in common beans and cassava.** In *Marker-assisted selection: Current status and future perspectives in crops, livestock, forestry and fish.* Edited by Guimaraes EP, Ruane J, Scherf BD, Sonnino A, Dargie JD. Rome: FAO; 2007:81–115.

35. Maxted N: Ford-Lloyd BV, Kell SP, Iriondo JM, Dulloo ME, Turok J (eds): *Crop Wild Relative Conservation and Use.* Wallingford: CABI Publishing; 2008.

36. The International Treaty on Plant Genetic Resources for Food and Agriculture [http://www.planttreaty.org/].

37. Food and Agriculture Organization of the United Nations (FAO): *The International Treaty on Plant Genetic Resources for Food and Agriculture.* Rome: FAO; 2009.

38. Food and Agriculture Organization of the United Nations: *Second Report on the State of the World's Plant Genetic Resources for Food and Agriculture.* Rome: FAO; 2010.

39. The Global Crop Diversity Trust. [http://www.croptrust.org/main/].

40. **Pre-breeding for Effective Use of Plant Genetic Resources.** [http://km.fao.org/gipb/].

41. Food and Agriculture Organization of the United Nations: *Pre-breeding for effective use of plant genetic resources: an e-Learning Course.* Rome: FAO; 2011.

42. Global Partnership Initiative for Plant Breeding Capacity Building (GIPB) [http://km.fao.org/gipb/].

43. The FAO Global System on Plant Genetic Resources for Food and Agriculture [http://www.fao.org/agriculture/crops/core-themes/theme/ seeds-pgr/gpa/gsystem/es/].

44. The World Information and Early Warning System of Plant Genetic Resources for Food and Agriculture (WIEWS) [http://apps3.fao.org/wiews/wiews.jsp].

45. **The World Information Sharing Mechanism on the Implementation of the Global Plan of Action for the Conservation and Sustainable Use of Plant Genetic Resources for Food and Agriculture.** [http://www.pgrfa.org/gpa/selectcountry.jspx].

46. Food and Agriculture Organization of the United Nations: *Second Global Plan of Action on Plant Genetic Resources for Food and Agriculture.* Rome: FAO; 2011.

47. Maluszynski M, Nichterlein K, Van Zanten L, Ahloowalia BS: *Officially Released Mutant Varieties - The FAO/IAEA Database.* Vienna: International Atomic Energy Agency; 2000:84.

48. Ahloowalia BS, Maluszynski M, Nichterlein K: **Global impact of mutation-derived varieties.** *Euphytica* 2004, **135:**187–204.

49. Shu QY: *Proceedings of International Symposium on Induced Mutations in Plants: Induced Plant Mutations in the Genomics Era: 11-15 August 2008; Vienna, Austria.* Rome: FAO; 2009.

50. The Joint FAO/IAEA Mutant Varieties and Genetic Stocks Database [http://mvgs.iaea.org/AboutMutantVarieties.aspx].

51. McCallum CM, Comai L, Greene EA, Henikoff S: **Targeting induced local lesions IN genomes (TILLING) for plant functional genomics.** *Plant Physiol* 2000, **123:**439–442.

52. Colbert T, Till BJ, Tompa R, Reynolds S, Steine MN, Yeung AT, McCallum CM, Comai L, Henikoff S: **High-throughput screening for induced point mutations.** *Plant Physiol* 2001, **126:**480–484.

53. Greene EA, Codomo CA, Taylor NE, Henikoff JG, Till BJ, Reynolds SH, Enns LC, Burtner C, Johnson JE, Odden AR, Comai L, Henikoff S: **Spectrum of chemically induced mutations from a large-scale reverse-genetic screen in *Arabidopsis*.** *Genetics* 2003, **164:**731–740.

54. Caldwell DG, McCallum N, Shaw P, Muehlbauer GJ, Marshall DF, Waugh R: **A structured mutant population for forward and reverse genetics in barley (*Hordeum vulgare* L.).** *Plant J* 2004, **40:**143–150.

55. Till BJ, Reynolds SH, Weil C, Springer N, Burtner C, Young K, Bowers E, Codomo CA, Enns LC, Odden AR, Greene EA, Comai L, Henikoff S: **Discovery of induced point mutations in maize genes by TILLING.** *BMC Plant Biol* 2004, **4:**12.

56. Triques K, Sturbois B, Gallais S, Dalmais M, Chauvin S, Clepet C, Aubourg S, Rameau C, Caboche M, Bendahmane A: **Characterization of *Arabidopsis thaliana* mismatch specific endonucleases: application to mutation discovery by TILLING in pea.** *Plant J* 2007, **51:**1116–25.

57. Till BJ, Cooper J, Tai TH, Colowit P, Greene EA, Henikoff S, Comai L: **Discovery of chemically induced mutations in rice by TILLING.** *BMC Plant Biol* 2007, **7:**19.

58. Slade AJ, Fuerstenberg SI, Loeffler D, Steine MN, Facciotti D: **A reverse genetic, nontransgenic approach to wheat crop improvement by TILLING.** *Nat Biotechnol* 2005, **23:**75–81.

59. Sato Y, Shirasawa K, Takahashi Y, Nishimura M, Nishio T: **Mutant selection from progeny of gamma-ray-irradiated rice by DNA heteroduplex cleavage using *Brassica* petiole extract.** *Breeding Science* 2006, **56:**179–183.

60. Cooper JL, Till BJ, Laport RG, Darlow MC, Kleffner JM, Jamai A, El-Mellouki T, Liu S, Ritchie R, Nielsen N, Bilyeu KD, Meksem K, Comai L, Henikoff S: **TILLING to detect induced mutations in soybean.** *BMC Plant Biology* 2008, **8:**9. doi:10.1186/1471-2229-8-9.

61. Suzuki T, Eiguchi M, Kumamaru T, Satoh H, Matsusaka H, Moriguchi K, Nagato Y, Kurata N: **MNU-induced mutant pools and high performance TILLING enable finding of any gene mutation in rice.** *Mol Genet Genomics* 2008, **279:**213–223.

62. Tadele Z, Mba C, Till BJ: **TILLING for mutations in model plants and crops.** In *Molecular Techniques in Crop Improvement.* 2nd edition. Edited by Jain SM, Brar DS. Dordrecht: Springer Publishing, Inc; 2009:307–322.

63. Till BJ, Afza R, Bado S, Huynh OA, Jankowicz-Cieslak J, Matijevic M, Mba C: **Global TILLING Projects.** In *Proceedings of International Symposium on Induced Mutations in Plants: Induced Plant Mutations in the Genomics Era: 11-15 August 2008.* Edited by Shu QY. Rome: FAO; 2009:237–239.

64. Forster BP, Heberle-Bors E, Kasha KJ, Touraev A: **The resurgence of haploids in higher plants.** *Trends Plant Sci* 2007, **12:**368–375.

65. Szarejko I, Forster BP: **Doubled haploidy and induced mutation.** *Euphytica* 2007, **158:**359–370.

66. Mba C, Afza R, Jankowicz-Cieslak J, Bado S, Matijevic M, Huynh A, Till BJ: **Enhancing genetic diversity through induced mutagenesis in vegetatively propagated plants.** In *Proceedings of International Symposium on Induced Mutations in Plants: Induced Plant Mutations in the Genomics Era: 11-15 August 2008.* Edited by Shu QY. Rome: FAO; 2009:293–296.

67. Sperling L, Ashby JA, Smith ME, Weltzien E, McGuire S: **A framework for analyzing participatory plant breeding approaches and results.** *Euphytica* 2001, **122:**439–450.

68. Ashby JA: **The Impact of Participatory Plant Breeding.** In *Plant Breeding and Farmer Participation.* Edited by Ceccarelli S, Guimaraes EP, Weltzien E. Rome: FAO; 2009:649–671.

69. Efisue A, Tongoona P, Derera J, Langyintuo A, Laing M, Ubi B: **Farmers' perceptions on rice varieties in Sikasso region of Mali and their implications for rice breeding.** *J Agronomy & Crop Science* 2008, **194:**393–400.

70. Witcombe JR, Joshi A, Joshi KD, Sthapit BR: **Farmer participatory crop improvement. I. Varietal selection and breeding methods and their impact on biodiversity.** *Experimental Agriculture* 1996, **32:**445–460.

71. Technical Advisory Committee of the Consultative Group on International Agricultural Research (TAC): *Systemwide Review of Breeding Methodologies in the CGIAR.*: Science Council of the CGIAR; 2001.

72. Suslow TV, Thomas B, Bradford KJ: *Biotechnology provides new tools for plant breeding.* Davis, CA: Seed Biotechnology Center, University of California Davis; 2002:19.

73. Moose SP, Mumm RH: **Molecular plant breeding as the foundation for 21st century crop improvement.** *Plant Physiology* 2008, **147:**969–977.

74. Stam P, Marker-assisted introgression: speed at any cost?: In *Eucarpia Leafy Vegetables 2003: Proceedings of the EUCARPIA Meeting on Leafy Vegetables Genetics and Breeding: 19-21 March 2003.* Edited by Van Hintum TJL, Lebeda A, Pink D, Schut JW. Wageningen: Centre for Genetic Resources (CGN); 2003:117–124.

75. Burstin J, Charcosset A: **Relationship between phenotypic and marker distances: theoretical and experimental investigations.** *Heredity* 1997, **79:**477–483.

76. Spelman RJ, Bovenhuis H: **Moving from QTL experimental results to the utilization of QTL in breeding programmes.** *Animal Genetics* 1998, **29:**77–84.

77. Hospital F, Moreau L, Lacourde F, Charcosset A, Gallais A: **More on the efficiency of marker-assisted selection.** *Theor Appl Genet* 1997, **95:**1181–1189.

78. Hospital F, Charcosset A: **Marker-assisted introgression of quantitative loci.** *Genetics* 1997, **147:**1469–1485.

79. Tanksley SD, McCouch SR: **Seed banks and molecular maps: Unlocking genetic potential from the wild.** *Science* 1997, **277:**1063–1066.

80. Bernacchi D, Beck-Bunn T, Emmatty D, Inai S: **Advanced backcross QTL analysis of tomato II. Evaluation of near-isogenic lines carrying single-donor introgressions for desireable wild QTL alleles derived from Lycopersicon hirsutum and L. pimpinellifolium.** *Theor Appl Genet* 1998, **97:**170–180.

81. Bernacchi D, Beck-Bunn T, Eshed Y, Lopez J, Petiard V, Uhlig J, Zamir D, Tanksley S: **Advanced backcross QTL analysis in tomato I. Identification of QTLs for traits of agronomic importance from Lycopersicon hirsutum.** *Theor Appl Genet* 1998, **97:**381–397.

82. Stuber CW, Polacco M, Senior ML: **Synergy of Empirical breeding, marker-assisted selection, and genomics to increase crop yield potential.** *Crop Sci* 1999, **39:**1571–1583.

83. Moreau L, Lemarie S, Charcosset A, Gallais A: **Economic efficiency of one cycle of marker-assisted selection.** *Crop Sci* 2000, **40:**329–337.

84. Meuwissen THE, Hayes BJ, Goddard ME: **Prediction of total genetic value using genome-wide dense marker maps.** *Genetics* 2001, **157:**1819–1829.

85. Charcosset A, Moreau L: **Use of molecular markers for the development of new cultivars and the evaluation of genetic diversity.** *Euphytica* 2004, **137:**81–94.

86. Lecomte L, Duffé P, Buret M, Servin B, Hospital F, Causse M: **Marker-assisted introgression of five QTLs controlling fruit quality traits into three tomato lines revealed interactions between QTLs and genetic backgrounds.** *Theor Appl Genet* 2004, **109:**658–668.

87. Collard BCY, Jahufer MZZ, Brouwer JB, Pang ECK: **An introduction to markers, Quantitative trait loci (QTL) mapping and marker-assisted selection for crop improvement: The basic concepts.** *Euphytica* 2005, **142:**169–196.

88. Tuberosa R, Salvi S: **Genomics-based approaches to improve drought tolerance of crops.** *TRENDS in Plant Science* 2006, **11:**405–412.

89. Brady SM, Provart NJ: **Extreme breeding: Leveraging genomics for crop improvement.** *J Sci Food Agric* 2007, **87:**925–929.

90. Ribaut J-M, Ragot M: **Marker-assisted selection to improve drought adaptation in maize: the backcross approach, perspectives, limitations, and alternatives.** *J Exp Bot* 2007, **58:**351–360.

91. Collard BCY, Mackill DJ: **Marker-assisted selection: an approach for precision plant breeding in the twenty-first century.** *Phil Trans R Soc B* 2008, **363:**557–572.

92. Choudhary K, Choudhary OP, Shekhawat NS: **Marker assisted selection: a novel approach for crop improvement.** *American-Eurasian Journal of Agronomy* 2008, **1:**26–30.

93. Eathington SR, Crosbie TM, Reiter RS, Bull JK: **Molecular markers in a commercial breeding program.** *Crop Sci* 2007, **(Suppl 3):**S154–S163.

94. **Pioneer Hi-Bred.** [http://www.pioneer.com/home/site/about/research/plant-breeding/].

95. Syngenta [http://www.syngentafoundation.org/index.cfm?pageID = 607].

96. Thro AM, Parrott W, Udall JA, Beavis WD: *Proceedings of the Symposium on Genomics and Plant Breeding: The Experience of the Initiative for Future*

Agricultural and Food Systems: 13 November 2002. Indianapolis, IN, USA: Crop Sci; 2004:1893–1919. Crop Science Society of America; 2004.

97. Ikeda N, Bautista NS, Yamada T, Kamijima O, Ishii T: **Ultra-simple DNA extraction method for marker-assisted selection using microsatellite markers in rice.** *Plant Molecular Biology Reporter* 2001, **19**:27–32.

98. **The Integrated Breeding Platform of the Generation Challenge Program of the Consultative Group on International Agricultural Research.** [http://www.generationcp.org/ibp].

99. James C: **Executive Summary Global Status of Commercialized Biotech/ GM Crops: 2011.** In *ISAAA Brief 2011*. Ithaca, NY: ISAAA; 2011:30.

100. Glover D: *Monsanto and smallholder farmers: A case-study on corporate accountability. IDS Working Paper 277.* Brighton: University of Sussex, UK, Institute of Development Studies; 2007:53.

101. Glover D: *Made by Monsanto: The corporate shaping of GM crops as a technology for the poor. STEPS Working Paper 11.* Brighton: STEPS Centre; 2008:45.

102. Convention on Biological Diversity (CBD): *Cartagena Protocol on Biosafety to the Convention on Biological Diversity.* Montreal: Secretariat of the Convention on Biological Diversity; 1992.

103. The African Agricultural Technology Foundation [http://www.aatf-africa.org/].

104. Boadi RY, Bokanga M, bioDevelopments-International Institute: **The African Agricultural Technology Foundation Approach to IP Management.** In *Intellectual Property Management in Health and Agricultural Innovation: A Handbook of Best Practices.* Edited by Krattiger A, Mahoney RT, Nelsen L, Thomson JA, Bennett AB, Satyanarayana K, Graff GD, Fernandez C, Kowalski SP. Ithaca, NY, USA: MIHR; Davis, CA, USA: PIPRA; Fiocruz, Rio de Janeiro, Brazil: Oswaldo Cruz Foundation; 2007:1765-1774.

105. Atkinson RC, Beachy RN, Conway G, Cordova FA, Fox MA, Holbrook KA, Klessig DF, McCormick RL, McPherson PM, Rawlings HR III, Rapson R, Vanderhoef LN, Wiley JD, Young CE: **Public sector collaboration for agricultural IP management.** *Science* 2003, **301**:174–175.

106. Delmer DP, Nottenburg C, Graff GD, Bennett AB: **Intellectual property resources for international development in agriculture.** *Plant Physiology* 2003, **133**:1666–1670.

107. Lusser M, Parisi C, Plan D, Rodríguez-Cerezo E: *New plant breeding techniques.* State-of-the-art and prospects for commercial development. Luxembourg: Publications Office of the European Union; 2011.

108. Blakeney M: **Patenting of plant varieties and plant breeding methods.** *J Exp Bot* 2012, **63**:1069–1074.

109. Xu Y, Crouch JH: **Marker-assisted selection in plant breeding: from publications to practice.** *Crop Sci* 2008, **48**:391–407.

110. Houle D, Govindaraju DR, Omholt S: **Phenomics: the next challenge.** *Nature Reviews Genetics* 2010, **11**:855–866.

111. Finkel E: **With 'Phenomics', plant scientists hope to shift breeding into overdrive.** *Science* 2009, **325**:380–381.

112. The Australian Plant Phenomics Facility [http://www.plantphenomics.org.au/].

113. Lemna Tec [http://www.lemnatec.com/].

114. Jülich Plant Phenotyping Centre [http://www2.fz-juelich.de/icg/icg-3/jppc].

115. Ecotron. [http://www.ecotron.cnrs.fr/].

116. The Ecophysiology Laboratory of Plant under Environmental Stress:, [http://www1.montpellier.inra.fr/ibip/lepse/english/].

117. The Biotron Experimental Climate Change Research Centre [http://www.thebiotron.ca/].

118. Mba C, Guimaraes EP, Guei GR, Hershey C, Paganini M, Pick B, Ghosh K: **Mainstreaming the continuum approach to the management of plant genetic resources for food and agriculture through national strategy.** *Plant Genetic Resources* 2011. doi:10.1017/S1479262111000943.

119. Food and Agriculture Organization of the United Nations: *Crop Improvement for the 21st Century.* Rome: FAO; Information Document for 23rd Session of the Committee on Agriculture; 2012.

120. Anthony VM, Ferroni M: **Agricultural biotechnology and smallholder farmers in developing countries.** *Curr Opin Biotechnol* 2011, **23**:1–8.

121. Ferroni M, Castle P: **Public-private partnerships and sustainable agricultural development.** *Sustainability* 2011, **3**:1064–1073.

122. **Plant Breeding and Related Biotechnology Capacity Assessment.** [http://gipb.fao.org/Web-FAO-PBBC/index.cfm?where = 02].

123. **Plant Breeding Studies.** [http://km.fao.org/gipb/capacity-assessments/ country-case-studies/en/].

124. Guimaraes EP, Kueneman E, Carena MJ: **Assessment of national plant breeding and biotechnology capacity in Africa and recommendations for future capacity building.** *HortScience* 2006, **41**:50–52.

125. Bliss FA: **Education and preparation of plant breeders for careers in global crop improvement.** *Crop Sci* 2007, **47**:S250–S261.

126. Gepts P, Hancock J: **The future of plant breeding.** *Crop Sci* 2006, **46**:1630–1634.

127. Repinski SL, Hayes KN, Miller JK, Trexler CJ, Bliss FA: **Plant Breeding Graduate Education: Opinions about Critical Knowledge, Experience, and Skill Requirements from Public and Private Stakeholders Worldwide.** *Crop Sci* 2011, **2011**:2325–2336.

128. Guner N, Wehner TC: **Survey of U.S. land-grant universities for training of plant breeding students.** *Crop Sci* 2003, **43**:1938–1944.

129. Knight J: *A Dying Breed. Nature* 2003, **421**:568–570.

130. Hancock JF: **Introduction to the symposium.** *HortScience* 2006, **41**:28–29.

131. Morris M, Edmeades G, Pehu E: **The global need for plant breeding capacity: What roles for the public and private sectors?** *HortScience* 2006, **41**:30–39.

132. Baenziger PS: **Plant breeding training in the U.S.** *HortScience* 2006, **41**:40–44.

133. Bliss F: **Plant breeding in the U.S. private sector.** *HortScience* 2006, **41**:45–47.

134. Khush GS: **Impact of international centers on plant breeding training.** *HortScience* 2006, **41**:48–49.

135. Ransom C, Particka C, Ando K, Olmstead J: **Report of breakout group 1. What kind of training do plant breeders need, and how can we most effectively provide that training?** *HortScience* 2006, **41**:53–54.

136. Weebadde C, Mensah C: **Report of breakout group 2. How will we provide improved varieties of specialty, minor and subsistence crops in the future?** *HortScience* 2006, **41**:55.

137. Terpstra K, Oraby H, Vallejo V: **Report of breakout group 3. How can the public and private sectors most effectively partner to train new generations of plant breeders?** *HortScience* 2006, **41**:56.

138. The African Centre for Crop Improvement [http://www.acci.org.za/main.asp?nav = 13].

139. The West Africa Centre for Crop Improvement [http://www.wacci.edu.gh/index.html].

2

Climate-smart agriculture global research agenda: scientific basis for action

Kerri L Steenwerth[1*], Amanda K Hodson[2], Arnold J Bloom[3], Michael R Carter[4], Andrea Cattaneo[5], Colin J Chartres[6], Jerry L Hatfield[7], Kevin Henry[8,9], Jan W Hopmans[2], William R Horwath[2], Bryan M Jenkins[10], Ermias Kebreab[11], Rik Leemans[12], Leslie Lipper[13], Mark N Lubell[14], Siwa Msangi[15], Ravi Prabhu[16], Matthew P Reynolds[17], Samuel Sandoval Solis[2], William M Sischo[18], Michael Springborn[19], Pablo Tittonell[20], Stephen M Wheeler[21], Sonja J Vermeulen[22], Eva K Wollenberg[23], Lovell S Jarvis[24] and Louise E Jackson[2]

Abstract

Background: Climate-smart agriculture (CSA) addresses the challenge of meeting the growing demand for food, fibre and fuel, despite the changing climate and fewer opportunities for agricultural expansion on additional lands. CSA focuses on contributing to economic development, poverty reduction and food security; maintaining and enhancing the productivity and resilience of natural and agricultural ecosystem functions, thus building natural capital; and reducing trade-offs involved in meeting these goals. Current gaps in knowledge, work within CSA, and agendas for interdisciplinary research and science-based actions identified at the 2013 Global Science Conference on Climate-Smart Agriculture (Davis, CA, USA) are described here within three themes: (1) farm and food systems, (2) landscape and regional issues and (3) institutional and policy aspects. The first two themes comprise crop physiology and genetics, mitigation and adaptation for livestock and agriculture, barriers to adoption of CSA practices, climate risk management and energy and biofuels (theme 1); and modelling adaptation and uncertainty, achieving multifunctionality, food and fishery systems, forest biodiversity and ecosystem services, rural migration from climate change and metrics (theme 2). Theme 3 comprises designing research that bridges disciplines, integrating stakeholder input to directly link science, action and governance.

Outcomes: In addition to interdisciplinary research among these themes, imperatives include developing (1) models that include adaptation and transformation at either the farm or landscape level; (2) capacity approaches to examine multifunctional solutions for agronomic, ecological and socioeconomic challenges; (3) scenarios that are validated by direct evidence and metrics to support behaviours that foster resilience and natural capital; (4) reductions in the risk that can present formidable barriers for farmers during adoption of new technology and practices; and (5) an understanding of how climate affects the rural labour force, land tenure and cultural integrity, and thus the stability of food production. Effective work in CSA will involve stakeholders, address governance issues, examine uncertainties, incorporate social benefits with technological change, and establish climate finance within a green development framework. Here, the socioecological approach is intended to reduce development controversies associated with CSA and to identify technologies, policies and approaches leading to sustainable food production and consumption patterns in a changing climate.

* Correspondence: kerri.steenwerth@ars.usda.gov
[1]Crops Pathology and Genetics Research Unit, Agricultural Research Service, United States Department of Agriculture (ARS/USDA), c/o Department of Viticulture and Enology, RMI North, Rm. 1151, 595 Hilgard Lane, Davis, CA 95616, USA
Full list of author information is available at the end of the article

Introduction

Globally, agricultural and forestry systems are expected to change significantly in response to future climate change, manifesting as major transitions in livelihoods and landscapes [1-4]. During the few past decades, crop yields have been reduced because of warming [5], and the results of modelling studies suggest that climate change will reduce food crop yield potential, particularly in many tropical and midlatitude countries [6-9]. Rising atmospheric CO_2 concentrations will decrease food and forage quality [10]. Price and yield volatility likely will continue to rise as extreme weather continues, further harming livelihoods and putting food security at risk [11]. Global demand for agricultural products, be they food, fibre or fuel, continues to increase because of population growth, changes in diet related to increases in per capita income and the need for alternative energy sources while there is less and less additional land available for agricultural expansion. Agriculture thus needs to produce more on the same amount of land while adapting to a changing climate and must become more resilient to risk derived from extreme weather events such as droughts and floods.

The term *climate-smart agriculture* (CSA) has developed to represent a set of strategies that can help to meet these challenges by increasing resilience to weather extremes, adapting to climate change and decreasing agriculture's greenhouse gas (GHG) emissions that contribute to global warming (Figures 1 and 2). CSA also aims to support sustainable and equitable transitions for agricultural systems and livelihoods across scales, ranging from smallholders to transnational coalitions. Forming a core part of the broader green development agenda for agriculture [12-14], CSA focuses on meeting the needs of people for food, fuel, timber and fibre through science-based actions; contributing to economic development, poverty reduction and food security; maintaining and enhancing the productivity and resilience of both natural and agricultural ecosystem functions, thus building natural capital; and reducing the trade-offs involved in meeting these goals. It invokes a continuous, iterative process for stakeholders, researchers and policymakers to meet the challenges presented by climate change and collectively transform agricultural and food systems towards sustainability goals [15]. Increased awareness and adaptive management are essential components of the CSA strategy. Yet, CSA is controversial. Such a broad agenda can be appropriated to support conflicting agendas or promote specific ecosystem services [16]. GHG emission mitigation by resource-poor farmers raises equity as an issue in developing countries because it may bring farmers little benefit unless it directly provides them with adaptive capacity. Setting CSA in the context of a safe operating space for humanity with socioecological systems

that support adaptive management and governance will require scientific metrics and science–policymaking dialogues [16] that depend on strong engagement of the scientific community.

At the 2013 Global Science Conference on Climate-Smart Agriculture (Davis, CA, USA), participants examined the state of global science and best practices concerning climate and agriculture worldwide. Participants built on the consensus achieved at the 2011 Global Science Conference on Climate-Smart Agriculture conference (Wageningen, the Netherlands), agreeing on a broad strategy for science and policymaking to strengthen food security, mitigation and adaptation [17]. Participants further examined current gaps in knowledge, identified existing and promising work within CSA and formulated agendas for interdisciplinary research and science-based actions to support CSA.

The relationship between vulnerability, resilience and adaptation was an overarching theme echoed across the conference and is crucial to CSA. *Vulnerability* describes exposure, sensitivity and capacity to respond to negative impacts of climate change, and *adaptation* is the means by which to reduce the vulnerability. Here *resilience* is regarded as the capacity to tolerate disturbance, undergo change and retain the same essential functions, structure, identity and feedback and is not indicative solely of returning to the same state that existed prior to a perturbation or disturbance [18-20]. Resilience focuses on factors that enable functioning despite adverse conditions [21,22], provides a means of framing the dynamic relationships between humans and the environment (socioecological systems) and considers society's capacity to manage change [23]. Thus, the principle of resilience can guide transformative change needed to meet the demands of food security, natural resource protection, and development, as well as to diminish vulnerability and promote adaptation (or adaptive capacity).

The recent increase in extreme weather events (climate shocks) threatens disruptive impacts on agriculture [24,25]. Projected adaptive actions include improving plant performance (for example, nutrition, yields, food quality) in response to elevated CO_2 and rising temperatures [26-28]; avoiding pest damage and food waste [28,29]; developing forecasting, management and insurance options to decrease the risk due to unexpected rainfall patterns, higher temperatures and shifting length in growth seasons [14,28,30]; and managing natural resources at the landscape and regional levels to assure the environmental quality and ecosystem services upon which agriculture depends [31-33]. Solutions involve trade-offs. For instance, planning now for higher temperatures and declining precipitation in arid zones may reduce water deficits for agriculture, but it will require institutional investment to support both the intensified demand for ground and

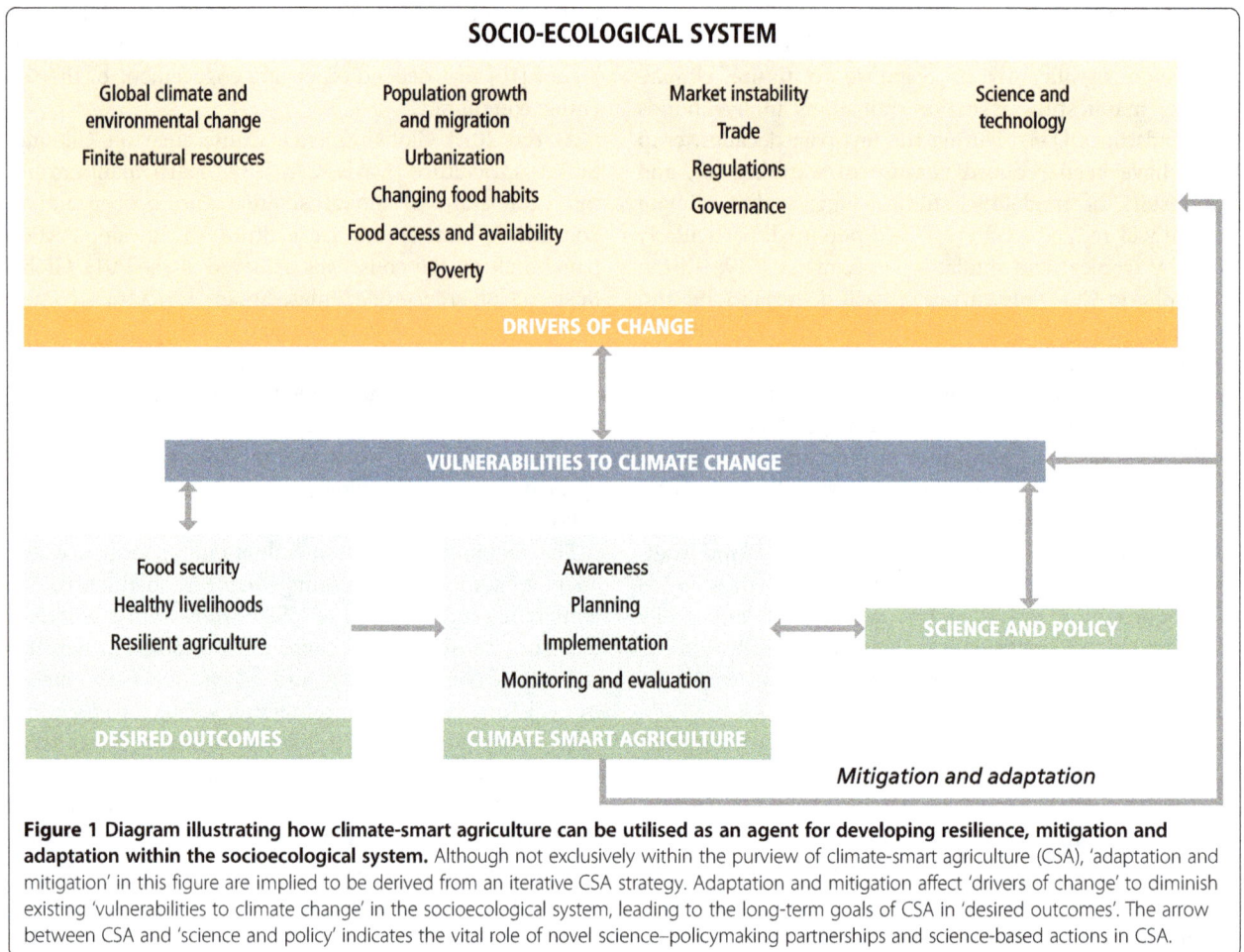

Figure 1 Diagram illustrating how climate-smart agriculture can be utilised as an agent for developing resilience, mitigation and adaptation within the socioecological system. Although not exclusively within the purview of climate-smart agriculture (CSA), 'adaptation and mitigation' in this figure are implied to be derived from an iterative CSA strategy. Adaptation and mitigation affect 'drivers of change' to diminish existing 'vulnerabilities to climate change' in the socioecological system, leading to the long-term goals of CSA in 'desired outcomes'. The arrow between CSA and 'science and policy' indicates the vital role of novel science–policymaking partnerships and science-based actions in CSA.

surface waters [34,35] and the necessary improvements in irrigation efficiencies [36]. Along with these adaptive actions, CSA seeks to contribute to the mitigation and reduction of GHG, mainly nitrous oxide (N_2O) and methane (CH_4) emissions, and to balance trade-offs with food security and livelihoods [7,37,38]. For example, combining agroforestry, afforestation and conservation efforts with agriculture to meet global food demand will help to mitigate GHG emissions, support biodiversity and concomitantly preserve ecosystem services [39,40]. Other trade-offs that occur when abrupt environmental changes stress agricultural systems include changes in rural and urban human migration patterns, as well as loss of cultural resources, which reduces the ability to manage land use effectively [41-43].

Without doubt, the development status of a country or region will influence the approach to mitigating and adapting to climate uncertainty and will affect the implementation and focus of the CSA strategy. For example, industrialized nations focus more strongly on mitigation of climate change through reduction of agriculture's environmental impacts, whereas developing countries'

approaches to climate uncertainty emphasize stabilizing and boosting food production, improving incomes and building adaptive capacity [7,15,44]. Gender can also influence decisions and capacity for mitigation and adaptation. Women in some regions in Africa have experienced greater exposure and vulnerability, especially to extreme events, than men, but they also have demonstrated greater collective action in farming decisions linked to social networking [45,46].

Crucial science questions and challenges for food systems in the face of climate change and uncertainty require comprehensive, collaborative investments and science-based actions. In the past few years, policies and programmes have included landscape-scale research on food security and natural resources, policy and governance to achieve agricultural resilience to climate change and capacity building [47]. Under CSA, transformative changes to achieve food security, poverty relief, mitigation and adaptation target novel types of science–policymaking partnerships and involve stakeholders and decision-makers in the public and private sectors to gain long-term commitment and investment to carry the new actions

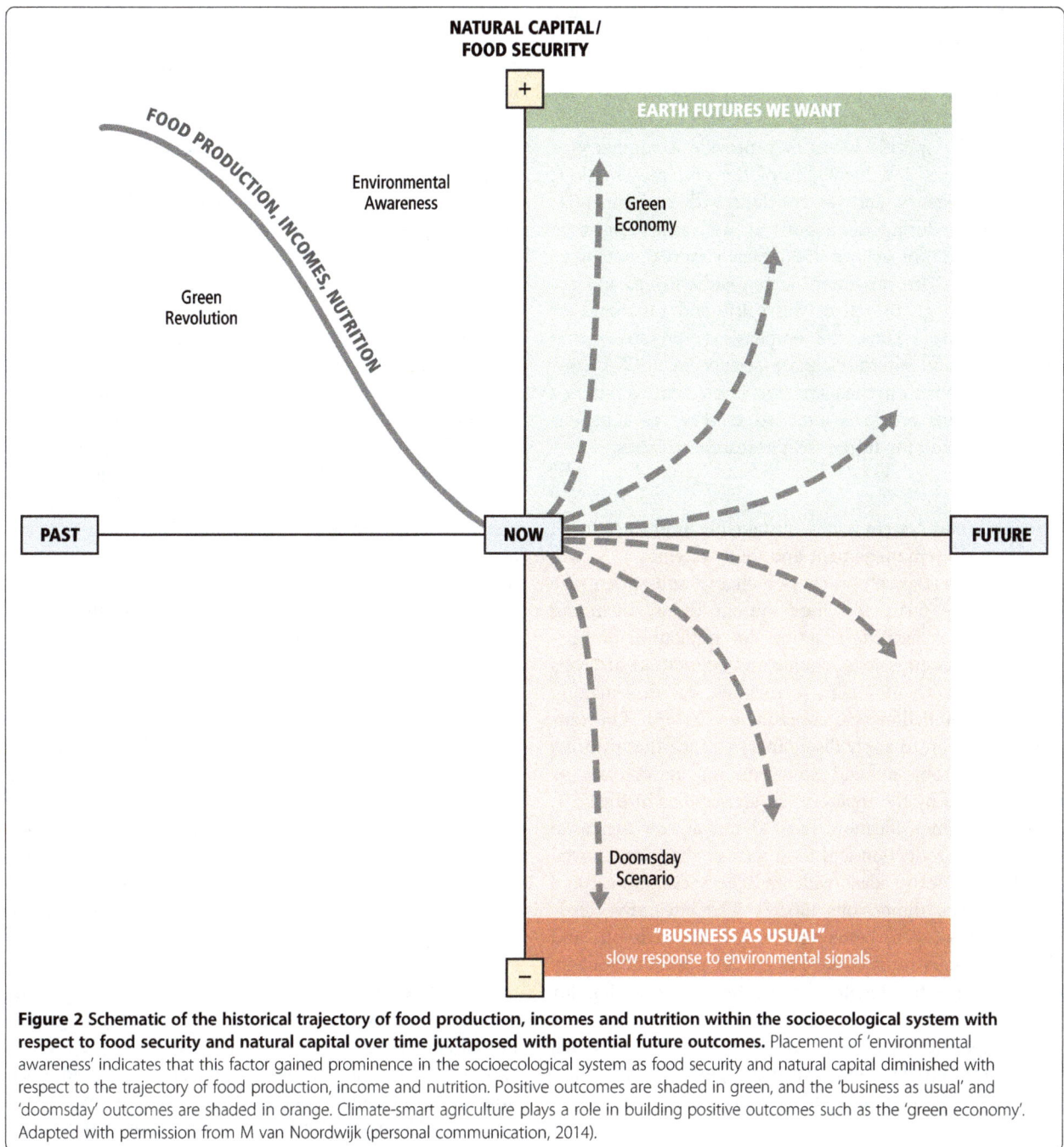

Figure 2 Schematic of the historical trajectory of food production, incomes and nutrition within the socioecological system with respect to food security and natural capital over time juxtaposed with potential future outcomes. Placement of 'environmental awareness' indicates that this factor gained prominence in the socioecological system as food security and natural capital diminished with respect to the trajectory of food production, income and nutrition. Positive outcomes are shaded in green, and the 'business as usual' and 'doomsday' outcomes are shaded in orange. Climate-smart agriculture plays a role in building positive outcomes such as the 'green economy'. Adapted with permission from M van Noordwijk (personal communication, 2014).

to fruition. CSA emphasizes the involvement of scientists with farmers, land managers, agroforesters, livestock keepers, fishers, resource managers and policymakers (stakeholders) to empower them in the formation of palatable choices to enact adaptive capacity and resilience 'on the ground' and within broader policies [14,15]. Farmer-led innovative approaches and social learning are crucial parts of this process, where social learning represents a 'change in understanding that goes beyond the individual to become situated within wide social units or communities of

practice through social interactions between actors within social networks' [48,49].

In this article, we summarize and synthesize the discussions and ideas presented at the 2013 CSA conference by an international community of scientists, growers, policymakers, research scientists, government officials, nonprofit entities and students who are working to achieve food security, poverty reduction, mitigation and adaptation within the CSA context. The three sections of this article reflect the scientific themes presented at the conference: (1) farm and food systems, (2) landscape and regional issues and (3) institutional and policy-related aspects. Within the first and second themes, parallel sessions at the conference charged participants to identify knowledge gaps, research initiatives and transformative actions required to address these specific issues. We provide a summary of the 12 sessions and highlights of the oral presentations by subject experts, and we conclude with recommendations offered during discussions as well as a consensus agenda for future actions [50]. Finally, broad outcomes and messages are presented, largely adhering to the actual proceedings to reflect the spirit and outcomes of this conference. Thus, the emphasis is on structuring disciplinary and interdisciplinary science in a CSA context rather than mechanisms for implementing science in action. This article is intended to serve as a benchmark and guide for future CSA research activities.

Theme 1
Farm and food system issues: sustainable intensification, agroecosystem management and food systems

Considerable research on climate change and agriculture exists at the farm and food system levels, including topics such as farming practices for mitigation of agricultural GHG emissions, choice and adaptation of crops and livestock to new climate regimes, decision-making by farmers and life-cycle assessments [51-55]. The tendency has been to apply disciplinary science that informs particular problems and solutions for agriculture, as demonstrated by the topics of the six sessions in theme 1. Sustainable intensification, focused initially on increased agricultural production and food security, has now moved to a broader set of goals with multiple social, ethical and environmental dimensions [56,57]. The integrative challenge for CSA is to better understand the trade-offs and choices farmers must make for greater multifunctionality and resilience to climate change. Because planning for climate change can be highly farm-, commodity- and context-specific, especially in response to extreme events, CSA is committed to new ways of engaging in participatory research and partnerships with producers [14].

Crop physiology and genetics under climate change

Responding to effects of climate change (for example, changes in nutrient availability and plant nutrient acquisition, higher CO_2 concentrations and temperatures, water deficits and flooding) that influence the closure of the yield gap between potential and actual production will require continuation of existing 'best management practices' coupled with improvements in agronomic management practices and crop-breeding [58,59]. Uncertain is the degree to which advances in crop physiology and genetics will continue to support higher agricultural production in response to more frequent

climate shocks. Whereas successful crop adaptation to new production locations may be a good predictor of future outcomes, much higher CO_2 concentrations and temperatures are conditions beyond our current set of experiences [21,60]. Molecular approaches and genetic engineering will foster better understanding and manipulation of physiological mechanisms responsible for crop growth and development, as well as the breeding of stress-adapted genotypes [61-63], but there are social controversies surrounding the use of some of these technologies. High-throughput phenotyping platforms and comprehensive crop models will lead to more rapid exploration of genetic resources, enabling both gene discovery and better physiological understanding of how crop improvement can increase tolerance to environmental stress [64-68]. Development of new crop genotypes to meet the need to thrive under future management and climate conditions, the expected increases in the frequency of climate shocks and the uncertainty of rates of climate change presents a challenge. The specific examples set forth in the following paragraphs demonstrate how greater understanding of biochemical pathways, plant traits and phenotypes and germplasm evaluation could help overcome bottlenecks in both yield and development of physiological resilience to environmental stresses.

Molecular approaches provide opportunities to establish linkages between biochemical pathways and physiological responses. In cereals such as rice, grain yield is highly dependent on the carbohydrate source (top leaves) and sink (florets) relationship, which is strongly influenced by the plant hormone cytokinin [69]. Cytokinin production also affects drought tolerance and senescence, and isopentenyl transferase (IPT) expression controls upregulation of pathways for cytokinin degradation. Therefore, it follows that tolerance of abiotic stress by delaying stress-induced senescence through manipulation of IPT expression in transgenic lines could maintain optimal levels of cytokinin, resulting in greater fitness and more seed and grain production [62]. When exposed to varying drought intensities pre- and post-flowering, transgenic rice with higher IPT expression maintained consistently higher grain yields and concentrations of sucrose and starch compared to the wild-type genotype. The delayed onset of drought-related symptoms in the transgenic lines caused positive source–sink relationships for a relatively longer period with higher photosynthetic rates than the wild type.

Combinations of multiple plant traits to survive stress, however, may produce more resilient crop production in the face of climate change [64]. Survival strategies employed by plants include early flowering to escape drought periods, stomatal control to prevent water loss, enhanced root growth in deeper soil layers to access water [70] and reduced leaf growth to minimize the transpiring surface [71]. These adaptations come at a cost, where reductions in the growth cycle, light interception and carbon (C) assimilation by photosynthesis are often accompanied by a higher C requirement to build additional plant roots, especially under nutrient stress [72]. Thus, the trade-offs of introducing new plant traits must be considered for specific types of environmental stress [65].

By examining the genetic basis of physiological mechanisms and environmentally induced stress responses, crops such as maize, wheat and other cereals can be bred to produce better yields and tolerances through targeted accumulation of alleles that confer robust responses to environmental stressors such as drought [73] (Figure 3). This approach is used by the International Maize and Wheat Improvement Center for the discovery and accumulation of drought-adaptive traits in wheat and maize germplasm from wild-type crop relatives and cultivars from a wide range of climates and growing conditions [65,67,74]. Screening for physiological traits can be highly effective in selecting such lines for a breeding programme. Canopy temperature (CT) is an example of a widely used, high-throughput germplasm screening tool. CT is linked to stomatal conductance, an indirect indicator of water uptake by roots, especially under drought and heat stress [75,76]. In one study, researchers found that 60% of variation in yield from recombinant inbred lines grown under drought conditions was explained by CT [77]. Screening for physiological traits in candidate genotypes as an initial step may thus accelerate the search for novel genes [75] and genotypes that will be needed to deal with rapid changes in climate, such as the greater intensity and frequency of drought. Trait-based breeding programmes will be most effective when approaches are developed to simultaneously screen a broad array of genotypes for phenotypic responses to environmental stresses quickly (for example see, [78]).

Complementary approaches are necessary for solving complex physiological plant responses to climate and management. Changes in temperature, precipitation, water delivery, salinity and CO_2 concentrations will occur simultaneously. Direct experimentation, high-throughput screening platforms using molecular-based techniques and predictive modelling are a set of tools for achieving multiple goals [79-81], which include exploration of genetic resources for broader use and dissemination, gene pool expansion and yield stability in the face of interannual weather variation. In addition, these tools can help with other crop selection criteria, including quality of food and feed, source–sink relationships, pest and disease resistance, plant–microbe interactions that reduce CH_4 and N_2O emissions, and postharvest storage [60,81]. Regional networks that examine environmental and physiological tolerances and yield potentials, as well as their coalescence into global crop improvement networks [82], will provide large-scale screening approaches to assessing both germplasm and phenotypic responses of crop plants. These networks already exist in representative target environments, such as the Network for the Genetic Improvement of Cowpea for Africa, Sorghum and Millet Networks, International Wheat Improvement Network, International Maize Improvement Network; and other regional networks linked to CGIAR that focus on grains and legumes in Africa, Latin America, the Caribbean and Asia. They also include networks for research and extension supported by Association for Strengthening Agricultural Research in Eastern and Central Africa. Participatory breeding by farmers and other stakeholders will eventually be an essential way to advance this agenda [83,84].

Livestock management and animal health
Livestock production not only contributes to climate change via GHG emissions (see [85]) but also suffers due to extreme weather events and disease related to

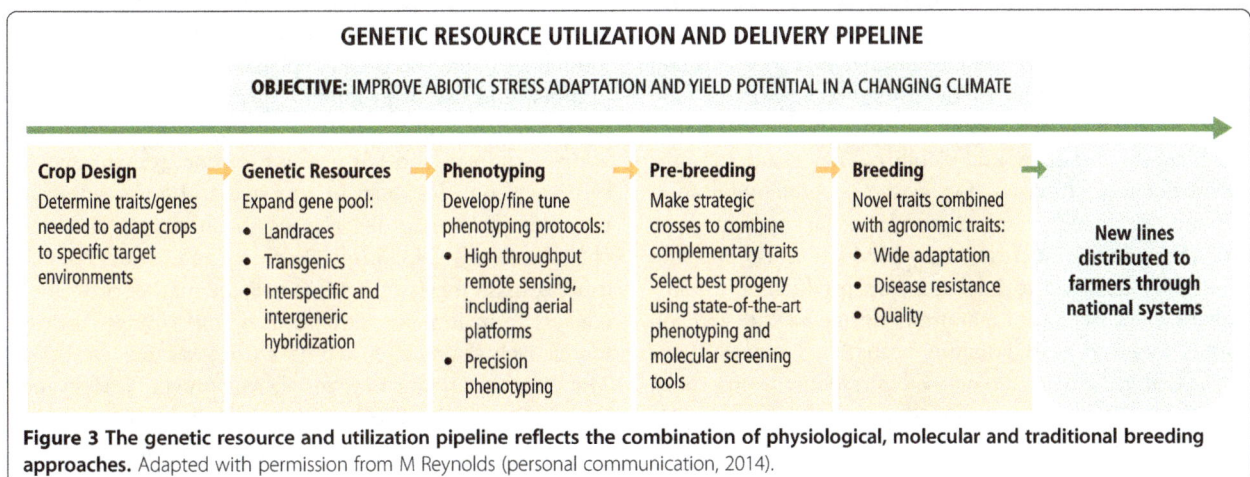

GENETIC RESOURCE UTILIZATION AND DELIVERY PIPELINE

OBJECTIVE: IMPROVE ABIOTIC STRESS ADAPTATION AND YIELD POTENTIAL IN A CHANGING CLIMATE

Crop Design
Determine traits/genes needed to adapt crops to specific target environments

Genetic Resources
Expand gene pool:
- Landraces
- Transgenics
- Interspecific and intergeneric hybridization

Phenotyping
Develop/fine tune phenotyping protocols:
- High throughput remote sensing, including aerial platforms
- Precision phenotyping

Pre-breeding
Make strategic crosses to combine complementary traits
Select best progeny using state-of-the-art phenotyping and molecular screening tools

Breeding
Novel traits combined with agronomic traits:
- Wide adaptation
- Disease resistance
- Quality

New lines distributed to farmers through national systems

Figure 3 The genetic resource and utilization pipeline reflects the combination of physiological, molecular and traditional breeding approaches. Adapted with permission from M Reynolds (personal communication, 2014).

climate change. Direct and indirect challenges in both mitigation and adaptation include fluctuating feed prices, habitat changes, expansion of vector-borne diseases in warm climates, impaired reproduction, pasture quality and availability and physiological heat stress [86,87]. Opportunities for mitigating emissions include dietary manipulation, genetic improvement and mortality reduction to enhance overall production potential; manure management; and reduction of deforestation and pasture burning through payments for ecosystem services [88,89]. Adaptation strategies include income and livelihood diversification by mixing crop and livestock production; sustainable intensification through pasture regeneration or destocking; diversifying livestock feeds; manipulation of rumen microbial composition; matching animal breeds to local environments and moving animals to other sites; and better risk management and transformative change (for example, exit from or entry into animal agriculture) [88,90-92]. These strategies rely heavily on sustainable intensification, as in the improvement of productivity and efficiency that exists in conjunction with incentives and investments that allow systems to intensify and in the development of regulations and limits on intensifying systems, among other aspects [93]. Access to credit or savings, land and resource inputs, and livelihood diversification are other potential pathways towards adaptation and food security [94,95]. Technology, supporting policies and investments will require varied mixtures of strategies, as shown by the examples described in the following paragraphs.

Flexibility in livelihood options for pastoralist, agropastoralist and ranching communities can increase a household's capacity to manage risk and adapt in the face of burgeoning external stress [96]. Adaptation options depend on household objectives and attitudes; local access to natural resources, inputs and output markets; and sustainable intensification. Nutrient management is fundamental to maintaining a livelihood in livestock production. In Madagascar, external nitrogen (N) inputs are not commonly used to replenish the N losses that occur through erosion, leaching, GHG emissions and harvest. Hence, Alvarez et al. [97] examined N flows through crop-livestock systems to determine management scenarios leading to improvement in their N use efficiency, productivity and economic viability. They evaluated four intensification scenarios for system productivity, food self-sufficiency and gross margins: (1) using supplementary feed (N inputs) to increase dairy production; (2) applying mineral N fertilizer to increase crop production; (3) improving conservation of manure N during storage and soil application; and (4) combining scenarios 1 and 3. They found that gross margin increased in response to improved retention of manure N and that increased N supply through supplementary feeding (scenario 4) across farm types led to increases in whole-farm N use efficiencies

from 2% to 50%, in N cycling from 9% to 68% and in food self-sufficiency from 12% to 37%. An example of adaptation to manage risk in East Africa is pastoralists who have shifted from cows to camels, which are better-adapted to survive periods of water scarcity and able to consistently provide more milk [98]. Risk adaptation by farmers may also involve changing from cultivated crops to livestock, as crops may be more environmentally and spatially constrained in the pastoralists' home regions [99].

Mitigation options at farm to regional scales form a large part of Brazil's multifaceted approach to managing direct and indirect GHG emissions from livestock. Brazil's commercial cattle industry is the largest in the world (more than 170 million head in 2006), and emissions from raising cattle are responsible for about half of the country's total emissions [100]. The principal targets for mitigating GHG emissions associated with cattle production in Brazil are reduction in deforestation and enteric fermentation, regeneration of secondary forest, recuperation of degraded pasture and soils and elimination of fire in pasture management. Maintenance of grazing productivity and high stocking rates through pasture reclamation and adoption of integrated crop-livestock systems, such as rotational grazing and introduction of legumes in pastures, buffers pressure on deforestation. Such pasture regeneration creates a potential for increasing soil C storage, with increases of up to 0.72 Mg of $C \cdot ha^{-1} \cdot yr^{-1}$ reported under improved management [101]. However, other pasture maintenance practices increase emissions. For example, burning accounted for 1.69 CO_2eq (Mt from total biome) in the Cerrado ecosystem from 2003 to 2008. Key mitigation efforts include reduction in enteric CH_4 emissions by genetic stock improvement and dietary manipulation [91]. This dietary manipulation through grain supplementation increases forage digestibility and reduces enteric fermentation, but it leads to greater emissions of N_2O through the use of fertilizers to grow the grain [100]. Several other promising technologies include grass and legume species with lower GHG emission potential, additives (for example, ionophores and secondary plant compounds such as tannins) and use of propionate precursors in feed to reduce methanogenesis [102]. To complement farm-based efforts, uniform and fair economic procurement and incentivized policies must be in place and enforced across the supply chain in order to establish supply and trade chains with low C footprints. Regional and national policies must contain mechanisms that balance market pressure to convert from low-impact land uses (for example, forests) to relatively more intensive uses (for example, ranching). The Norwegian Agency for Development Cooperation (NORAD) and the Brazilian organization Aliança da Terra, which includes farmers, researchers and agribusiness entrepreneurs, are partnering to increase contributions by private landowners throughout the Brazilian Amazon and Cerrado to the goals of the Reducing Emissions from Deforestation and Forest Degradation (REDD+) programme by combining sustainable rangeland economic development with forest and water protection [103].

Effectively managing emerging zoonotic diseases and outbreaks due to climate change is a strong component of maintaining agropastoralist livelihoods. Increased temperatures due to climate change will affect the survival of pests in the winter and thus distribution of pests and diseases (for example, zoonotic, endemic, emerging, foodborne and noninfectious diseases), though some regions may find relief from these existing pressures in a changing climate. The Emerging Pandemic Threats Program (PREDICT) addresses the broad geographic issues in disease emergence from farm to national to global scales related to a shifting climate [104,105]. The programme, which is run within the US Agency for International Development, leverages existing networks within national and local governments as well as networks of scientists and specialists involved in outbreak reporting, microbial characterization and pathogen discovery [104]. The programme includes 20 developing countries where hotspots of emerging infectious diseases exist and is focused on surveillance of human–animal interfaces where transmission is more likely. PREDICT is focused on the prevention of pandemics by addressing underlying ecological, economic and social drivers of change, such as shifts in land use from forested systems to livestock. It is used to develop and deliver new technologies (such as information management and communication tools) to improve control efforts close to the pathogen's source. This type of interdisciplinary effort that moves from farm-level to broad spatial scales is considered necessary for creating comprehensive strategies for the control and prevention of emerging zoonotic diseases in a changing climate [104].

In support of agropastoral farming systems, models must integrate mitigation options, alternative intensification pathways, zoonotic disease and vector ecology (for example, genetic shifts, patterns of emergence); mechanisms of effecting behavioural change; and adaptation to future climate change scenarios. Some existing models for predicting regional GHG emissions from livestock production include BEEFGEM (Ireland; [106]), IFSM (USA; [107]) and SIMS$_{DAIRY}$ (UK; [108]). Reisinger et al. [109] recently evaluated different metrics on the integrated assessment model, MESSAGE, and the land-use model, the Global Biosphere Management Model (GLOBIOM), to examine the global costs of abatement strategies used to reduce the magnitude of climate change and subsequent effects on regional food production and supply prices for livestock products and other agricultural commodities. Other transformative approaches to livestock

production include identifying the value of a blend of market-orientated smallholders vs. large-scale farms, evaluating ecosystem services payments as a means of income diversification, forming institutional and market mechanisms for reaching smallholders to foster technological change, finding the best locations for both livestock production and marginal land rehabilitation, and creating new capacity of the livestock sector for mitigation and adaptation in the face of climate change [90,110] (Figure 4). The social and economic impacts (for example, income stability, human nutrition, product value chains, transaction and opportunity costs of other alternatives) of landsparing and reducing livestock consumption, two recently suggested mitigation options, merit further investigation, especially with respect to gender, region and income differentiation [92].

Nitrogen management: agricultural production, greenhouse gas mitigation, and adaptation

Future food security will continue to rely on N fertilizer inputs, but cropping systems must achieve yield potential (that is, close the yield gap) while minimizing trade-offs in air, water and soil quality [58,59]. The long-term ramifications of N-related GHG emissions; off-site movement of N on eutrophication, acidification and pollution of aquatic and terrestrial ecosystems; and human health problems have led to a recommendation that anthropogenic inputs of reactive N to terrestrial ecosystems be reduced by up to one-fourth of present quantities, or about 35 million tonnes of N per year [112]. Even if this reactive, anthropogenic N entering agroecosystems is emitted as N_2 rather than N_2O, the energy associated with the Haber-Bosch process and transport of fertilizers will still contribute to GHG emissions [113]. Cropping system diversification, careful selection of crop rotations to reduce nutrient loss, and improved soil organic matter content are means by which to promote sustainable intensification. Yet, this often involves a set of complex trade-offs for producers and their livelihoods [114], emphasizing the need for a CSA strategy that involves stakeholders from the beginning to develop viable scenarios that include both mitigation and adaptation to climate change. The examples presented here demonstrate how strategies for N fertilization practices provide both mitigation and adaptation benefits by decreasing GHG emissions, reducing reliance on synthetic mineral fertilizer and enhancing food security.

Enhanced-efficiency fertilizers (EEFs), such as slow-release fertilizers or those containing nitrification inhibitors and urease inhibitors, hold potential to mitigate GHG emissions. According to the Intergovernmental Panel on Climate Change (IPCC) Fourth Assessment Report [114], the mean mitigation potential of N_2O by nutrient management using nitrification inhibitors and slow-release fertilizers has been estimated to be 0.07 t

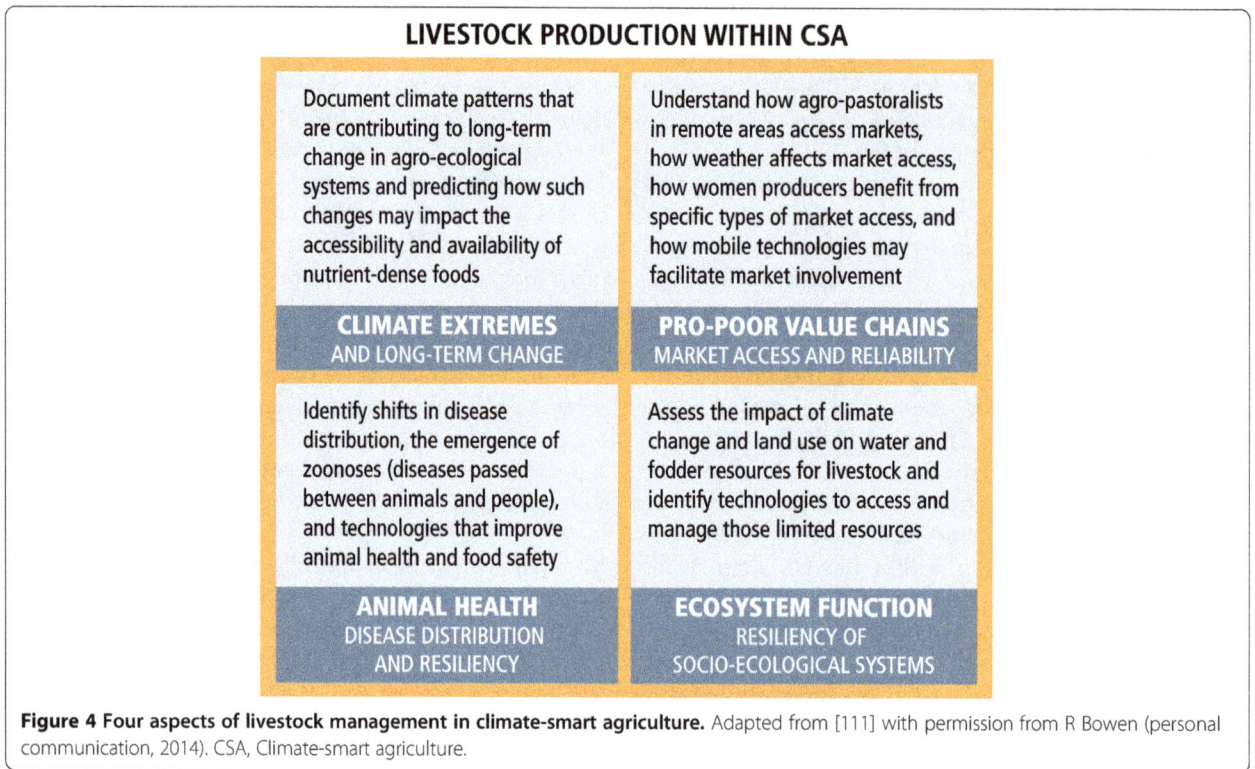

Figure 4 Four aspects of livestock management in climate-smart agriculture. Adapted from [111] with permission from R Bowen (personal communication, 2014). CSA, Climate-smart agriculture.

$CO_2\text{-eq}\cdot ha^{-1}\cdot yr^{-1}$ (as a reference, agriculture accounted for an estimated 5.1 to 6.1 $GtCO_2eq\cdot yr^{-1}$ in 2005, which amounts to 10% to 12% of total global anthropogenic emissions of GHGs). In practice, N_2O emissions decreased by 54% from a no-till corn–dry bean rotation receiving urea, urease and nitrification inhibitors in comparison to a urea-only application in Colorado (USA) [115]. According to a recent global meta-analysis of enhanced-efficiency fertilizers, nitrification inhibitors can reduce N_2O emissions by 38% and polymer-coated fertilizers by 35%, on average, compared to conventional fertilizer, but urease inhibitors alone are not as effective in reducing N_2O emissions [116]. Nitrification inhibitors are compatible with both chemical and organic fertilizers, making them a seemingly attractive mitigation option, but their efficacy varies with edaphic factors. For example, EEF materials were applied to rainfed corn in the central Corn Belt (Midwest region, USA), a more humid region than Colorado [117]. Although all EEF treatments had lower cumulative emissions than the treatment that did not include EEFs, episodic N_2O emissions from EEF treatments corresponded to rainfall patterns, and the relative effectiveness among EEF materials was similar. Together, these findings suggest that the impact of EEF materials may be diminished in rainfed agriculture systems compared to irrigated systems with regulated water availability. Although yield responses to EEF materials may also vary with respect to crops and location, consistent yield increases in corn (central

Corn Belt) grown with EEF materials were reported to occur as a result of the increased duration of photosynthetic leaf area during grain-filling [118]. In microirrigation systems, the results are less impressive, likely due to the increased efficiency of EEF where fertilization by fertigation matches crop needs more precisely, leaves less residual fertilizer and avoids its loss [119]. With this emerging evidence that use of EEF materials could have a positive impact on crop production and limit N_2O emissions, the results of research on understanding the conditions for which these materials are useful could underpin both the development of risk assessment tools and the feasibility of grower adoption of these technologies.

Mitigating GHG emissions through C sequestration depends on the stability of soil C pools. Declining productivity in the rice–wheat cropping systems of India's Indo-Gangetic plains has been attributed to reductions in soil C [120]. Mandal *et al.* [121] found that, to combat this, addition of NPK fertilizer during double-rotations of rice led to increases in soil organic C stocks compared to adding just N or NP alone [121]. When compost was applied during rice production, as much as 29% of compost-derived C was stabilized [121]. This was attributed to high lignin and polyphenol content in crop residue and compost and also to the diminished soil C decomposition stemming from anaerobic conditions due to soil submergence under rice cultivation. Crop residue management improves poor soil fertility through soil organic matter accumulation, leading to reductions in soil N loss by leaching and gaseous emissions; in many situations in developing countries, however, crop residues are used to feed animals, to provide fuel for cooking or are turned into biochar [122,123]. Developers of mitigation strategies for increasing soil C and decreasing N_2O emissions have to take into account the dynamics of crop residue, tillage and nutrient management, along with climate, in order to evaluate the efficacy of different practices across locations [124].

Legumes, a form of ecological intensification, offer both mitigation and adaptation options, especially to smallholder farms susceptible to deficits in soil fertility, climatic uncertainty and reduced economic access to agricultural inputs such as mineral fertilizer. The biologically fixed N from legumes is often tightly synchronized with plant N demand and has a much lower C footprint than industrially produced synthetic N fertilizers [125]. For instance, intercropping with N-fixing trees in Sub-Saharan Africa were found not only to reduce reliance on fertilizers but also to enhance soil C sequestration and reduced N_2O emissions [126]. In this intercropping system, 10.9 Mg $C \cdot ha^{-1} \cdot yr^{-1}$ were sequestered in the soil. The potential for N_2O mitigation was only 0.12 to 1.97 kg $N_2O \cdot ha^{-1} \cdot yr^{-1}$ [126]. However, the authors of a review of 71 site-years of pasture, cropping and agroforestry systems indicated that providing N additions via legumes can increase accumulation of soil C at rates greater than can be achieved with other crops, such as cereals or grasses, even when they are supplied with N fertilizer [125]. Furthermore, intercropped mixtures of peas and barley (*Hordeum vulgare* L.), compared to the respective sole crops, were found to lead to effective weed suppression in weed communities across sites in Western Europe [127]. Adaptation options that include legumes to reduce dependence on fossil-fuel derived fertilizers include integration of intercropped or rotational legumes into management regimes, development and facilitation of access to new legume cultivars with broader stress tolerance and removal of barriers to legume use and consumption in the food system (for example, competing uses, seed availability, labour).

The design of more efficient N management strategies will only be conducive to climate change solutions if based on knowledge systems and participatory research with stakeholders to ensure viable action and adaptive management. Although decision-making support tools and metrics are being developed to aid producers in tempering N inputs for the desired outcomes of higher crop production (for example, quantity and nutritional quality) and lower environmental impacts [128], adoption is a major obstacle. When extension agents are involved in troubleshooting with and training of participants, the new knowledge systems that are created begin to

delineate clear pathways that benefit farmers' livelihoods. In regions dominated by smallholder farmers who are already experiencing climate impacts such as increased drought, flooding or heat waves, the priority is on adaptive measures for reliable N availability to support food security and minimize vulnerability. Combining low inputs of synthetic N fertilizers with practices that increase soil quality through organic matter management and acquisition of N from biological N fixation allows adaptation measures to contribute to GHG mitigation. However, synthetic N sources are fraught with constraints such as high cost, price fluctuations and availability, whereas biological N sources are affected by constraints of labour, time and physiological tolerance. Future food security also will depend on a substantial rate of yield gains for major cereal crops. Maintaining these yield increases above a 1% annual growth rate will require constant improvement in crop yields, stress avoidance and agronomic management to achieve physiological yield potential [129]. However, maintaining a compounding rate of yield increases is not consistent with historical trends and likely is not achievable without great effort [130]. Therefore, the limits of current crop productivity need to be estimated using potential yield and water-limited yield levels as benchmarks. Determining and closing the yield gap, especially in developing countries, is fundamental to achieving food security because variety improvement through breeding and genetic modification might be insufficient [129,131,132].

Farmer decision-making and barriers to the adoption of climate-smart agriculture practices

Climate change challenges farmers' decisions by altering risks and uncertainty and incorporating new information into their traditional knowledge-processing systems. The unfolding of the decision-making process and its translation into action depends on the socioecological context in which farmers are embedded. How well innovation models apply to all climate-related behaviours is a major question, especially given that governance regimes at the national and international levels strongly influence farmers' actions [133]. The massive literature on innovation systems has established the basic hypothesis that farmers evaluate the costs and benefits of different practices in light of information accessed through social networks and other communication channels. The diffusion of innovation model can provide critical insights into adoption decisions. In this model, adoption of innovations follows a sequence of stages: knowledge, persuasion, decision, implementation and confirmation [134]. Innovations generated by agricultural research are communicated by extension agents to farmers. This approach may place too much emphasis on traditional socioeconomic variables and ignore how other social factors (for example, networks, gender, social norms, values, climate-change attitudes), and uncertainty may be implicated by practices that are ostensibly consistent with CSA priorities (for example, adoption of new crops and cultivars or changes in N fertilization) [135-137].

Effective outreach strategies will manifest with greater understanding of farmers' beliefs about climate change and their readiness to respond to climate change through mitigation and adaptation. Little is known about farmers' and their advisors' willingness to use outreach tools, their information needs with respect to climate change or their ability to incorporate this knowledge into existing decision-making processes. A survey of almost 5,000 farmers in 22 top corn-producing watersheds across the United States showed that farmers' climate change beliefs correlated with both their perceptions of climate risk and their willingness to respond and adapt to changing conditions [138]. Farmers who believed that climate change is occurring, and is due in large part to human activity, were significantly more likely to support both mitigation and adaptation actions and also more likely to support government- and farm-level GHG reduction efforts. Most farmers supported adaptive strategies, with two-thirds agreeing that they should take efforts to protect land from increased weather variability. Many (59%) expressed lower levels of support, however, for mitigation through GHG reduction. These farmers obtained much of their information through social networks that included professional advisors. A survey of corn grower advisors, including government, nonprofit, for-profit and agricultural extension personnel, found that advisors are more influenced by current weather conditions and 1- to 7-day forecasts than by longer-term climate outlooks [139]. The advice given to farmers has been based predominately on historical weather trends and focused on short-term operational decisions rather than on long-term strategies. For climatic data to be useful to such populations, designing outreach strategies that target extension agents and other professional advisors will increase the potential to influence beliefs and practices of farmers. Furthermore, though mitigation policies alone might not resonate with farmers, those that combine mitigation with adaptation could be effective. In general, adoption of best management practices can be promoted by focusing on implementation among farmers most likely to adopt them, followed by leveraging social networks to inform other farmers about the benefits of adoption [140].

The constraints that farmers face when making decisions, such as whether to use conservation agricultural techniques, may create barriers to practices that could improve resilience to climate change. Conservation agriculture includes practices such as minimum mechanical soil disturbance, permanent organic soil cover and crop rotation, all of which typically increase soil C storage, especially when applied in concert [141,142]. Cited benefits of conservation agriculture in Sub-Saharan Africa

include increased yields, reduced labour, improved soil fertility, reduced erosion and land-saving [141-143]. Reports of conservation agriculture's widespread adoption may be overrated, though, because many farmers seem to adopt technologies only while incentives are offered and the project is actively supported, and then they quickly return to their former crop management practices once project support ceases [144]. Constraints to adoption include strong competition for mulched crop residues for livestock feeding, increased labour demand for weeding (which often changes cultural gender divisions of agricultural work) and lack of access to and/or use of herbicides and other inputs [143,144]. Although there are some recognized factors that influence adoption (for example, larger farm size and more education), no universal variables seem to explain adoption [145], leading some to suggest that conservation farming may be successful only under certain agroecological conditions [144,146].

Recent work in Zambia may help to explain regional variation in farmer adoption and rejection of conservation agriculture practices. Analysis of surveys of rural incomes and livelihoods revealed that rates of rejection in Zambia were high (approximately 95%), and practice dropped from 13% to 5% of farmers between 2004 and 2008 [145]. Rainfall data reveal that, during the past 10 years, the onset of the first rains needed for planting have been progressively delayed. Although adoption decisions are not strongly or explicitly based on labour constraints, farmer age or education level, farmers in districts that experience more rainfall variability are more likely to adopt conservation agriculture practices and to implement those practices with greater intensity [147]. Because conservation agriculture allows planting to occur as soon as the rains begin, it offers an adaptive response to changing rainfall regimes [148].

Fundamentally, an existing lack of food security and farmers' concerns about poor health will counteract incentives to their adoption of new farming technology [149,150]. Although many farmers believe climate risk is real, they are less likely to believe it is caused by human behaviour. They have paid the most attention to climate variables that have traditionally constrained their operations and have relied on an existing suite of adaptive behaviours [53]. Thus, knowledge networks are especially critical to their understanding of trade-offs between the short-term costs and longer-term benefits of adopting new farming technology and practices that will help them mitigate and adapt to the effects of climate change as well as to increases in climate variability. Adaptation to climate change and the idea of climate change itself define and change human cultures. Indeed, cultural factors (for example, place attachment, value systems, individual and collective identities) shape how people support and respond to adaptation interventions and must be woven into climate change policies and programmes [151]. Key to this effort is linking science, technology and decision-making to the context of socioecological systems to better achieve balance between economic, cultural and social needs [152]. Systems that effectively leverage science and technology in support of sustainability efforts create salience, credibility and legitimacy across boundaries where boundaries exist between science and policy, disciplines, public and private sectors, and/or organizational hierarchies. Actions employed within these systems include convening (bringing all stakeholders in the CSA context together to foster communication and build trust), translation (defining a shared ontology and language), collaboration (actors working together to produce applied knowledge and specific outcomes, with specific mechanisms in place to facilitate interactions across multiple boundaries) and mediation. Specifically, mediation is 'a process by which different interests are represented and evaluated so that mutual gains can be crafted and value created in a way that leads to perceptions' of fairness and procedural justice by multiple parties' [152], p. 470. These components, as well as broad stakeholder engagement from the initiation of a project, are keys for linking science with action, developing knowledge networks and forming critical capacity to reach desired outcomes also see [135-137,152]. Other approaches for forming new knowledge networks and adaptive capacity in the socioecological system combine both back-casting and explorative scenarios [137]. Interactions between climate change and culture, as well as ideas regarding the ethics and morality involved with climate change and the role of these constructs in stakeholders' and the larger society's adoption of actions related to mitigation and adaptation, are outside the scope of this article, but they are discussed by Hayward [153] and Markowitz and Shariff [154].

Climate risk management: financial mechanisms, insurance and climate services for farmers

An alternative to emergency aid in the face of climate shocks is reliable programmes developed to minimize farmer risk, which could prove to be more effective by preventing the slide into poverty traps [155]. The uncertainty of climate change, especially extreme events, makes it difficult for individual farmers to incorporate risk into their decision-making [156,157]. Vulnerabilities to climate effects on production, pests, disease and price volatility depend on farmers' assets and natural resource base [158]. Appropriate risk management tools, such as improved forecasts and extension support, and appropriately designed safety nets or insurance instruments must revolve around the vulnerabilities in specific farming situations. Rural households in developing countries, limited in both resources and access to information, could

be disproportionately affected unless appropriate measures are introduced to manage the additional risk and uncertainty related to climate change [159-161]. Innovative management of risk and uncertainty employs financial mechanisms (for example risk transfer or insurance contracts) that use several types of methods to understand investment decisions, technology choices, and risk perceptions. These methods include remote-sensing technology, micro-level household data, analysis of diversification, and farm surveys. Implementation of such insurance instruments requires appropriate technical innovation, building awareness and trust, ensuring viable market demand, and enhancing local capacity building among local financial institutions [162,163].

Index insurance is one such instrument that effectively reduces farmers' risk under a changing climate and generally has many advantages. With index insurance, indemnity payments are decoupled from actual crop losses, instead of linking payments to changes in attributes that impact or reflect crop growth or survival over a given spatial extent. This then reduces transaction costs associated with verifying ownership and losses, removes the opportunity for individuals to change their risk behaviours to increase the likelihood of receiving a payout, and allays the problem of adverse selection, in which high risk individuals are disproportionately represented in the insured pool. Most vitally, the rural poor are no longer widely excluded from insurance by the need to demonstrate assets as a prerequisite to purchasing a policy [161]. For example, the Index Based Livestock Insurance (IBLI) programme recently launched by The Index Insurance Innovation Initiative seeks to accurately represent the insured's loss experience through the use of landscape-level data derived from measures such as the Normalized Difference Vegetation Index (NDVI) (Figure 5). NDVI is a satellite-derived indicator of photosynthetic activity or a proxy for plant production to feed livestock, which is available in real time every 10 days [165]. Livestock in Northern Kenya's arid and semi-arid lands account for more than two-thirds of average income, with most livestock mortality associated with severe drought [164]. Herd losses that push a household below a certain threshold tend to result in long-term consequences, including destitution, which can trap the household in poverty. The data derived from the developed index showed that the NDVI performed well when tested against other herd mortality data from the same region, and, when compared to drought experiences over the past 27 years, removed 25% to 40% of total livestock mortality risk in simulations. The IBLI programme has been implemented, with initial payouts issued to households in October 2011 [166]. Actions needed to facilitate establishment of the IBLI include identification of systematic criteria for end users to evaluate whether they

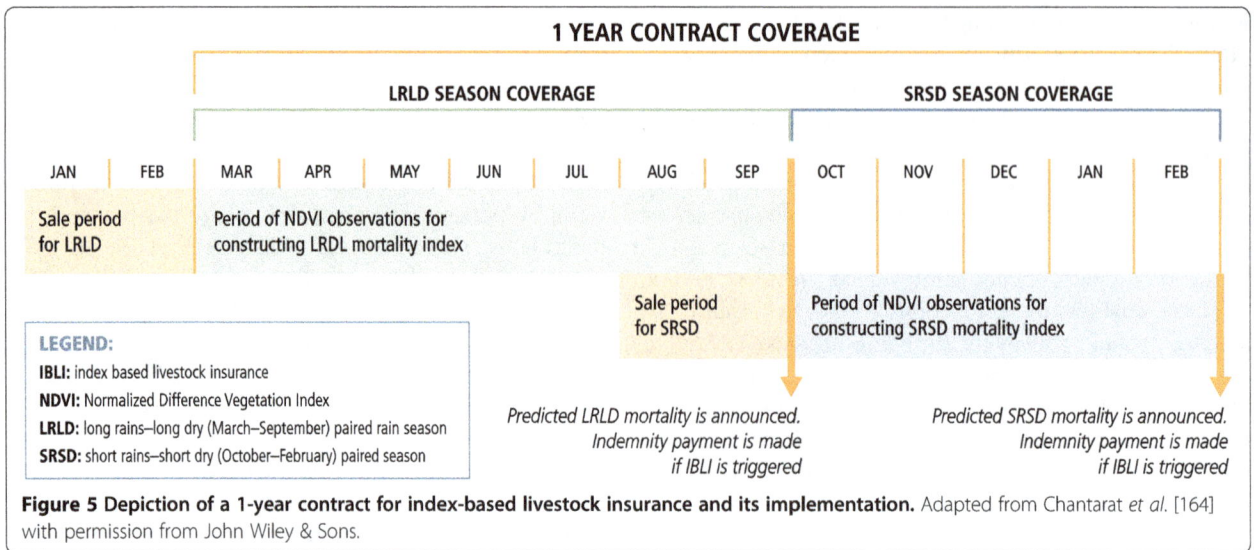

Figure 5 Depiction of a 1-year contract for index-based livestock insurance and its implementation. Adapted from Chantarat *et al.* [164] with permission from John Wiley & Sons.

need to purchase this insurance product [167] and development of programmes for client recruitment, low-cost marketing, and claim settlements.

To provide long-term farm and community security in support of CSA, bundling agronomic breeding programmes for drought tolerance and financial programmes with index-based drought insurance will maximize farms' resilience to financial shocks due to drought, especially as the drought tolerance of crops diminishes with more severe drought stress. Developing and planting crops with drought tolerance is primarily a more cost-effective risk management tool than index insurance in the face of less extreme climatic events; however, index insurance could complement both private and public crop improvement programmes by providing assistance when even drought-tolerant varieties fail during extreme climatic events. Demand for bundled strategies seems likely to be high [168,169], thus creating a sustainable market for both drought-tolerant varieties and index insurance. To assess how bundled strategies affect household welfare and operate in practice in a drought-prone region of Ecuador, Carter and Lybbert [170] estimated the underlying probability structure for traditional maize yields from yield data collected annually by the Ecuadorian government from random samples of producers in different regions of the country. The certainty equivalent of the drought-tolerant technology was 6% higher than that of traditional technology. Incomes were most stable under drought pressure when drought-tolerant and insurance index technologies were combined, but interactions of such bundled strategies with other risk management and safety net programmes remain to be determined.

Uncertainty influences individual farmers' expectations of yield and dramatically impacts their adaptation behaviour. For example, government policies to protect farmers

against climate-change risks, such as insurance programmes and direct *ex post facto* payments after extreme climate shocks, may reduce farmers' incentives to diversify farm production away from more climate-sensitive crops. Antón *et al.* [30,171] examined farmers' responses to agricultural risk management policies under conditions of climate change using a stochastic microeconomic simulation model calibrated with data derived from farming in Australia, Canada and Spain. They distinguished between farming risk and uncertainty with regard to climate and farmers' beliefs. They examined the impacts of *ex post facto* disaster payments and three types of crop insurance (individual yields, area-based yield and weather index) utilizing a combination of climate-change scenarios (no change, marginal change, change with an increase in extreme events) and farmers' behavioural response options (lack of adaptation due to misalignment of expectations, diversification, structural adaptation). Their model results indicated that farmers in Australia and Spain, in the absence of government policy, would respond by increasing diversification, assuming they correctly anticipated climate change. The introduction of risk management policies in these two countries tended to crowd out diversification, and this effect increased with climate change. The relative cost-effectiveness of policies depended strongly on the extent of extreme events and farmers' misperceptions of climate (that is, misalignment), which can greatly inflate a policy's budget. Reducing the uncertainty that farmers face, with regard to how climate change will affect them, by developing information strategies will aid in the design of robust risk management policies and will limit the excessive financial costs brought on by misperceptions [30].

The goal in using the risk management instruments described here is to promote resilience of rural households to weather shocks and climatic variability, a key premise of CSA. Although not addressed here, other index insurance products promote the integration of rural households into market production and often are used in concert with programmes aimed at promoting agricultural value chains and supply chain risk management [162]. These kinds of programmes consolidate and facilitate the participation in the agricultural value chain by specific populations in discrete regions, and they are intended to help increase farmers' access to credit and to encourage investment in appropriate technology to increase productivity.

Energy and biofuels: development of production methods and technologies to cut emissions without interfering with food production

Bioenergy is the native energy resource embedded within agriculture, but, more fundamentally, agriculture is itself an energy conversion process with the capacity to develop a rich portfolio of products for diverse markets, including markets for food and energy. The role of biofuels in achieving reduction goals (that is, mitigation) for GHG emissions and meeting future energy needs (that is, adaptation), as well as their impact on food commodity prices, remains a principally global issue [172,173]. Estimates of increases in food and commodity prices suggest that between 3% and 70% of retail food price increases can be attributed to biofuels; however, this wide range stems from differences in time periods, data sets using different price series (export, import, wholesale, retail) and different food products [174-176]. Global models used to predict mid- to long-term effects of biofuel production growth on prospective prices, production of feedstocks (for example, maize, sugar cane, oilseeds), mitigation and adaptation measures, and land-use change are general or partial equilibrium (PE) models. General equilibrium models encompass supply, demand and prices in the entire economy and take into account multiple markets and associated inputs; PE models are focused on equilibrium conditions in an individual market or sector of a national economy, in which prices, quantities under demand and product supply remain constant. Along with models used to assess land-use change in response to bioenergy production [177] are models such as the Asia-Pacific integrated model, which is used for analysis of global and national CO_2 emissions, mitigation costs and C taxes [178]; the Modular Applied General Equilibrium Tool which is used to examines links between agricultural markets, the general economy and agricultural policy issues [179]; the Global Change Assessment Model, which is an integrated assessment model of energy, agriculture and climate used extensively by IPCC and others [180,181]; GLOBIOM, which is used in analysis of mid- to long-term land-use change scenarios in agriculture, forestry and bioenergy

[182]; and the Model of Agricultural Production and its Impact on the Environment which is utilized in evaluating spatially explicit patterns of production, land-use change and water use in different global regions and linking economic development with food and energy demand [183,184]. These models can provide information regarding uncertainties, costs and trade-offs crucial to CSA for (1) climate policymaking, GHG mitigation and sustainable energy futures and (2) projections regarding agriculture, agricultural markets and the future of the world's food and feed supplies. The case studies described here are used to assess costs and trade-offs of biofuel expansion at the farm and global scales as well as the impacts of enacted policies in the European Union (EU) and the State of California in the United States.

Increased future demands for food, fibre and fuels from biomass can only be met if the available land and water resources on a global scale are used and managed much more efficiently than they are now. Therefore, developers of an integrated bioenergy framework must incorporate not only bioenergy's mitigation potential but also its costs and trade-offs with food, water security and land use. To assess the cost-effectiveness of bioenergy for climate change mitigation, Popp et al. [184] coupled global models of vegetation and hydrology [185,186], land-use optimization (MAgPIE) and the energy–economy–climate interface [187]. If all suitable land for agricultural production was made available, bioenergy from specialized grassy and woody bioenergy crops, such as Miscanthus (poplar), could produce 100 EJ globally by 2055 and up to 300 EJ by 2095. However, bioenergy cropland would grow from 1.52 billion ha to 1.83 billion ha, thereby increasing CO_2 emissions due to deforestation. Meeting bioenergy needs while preserving intact and frontier forests would require higher rates of technological change in agriculture (by 0.9% per year until 2095), thus leading to additional costs. The potential trade-offs of conserving forests and cultivating bioenergy crops on a large scale include conflicts with respect to food supply, food prices (especially in the tropics) and water resource management [188].

In the EU, market demand for biofuels and biomass will likely increase as the region becomes less reliant on fossil fuels and the EU implements targets for renewable energy, such as the Renewable Energy Directive and the ensuing national renewable energy action plans. This demand was first met with imported biomass sources from residue streams, such as palm kernel shells and wood pellets, and industrially produced biomass, such as palm oil and ethanol [189]. In an analysis conducted for the International Energy Agency, Hoefnagels et al. estimated future intra- and inter-European trade of solid bioenergy biomass by combining geographic information system models of transport routes with models of supply and demand for energy crops, forestry products and/or residues and agricultural residues [189]. They estimated that intra-European biomass trade could increase to 6,560 kilotonnes of oil equivalent (ktoe) by 2020 in the low-import scenario and to 5,640 ktoe in the high-import scenario. Transportation costs could contribute substantially to these totals (for example, up to 75% (9 €/GJ) of the total cost (12 €/GJ) in the case of forestry residues). However, they determined that the lower transportation costs of pelletized biomass would not make up for its high production costs. In both scenarios, the chief future exporting regions for inter-European biomass trade included Poland, Estonia, Hungary and Slovakia and the major importing regions included Germany, Italy, the United Kingdom and the Netherlands. Within the CSA strategy, these modelled outcomes can help in the identification of the issues and stakeholders that should be involved in the development of future energy use and policy.

Newly enacted low carbon fuel standard (LCFS) policies in California and the EU offer promising approaches to reducing the C footprint of transportation fuels. The LCFS applies to itself a direct life-cycle C intensity analysis that captures all GHGs emitted per unit of fuel energy during extraction, cultivation, land-use conversion, processing, transport and fuel use [190]. Both California's LCFS and the European Parliament's revised fuel-quality directive require a 10% reduction in GHG emissions by 2020, and both allow credit-trading. These standards differ from previous policies aimed at reducing petroleum fuels, which comprised volumetric mandates and only indirectly required reductions in GHG emissions. As a case in point, the US renewable fuels standard requires annual sales of 36 billion gallons of biofuels by 2022, 21 billion gallons of which must derive from advanced biofuels and achieve a 50% reduction from baseline life-cycle GHG emissions. The other 15 billion gallons must come from corn ethanol [190]. With this focus on total GHG emissions rather than on volume, biofuels under LCFS will not be forced into a small number of categories, and transformative innovation, a key part of the CSA strategy, will be promoted. The flexibility and performance-based nature of the LCFS allows industry, rather than government, to pick the likely biofuel winners [190]. If implemented on a global scale, such changes in biofuel policies will heavily influence agricultural markets and environmental outcomes. Tokgoz *et al.* [191] simulated a reduction in maize ethanol production of the magnitude suggested by the LCFS analysis by utilizing a modified version of the International Food Policy Research Institute's (IFPRI) PE model, or the International Model for Policy Analysis of Agricultural Commodities and Trade (IMPACT). IMPACT was developed to project future global food supply, demand and security in 115 country regions. Holding biodiesel production levels constant at 2010 levels in this model dramatically decreased rapeseed and soybean oil prices and increased the availability of food calories. Building future international policies upon the LCFS policies implemented in Europe and California will further the demonstrated benefits of reducing fuel C intensity rather than promoting policies that benefit biofuel producers who pursue ongoing profit-driven growth.

Policymakers and financial institutions have been hesitant to invest in bioenergy, owing to negative press and the resultant uncertainty about its long-term sustainability. In response, the scientific community must present a balanced perspective of how bioenergy can (or cannot) be managed as part of CSA (for example, see [172,173]). Models comprising the global impacts of bioenergy, along with agricultural productivity at local, regional and country scales, can be utilized to effectively assess the realization of environmental and economic objectives via policy and technology [192]. Separate consideration of bioenergy in the agricultural context will lead to suboptimization of the system with the likelihood of realizing lower environmental and economic benefits [193]. The viability of biofuels will be achieved when their cost is competitive with those of fossil fuels when it includes both the cost of the feedstock seed and the value of coproducts derived from the biofuel by-products, which can provide additional revenue. In some cases, large subsidies are required to make biofuels competitive with fossil fuels (for example, *Jatropha*-based oil in Senegal) and/or feedstock seeds must be imported to satisfy demand, suggesting that alternative feedstocks should be adopted [193]. A stable supply of feedstock, determination whether other industries strongly compete for the same feedstock and access to a well-functioning value chain for the product are all crucial to facilitating vertical integration of production, conversion and processing, as observed in Brazil's biofuel sector. Msangi and Evans [194] suggested that growing a biofuel feedstock that can serve as a food product with coproducts will create greater stability for the farmer and that solving problems of food security in developing countries will lead to a flourishing biofuel sector. Furthermore, increases in food crop production and efficiency underpin the success of increased reliance on bioenergy and the conservation of forested lands in lieu of expansion of agricultural lands [188]. Lignocellulosic biofuels also can be a strong component of GHG mitigation with small impacts on global food prices, especially if sufficient land for feedstock production exists and does not compete with land devoted to food production, as indicated by modelled outcomes [173]. It is imperative to engage producers and affiliated industries in research to better understand how markets for new development of bioenergy and nontraditional biological products can become an integral part of energy-efficient agriculture.

Theme 2

Landscape and regional issues: land use, ecosystem services and regional resilience

Recently, extensive research on climate impacts on landscape and regional scales has been stimulated in part by policies that require institutional action to mitigate and adapt to climate change [14,195]. Such research includes use of remote sensing to analyse land-use mosaics, inventory approaches to assessing C stocks and water resources, and models to examine the potential of land-use change in different climate scenarios [196-198]. These techniques are being combined with farm- and field-scale data on crop performance, soil biogeochemistry and irrigation use to analyse if and how mitigation and/or adaptation strategies build food security and ecosystem services [34,199-201]. Interdisciplinary science underpins an integrated landscape approach, along with involvement of stakeholders who hold key information for developing climate-change scenarios and innovation pathways [202,203]. Landscape approaches that expand beyond agriculture itself are needed to understand how extreme events trigger rural outmigration and create new types of rural–urban connections. The development of metrics and indicators to track responses of climate change and ecosystem services is accelerating with broader recognition of the need for greater accessibility of data, formation of more types of socioecological assessments [203-205] and charting of the progress of climate-change policies.

Climate change and food security: modelling adaptation and uncertainty

Determining the adaptive capacity of mitigation and adaptation scenarios that will evolve with CSA's participatory processes rely, in part, on biophysical models. Models that will be used to examine the limits to crop adaptation as well as the impacts of climate change on biodiversity, land use and ecosystem services are now available [2,206]. They still contain much uncertainty due to (1) the ability of process models to accurately simulate the growth and development of crops when exposed to very high temperatures and elevated CO_2 levels, (2) the rate and degree to which agricultural productivity and development can progress in concert with reductions in GHG emissions and (3) the ramifications of successful agricultural adaptation to climate change for land-use change and associated ecosystem services [207-209]. Despite these uncertainties, the use of models and scenario-building has led to the exploration of potential synergies and obstacles to coping strategies in agricultural that would not have been possible

with empirical data alone [210,211]. Here we present modelling approaches to evaluating adaptation scenarios across the EU, the Mediterranean region and the United States.

Modelling can be used to identify climate-change impacts and sensitivities as well as possible adaptation strategies. Rather than being focused solely on climate-change constructs, such vulnerability assessments also include changes in CO_2 concentrations, GHG emission management, N deposition, land use, and socioeconomic trends to manage vulnerability. The Advanced Terrestrial Ecosystem Analysis and Modelling (ATEAM) program produced a new set of climate scenarios for Europe in multiple global change scenarios and ecosystem models [212]. A dialogue among relevant stakeholders from the private sector, governmental and nongovernmental organizations and policymakers was conducted. Unlike global trends, European trends included moderate or no population increase, little urbanization, increased forest area and decreasing demand for agricultural land. The modelled outcomes allowed for changes in land management that could decrease vulnerability, such as C sequestration due to reforestation. Modelled outcomes indicated that the Mediterranean region could face increased risks of forest fires, water shortages, changes in tree species distribution and losses of agricultural potential. Under the different scenarios, which ranged from business as usual to greatly reduced GHG emissions, 20% to 38% of the population in the Mediterranean would live in watersheds under stress and experience water scarcity exacerbated by increased tourism and demand for irrigation. Mountain regions would be especially vulnerable because of less snow cover and subsequent changes in river runoff. These modelled outcomes provide opportunities for back-casting and identification of sensitivities where mitigation and adaptation efforts should be focused, as well as how subsequent research could inform policies around such efforts.

The participants in the EU SmartSOIL project [213] employ a CSA-like strategy that includes stakeholder involvement and is used to examine the implications of findings for economics and policy implementation. As of 2012, consultation with policymakers and advisors had begun in six case study regions [214]. The creators of SmartSOIL developed a framework of C flows and stocks informed by new data and meta-analysis of long-term European experiments that are relevant to short- and long-term CSA management decisions. This framework will be used to improve existing soil and crop simulation models out of which a simplified model will be derived to predict scenarios for future management systems to improve productivity and enhance C sequestration. As an example of modelling for C sequestration, Lugato *et al.* [197] used the CENTURY model to inform proposed European policies on the mitigation potential of agricultural soils through C sequestration and to assist in the evaluation of the agricultural sector's deployment of 'greening' measures for agriculture that benefit climate and environment as required in the EU's post-2013 Common Agricultural Policy. Nearly 16 soil–climate–land combinations in the EU and neighbouring countries (Serbia, Bosnia and Herzegovina, Croatia, Montenegro, Albania, former Yugoslav Republic of Macedonia and Norway) were used in calculations, including the main arable crops, orchards and pastures as well as management practices (for example, irrigation, mineral and organic fertilization, tillage) (nearly 164,000 scenarios). Testing modelled results against soil inventories collected using comprehensive and standardized approaches (the European Environment and Observation Network and the Land Use/Cover Statistical Area Frame Survey) strengthened the examination of the uncertainty of modelled outcomes. Consideration of a broad spatial extent (pan-EU scale) allowed for better evaluation of C sequestration, in which an estimated current stock of 17.63 Gt C is predicted to increase through 2100. Within the pan-EU region, stocks will diminish in the southern and eastern parts because of higher soil respiration, whereas these losses will be offset by increases in the central and northern regions due to increased CO_2 atmospheric concentration and favourable crop-growing conditions. Such survey and monitoring programmes support the need for further spatiotemporal analysis of climate trends and stakeholder dialogue in modelling efforts so that proposed adaptation strategies are relevant to economic and socioecological contexts such as local, national and EU-wide policies and regulations.

Many climate modelling studies are focused on yield variations in response to changes in mean climate conditions [215]; yet, this approach overlooks several key factors, such as the occurrence of extreme events in which variance is changing [216]. Empirical approaches that capture the effects of extreme temperatures can be used to more efficiently assess climate impacts and adaptation. For example, in Mediterranean sunflower and wheat, an increase in both mean temperatures and climate extremes modelled under A2 and B2 scenarios (year 2100 business-as-usual and reduced GHG emissions scenarios, respectively) would cause severe yield reductions by shortening growing seasons and intensifying heat stress [217]. In the United States, yield patterns of rainfed maize have been explained by accounting for extreme events using the process-based Agricultural Production Systems Simulator (APSIM). With APSIM, observed negative yield responses to extreme heat shocks (measured as accumulated extreme degree days) were best explained by increased vapour pressure deficit (VPD). VPD contributed to water stress by increasing plant demand for soil water and reducing future water supply as a

consequence of higher plant transpiration rates [218]. The ratio of water supply to demand, as modelled with APSIM, was three times more responsive to a 2°C mean warming than to a 20% reduction in rain. The results of these studies direct researchers, policymakers and extension agents to take science-based actions that rely on climate scenarios and predicted outcomes that are not based solely on the change in climatic means but include climate extremes. Despite incongruences between actual biological patterns and model simulations, model outputs provide an evolving information base for planning strategies and new research directions.

Quantitative assessments of adaptation to consider the effects of extreme events on agriculture can inform policymaking by providing a much wider set of outcomes than is possible with perceptions or projected impacts. Modelled outcomes evaluated in a socioecological context allow investigation into the limits of adaptation and related consequences for agricultural productivity, other economic sectors and land use (for example, an indicator-based, spatially explicit and scenario-driven adaptive capacity model [211]). Coordinated cycles of model improvement and projection across multiple spatial scales (global, regional, local) will facilitate model validation and calibration as well as effective use of studies with different geographical domains [219]. Challinor et al. [219] recommended that different model intercomparisons and improvement programmes (MIPs) form separate but linked strategies, that detailed modelling studies of response mechanisms (for example biophysical processes, crop yields) and robust experimental data (for example, see [208]) underpin the models and that systematic comparisons of impact studies and their outcomes be used to address sources of models' uncertainties. Involvement of stakeholders at the outset of model development also aids in development of relevant scenarios and tools [152]. Modelled outcomes form a key part of the climate policy and governance process necessary to attain the Copenhagen 2°C target, which requires 70% to 90% worldwide emission reduction targets and which has been questioned as being too weak [220]. A wider set of options for targets will be facilitated by examining options for climate governance (that is, institutional mechanisms to guide and direct societal policymaking) and their societal implications, as well as assessing the potential for success in achieving the target within existing political structures (for example, democracy, autocracy) [221,222]. For example, the 'mitigate for 2°C but adapt for 4°C' option implies that society will take steps to adapt to existing in a warmer world, but will maintain the goal of reaching the current 2°C target. This approach will diminish conflicts and trade-offs associated with the water–food–energy nexus. Yet, if it is perceived that the 2°C target is unattainable and investment is strongly supportive of adaptation, then acceptance of even higher target values and a greater need for adaptation-related burden-sharing will be a consequence [221].

Soil carbon and achieving multifunctionality through mitigation and adaptation

Soil resource degradation has led to loss of functions and ecosystem services, such as water availability, water-holding capacity, C storage, mitigation of GHG emissions and sustained agricultural productivity [223,224]. Soil degradation limits resilience to climate change and extreme events, such as drought, and therefore impacts food security and augments susceptibility to poverty, especially in vulnerable regions such as Sub-Saharan Africa. Better understanding of the biophysical capacity of agricultural landscapes to act as C sinks through capture and storage of atmospheric CO_2 in soils and perennial vegetation leads to strategic design and operational management for both mitigation and adaptation actions [122,225]. Improving biophysical capacity for desired functions such as GHG mitigation, food production and maintenance of soil and ecosystem biodiversity is a form of ecological intensification and is enhanced within a multifunctional landscape. Ecological intensification builds resilience by leveraging ecological processes to increase outputs from agricultural lands to promote (provisioning supporting, and regulatory ecosystem services) and decrease dependence on external inputs [93]. Balancing trade-offs between the different types of services can be facilitated by assessing indicators such as soil organic C (SOC). Trade-off analysis can employ simulation methods and modelling tools (for example, the Agricultural Model Intercomparison and Improvement Project, known as AgMIP; see [9]) to evaluate existing and alternative agricultural systems, changes in market conditions affecting supply and demand, and related policies in relation to climate change. The negative trade-offs can be minimized when landscapes are managed to achieve multifunctionality objectives, such as by a diverse set of land-use types, each providing a different combination of services [31]. The case studies below are focused on tools for accounting for GHG emissions and soil C storage, processes to enhance soil C storage and use of a paired economic-biophysical model to assess impacts of mitigation efforts within multifunctional landscapes.

Climate-change mitigation and adaptation within multifunctional landscapes depends on the multiple roles of SOC, which include a reservoir for plant nutrients (N and P) to support crop production and reduce external inputs, a substrate for soil organisms affecting their activity and diversity, and a promoter of soil physical structure leading to enhanced water quality and reduced erosion [223]. To maximize mitigation efforts, accurate GHG calculation can engage stakeholders and other end users to form a database with which to understand the C budget of their practices, such as SOC sequestration and storage and CO_2 emissions from fossil fuel combustion. Many models used to calculate GHG emissions and SOC are designed for specific geographical areas to meet distinct needs. Colomb et al. [226] provided information on the features of 18 available calculators and created a framework for choosing the most suitable GHG and C calculators for a given situation. They found that major sources of GHG emissions were usually well-identified, but that the calculators used failed to account for landscape effects due to land-use change. Few calculators accounted well for emissions from the loss of previous biomass, which is especially crucial in cases of deforestation–reforestation or rehabilitating and restoring grazed and ungrazed grasslands. To illustrate this point, Colomb et al. [226] used seven calculators to assess the GHG balance of replacing grassland by wheat, a case where the average emissions due to land-use change were greater than those that occurred during the production of wheat itself. Owing to differences in reporting units, measurement of emissions and scope, the results obtained with different calculators could not be directly compared and uncertainty levels were very high. Minimizing uncertainty in C and GHG accounting methods will provide reliable data to aid global markets and agencies for use in developing GHG- and C-footprinting and life-cycle assessment criteria. Greater standardization of metrics will also help in the enumeration of trade-offs in balancing between crop management and land use.

The design of multifunctional, ecologically intensive landscapes when providing ecosystem services of local and global interest is informed by analysing synergies between agricultural practices and landscape attributes [58]. For example, an analysis of carbon stocks and flows in smallholder farms in Kenya revealed positive synergies between agricultural production, on-farm biodiversity and above-ground C storage [227]. Dominant land-use types considered included home gardens, food-crop plots, cash-crop plots, pasture plots and woodlots. Close to the homestead, home gardens received the most organic nutrients in the form of compost, kitchen waste and manure, and downslope and farthest away maize, vegetables and eucalyptus woodlots were planted. Tree species diversity was highest in home gardens and near crop fields. Although such trees contributed up to 39% of total aboveground C storage, the greatest contribution came from monospecific woodlots dominated by *Eucalyptus saligna* (which contributed up to 81% of total aboveground farm C). In a landscape survey of 250 farms across 6 regions in Kenya, SOC, available P and exchangeable K^+ varied widely but generally varied by management practice and reflected diminished soil fertility with greater distance from the homestead [58].

Thus, a combination of land-use practices contributed to C storage below and above ground as well as to multiple functions on the farm (Figure 6). Including the diverse agricultural landscapes in such studies leads to understanding of how management practices support ecological processes for C storage, and farmer participation supports identification of economically viable options for smallholder farmers [58].

Trade-offs between mitigation and adaptation occur often in agricultural systems, notably in the allocation of scarce resources between competing activities. The Trade-off Analysis model for Multi-Dimensional Impact Assessment is used to evaluate climate-change impacts and the viability of adaptation strategies by combining survey, experimental and modelling data [229]. Its next step is calculation of future land use, output, output price, cost of production and farm and household sizes for different climate-change and socioeconomic scenarios. The authors applied the model to the Vihiga and Machakos districts in Kenya to simulate changes in crop and livestock productivity and the effects of climate change to 2030. Climate change was projected to have a negative economic impact for 62% of farmers in Machakos and 76% in Vihiga, but these modelled effects could be partially offset by specific adaptation strategies. The most viable adaptation strategies included introduction of an improved maize variety or low-yielding,

dual-purpose sweet potatoes in Machakos and improved feed quantity and quality combined with livestock breeds adapted to increased drought and high temperatures in Vihiga. In some cases, mitigation activities result in negative trade-offs, such as organic practices that increase SOC offset net GHG emissions, leading to competition for feed for livestock or fuel, or even to decreases in average yields, thereby exacerbating forest conversion to agricultural land [122]. Agroforestry, however, contributes to multifunctional landscapes that support mitigation and adaptation and can lead to improvements in livelihoods, whereby provision of fuel wood, timber, fruits and/or fodder is often associated with the cobenefits of improved soil fertility, water infiltration and below- and aboveground C sequestration [40,150].

Currently, agricultural decision-makers and policymakers rarely consider SOC to be a major factor in agricultural management or land-use change, and the concept of multifunctional landscapes is an emerging idea in the science-based policymaking realm. Yet, the study of SOC formation, its functions, its physical and chemical protection and identification of those fractions most susceptible to degradation is an area of active research. Through various international conventions, this scientific knowledge is slowly becoming part of the science–policymaking interface relevant to climate-change mitigation and adaptation (for example, the United Nations

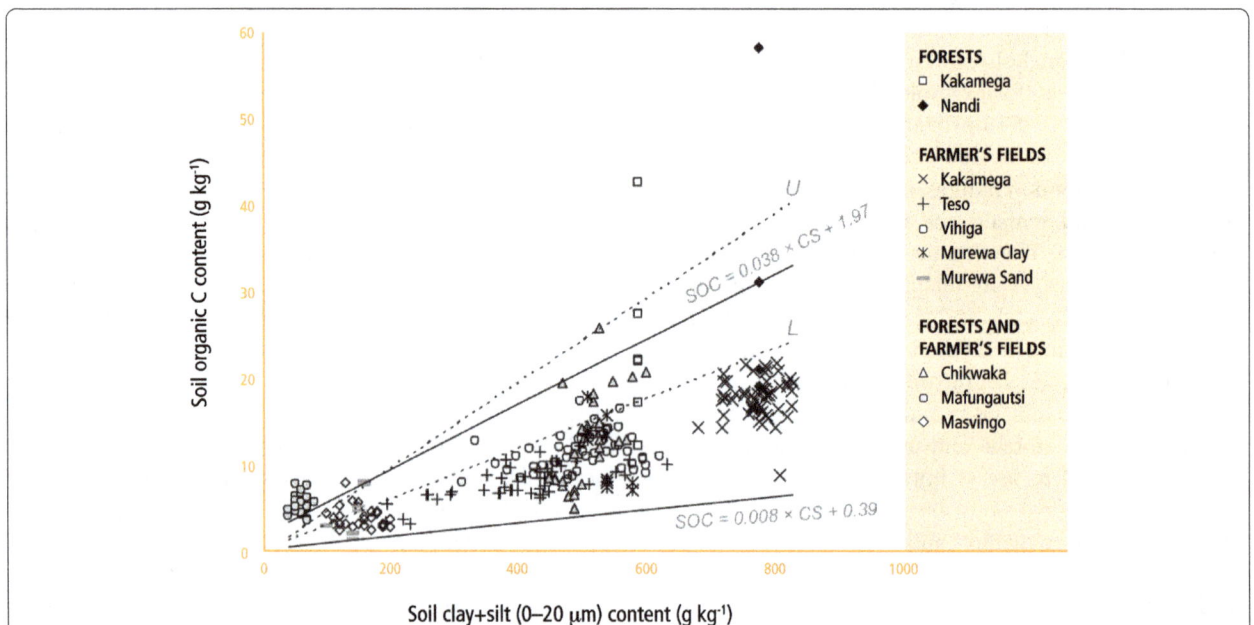

Figure 6 Soil organic C from the upper 0 to 20 cm as a function of clay plus silt. The tropical forests (Forests) and cultivated fields converted from the forests (Farmer's Fields) represented here were in Zimbabwe (Chikwaka, Mafungautsi, Masvingo, Murewa) and Kenya (Kakemega, Nandi, Teso, Vihiga). The Zimbabwean forests were the Miombo woodlands with unimodal rainfall of 800 mm. The Kenyan forests were rain forests with bimodal rainfall of 1,800 mm. Upper (U) and lower (L) boundary lines were fitted to the 95th and 5th quantiles, respectively. Samples collected at Chikwaka, Mafungautsi and Masvinga (Forests and Farmer's Fields) were collected along a temporal gradient from 0 to 60 years after conversion from tropical forests, but were not differentiated by land use in this figure. Adapted from Tittonell [228] with permission from the author.

Framework Convention on Climate Change, the United Nations Convention to Combat Desertification, and the Global Soil Partnership). However, the complex trade-offs in land-use decision-making regarding provision of multiple ecosystem services in a given landscape are usually local, so new interdisciplinary and socioecological research approaches are needed in order to downscale information and options regarding how to best manage soil C in relation to other ecosystem services and farmers' livelihoods [40,122,150,230-232].

Water management for food and fishery systems

The effects of climate change on hydrology are far more uncertain than temperature change, and yet, global irrigation water demand will likely increase by approximately 10% by midcentury [233]. IFPRI models indicate that calorie availability in developing countries could potentially reach almost 85% of that in developed countries by 2050, but in more pessimistic scenarios, calorie availability will decline in all regions, due in part to lesser water availability [36]. IPCC models for irrigated areas within this same time frame indicate that the gap between potential evapotranspiration and effective rainfall will be about 17% by 2050 under a high-emission scenario, placing extra stress on demand for irrigation water [234]. Taylor *et al.* [235] asserted that land-use change may have even more noticeable impacts on the hydrological cycle than climate change itself, but that, given the strong focus of mitigation and adaptation planning on land-use change, the two will remain intimately linked. For example, following conversion of forests and grasslands to agriculture in the West African Sahel [236], Southeastern Australia [237], New Zealand [238] and Southwest USA [239], runoff and/or groundwater recharge increased up to two orders of magnitude. Such increases are not always sustained, owing to a range of vegetation cover and hydrological response factors [240]. Forests and woodland cover can also support water quality and, in some cases, can assist in reducing dryland salinization and water-quality decline in semiarid environments [241-243]. Massive abstraction of groundwater and redistribution to agricultural land (nearly 70% of global freshwater withdrawal and 90% of consumptive water use for irrigation) has led to groundwater depletion in regions with primarily groundwater-fed irrigation (for example, regions of China and in the Ogallala Aquifer region in the United States). With projected increases in drought incidence and severity, changes in rainfall patterns and intensification, and decreases in snowpack, agricultural areas that are currently irrigated with surface water will become heavily reliant on groundwater. In Mediterranean-type climate regions in California's Central Valley and in southern Europe, groundwater recharge will be highly dependent on uncertain changes in precipitation patterns

[235]. Aquifer salinization is also predicted to increase, at least in California's Central Valley [244]. Sea-level rise also threatens groundwater and surface water with saltwater inundation [245]. The case studies here depict adaptation measures that have been employed to meet the challenge of water management in the face of climate change across a range of spatial scales.

In the Central Valley of Chile, multidisciplinary teams have enacted a CSA-like strategy to address climate-related changes in water [245]. In Chile, farmers' permanent water rights are determined by estimates of minimum stream flow. In a high-emissions scenario, the Central Valley may experience temperature increases of 4°C by the end of this century [4], which would lead to decreases in water supply and thus challenge the existing system of determining water rights and their allocation. In the Maipo basin of Chile, snowmelt from the mountains will be reduced, affecting both river discharge and water demand. In a moderate climate-change scenario (B2), modelled reference evapotranspiration, an indicator metric of irrigation demand, was discovered to potentially increase by 10% to 15%, whereas under the high-emissions scenario (A2), increases ranged from 14% to almost 20% [31]. Permanent water rights vulnerability under the two scenarios, on the basis of data for monthly mean river flow and an agricultural census, indicated that water demands would be inadequately met in 40% to 50% of years under the more severe climate-change scenario. In response, farmers could change crops and/or cultivars, increase irrigation or sell their land and water rights. Even under current climatic conditions, farmers' existing water rights have been questioned because of increasing demand by urban users [245]. To address this issue of failing water rights and limited availability in future climate scenarios, a 'science-policy' strategy has been employed that involves civil society, scientists and policymakers in an iterative dialogue to identify the challenge and its solutions (Figure 7). Since 2008, annual meetings have been conducted with researchers and stakeholders from the national water service, irrigation commission, and environment ministry in Chile). The result has been increased inclusivity and quality of overall participation in topics such as climate-change impact assessment, water-allocation system reliability and water-sector adaptation evaluation, leading to improvements in decision-makers' support of studies on uncertainty in evaluating irrigation projects and future reservoir operations. The science-policy approach supports dissemination of information and projects to strengthen vulnerability assessment tools and coping strategies for irrigated agriculture.

In the Mekong River Delta in Vietnam, more than 700,000 ha of coastal habitats used for aquaculture are threatened by rising sea levels due to climate change.

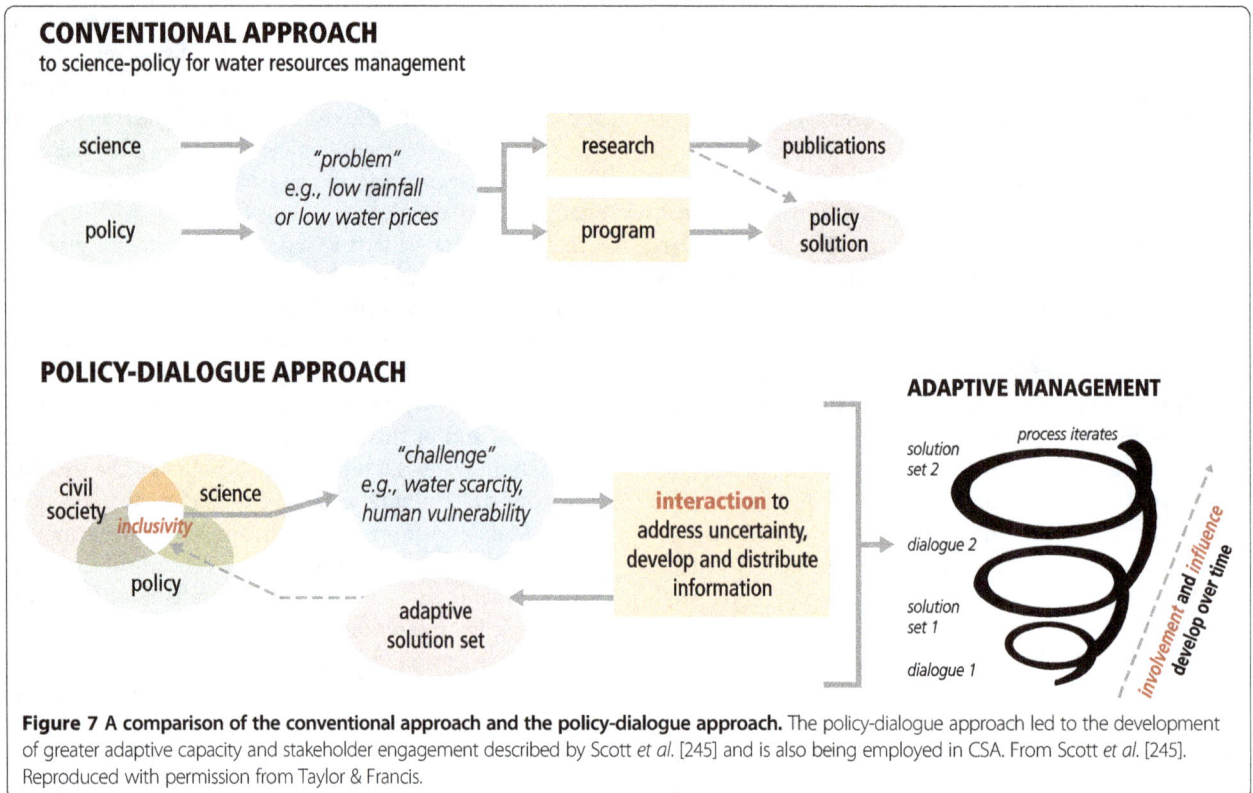

Figure 7 A comparison of the conventional approach and the policy-dialogue approach. The policy-dialogue approach led to the development of greater adaptive capacity and stakeholder engagement described by Scott *et al.* [245] and is also being employed in CSA. From Scott *et al.* [245]. Reproduced with permission from Taylor & Francis.

Kam *et al.* [246] analysed the farm-level economic costs and benefits of several alternatives: (1) autonomous adaptation, that is, spontaneous adoption or response, to climate change; (2) no climate change; and (3) planned, or policy-driven, adaptive strategies in which costs are distributed more equitably across the supply chain or are borne by government and other entities. Here 'autonomous adaptation' includes farmers' responses to changes in land and water availability, commodity prices, market incentives, and climate variability. Such responses incur incremental capital costs and include using different levels and combinations of inputs, altering species and production systems, adjusting the height of pond dikes, and increasing water volumes pumped into ponds. Shrimp farmers will be better able to bear the cost of autonomous adaptation than catfish farmers because they sustain relatively higher profit margins and require lower capital investments than catfish farmers. However, without government intervention to prevent flooding and salinity intrusion, the shrimp industry in aggregate will likely experience higher adaptation costs, as it covers more area. Planned adaptive strategies include genetic improvement of breeding stock and pathogen control. Although constructing dikes would reduce river and coastal flooding and salinity intrusion in support of fish production (a provisioning service), opportunities for expansion in both brackish-water and mangrove aquaculture

systems that are key to coastal preservation (supporting service) will be lost. In general, evaluating adaptive planning with many types of metrics, including those for ecosystem services through restoration of coastal and intertidal vegetation, were found to provide more data to inform the final choices made by stakeholders [247].

Recently, the concept of rainbow water, or terrestrial and oceanic evaporation as a source of atmospheric moisture and subsequent precipitation, has emerged. This conceptualization frames how to harmonize the interests of all users of the hydrologic cycle [248]. Available blue water sources—water used for irrigation, industrial or domestic use—and grey water sources cannot support the rate of agricultural intensification, so interest in green water—rainfall used by forests and other vegetation—has grown. Although controversial, passage of air over vegetation with a specific leaf index of 1 in the 10 days preceding rainfall was observed to lead to increased precipitation in Africa [249]. It follows that assessments of climate must take into consideration whether, where and how landscape changes alter large-scale atmospheric circulation patterns of water far from where the land use and cover changes occur to avoid misalignment of investment in climate mitigation and adaptation [248].

Given that climate change is likely to reduce water availability across many agricultural regions, it is critical that water policy and management practices focus on efficient and equitable water rights and allocation policies; increasing water productivity via more and better irrigation storage, conveyance and delivery systems that reduce evaporative losses; in-field water-use efficiency improvements; and technologies that reduce seawater intrusion in coastal environments. These challenges are equally important in the quest to increase agricultural productivity to feed a growing global population, irrespective of the degree of climate-change impact. Responses to the spatial and temporal shifts in water quantity and quality due to climate change involve many scales and stakeholders, and the need for coordinated planning at regional and national scales will increase with growth in the urban and industrial sectors. Approaches to increasing the efficiency of water used for food supply must employ drought-tolerant crops and irrigation technology (for example, water-conserving irrigation systems, crop coefficients and surface renewal [250,251]). They also need to address both consumptive behaviour (that is, overconsumption and resource-intensive food selection) and waste incurred during postharvest and along the supply chain (for example, threshing, transport, storage) [252]. Other adaptive strategies include the involvement of communities and government agencies in increasing storage capacity via small-scale reservoir projects, rainwater harvesting, groundwater banking through artificial and/or natural aquifer recharge and flood harvesting (that is, directed capture of floods in floodplains) and restoration of coastal vegetation to promote opportunities for aquaculture [242,244,252]. Additional adaptation options include reduction in end-user demand, deengineering and reoperation of water systems to create adequate supply and distribution, improved wastewater treatment plants to facilitate wastewater reuse, desalination plants and targeted water-conservation projects [253].

Managing forest biodiversity to increase ecosystem services and resilience

Forest loss and degradation cause GHG emissions and loss of C stocks, biodiversity and ecosystem services. Trees and forests buffer microclimates, regulate water quality and flows, store C and provide habitat for plants and animals in protected areas and corridors [248,254,255]. When landscapes are managed to contain a mosaic of forestry and agroforestry ecosystems, the diversification of food, feed and timber production, income sources, and markets promotes greater resilience to environmental uncertainty [149,256]. REDD + programmes to pay developing countries for conservation and sustainable use of forests have evolved over the past decade toward greater attention on (1) increased interactions between institutional networks and (2) achieving reduced

GHG emissions along with improvement of livelihoods of local communities and biodiversity conservation [257]. A systems approach involving biophysical and social sciences, as well as indigenous knowledge, is fundamental to demonstrating that REDD + projects are performance-based, fair and equitable [33]. Although afforestation and reforestation are often considered in REDD + projects, trees on farms are usually not included, owing to strict 'forest' definitions. Yet, agroforestry systems offer many REDD + -related benefits. Intentional integration of trees on farms and in agricultural landscapes increases C sequestration, along with greater food security and resilience [40,229] (for example, see Figure 8). Assessing such multifaceted trade-offs across an agricultural landscape is relevant to the CSA strategy, but will require greater coordination on local, regional and international levels to be incorporated into REDD +.

Examples of agroforestry types in agricultural landscapes include remnant forest or savanna, agroforests, tree crops, home gardens and boundary plantings [258]. Tree species and densities for each type are selected by desired ecological processes, farmers' criteria and land-use policies. An integrated landscape approach allows valuation of the ecosystem services derived from these management options and can be used to determine potential trajectories of tree-cover transitions [31,149]. It permits the nesting and spanning of spatial scales of different agroforestry types, the confrontation of biases for C benefits versus livelihood choices, and the optimization of tree-diversity exploration. It also opens opportunities to identify synergies and trade-offs and helps sidestep definitional challenges that result in negotiation platforms for proactive actions that reduce vulnerability and increase benefits (for example, see [259]). The landscape perspective is useful for scenario-building, such as comparing financial incentives that emphasize economic efficiency for agricultural and timber purposes versus socially 'green' and 'rights-based' approaches that support resilient livelihoods and broader sustainable development goals. The current scientific literature does not adequately detail these socioecological and community-based processes or how they underpin decision-making.

Examining trade-offs in REDD + can provide scientific information to enable science-based policies and decision-making, as well as coordination and standardization of REDD + practices. Many of the trade-offs involve livelihood issues that increase productivity and wealth, thereby encouraging land tenure and sustainable intensification through agroforestry. The results of household surveys and farm inventories have shown that agroforestry can help farmers deal with drought, flood and rain variability by reducing the need to sell land and livestock at low prices and instead sell seedlings, timber and firewood and consume tree fruit during the 'hunger gap' [33,40,260]. Sequestering

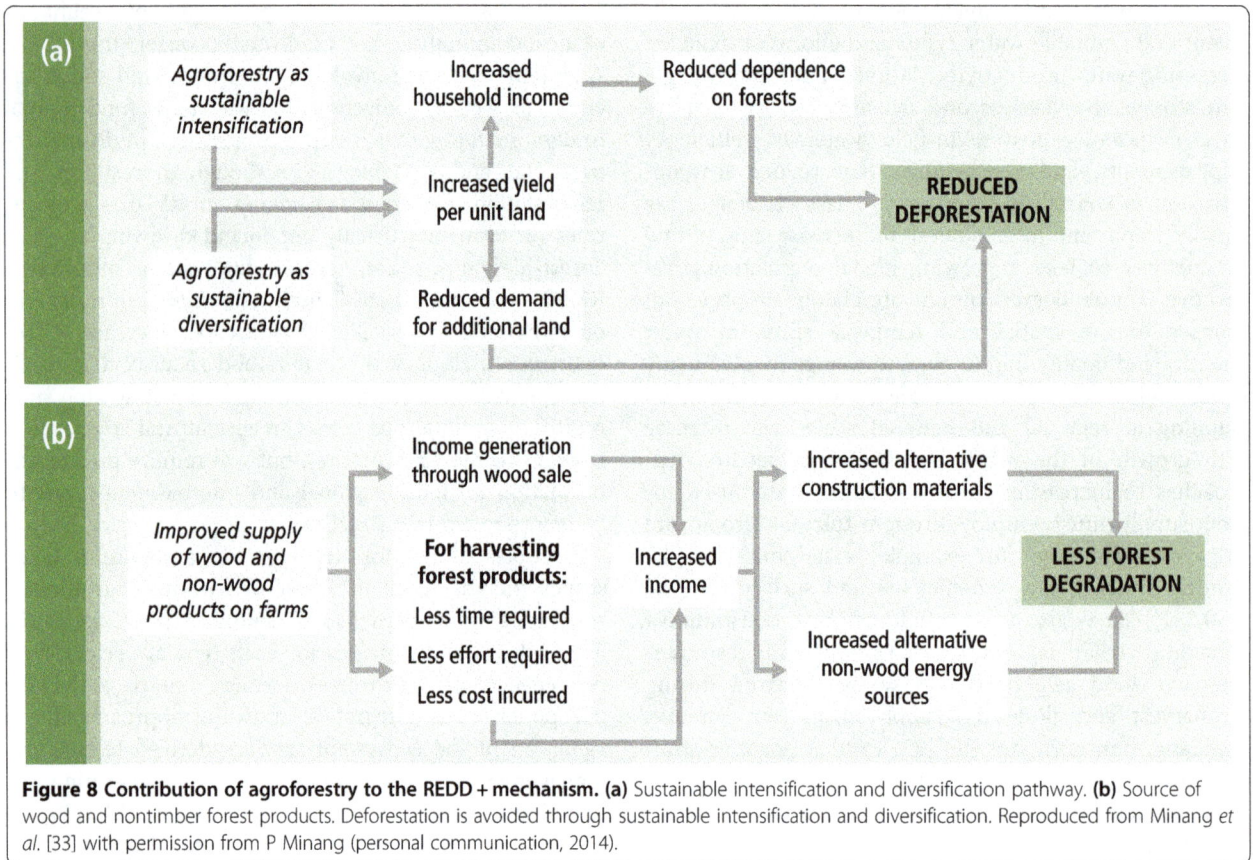

Figure 8 Contribution of agroforestry to the REDD + mechanism. (a) Sustainable intensification and diversification pathway. **(b)** Source of wood and nontimber forest products. Deforestation is avoided through sustainable intensification and diversification. Reproduced from Minang *et al.* [33] with permission from P Minang (personal communication, 2014).

C on farms for climate-change mitigation will only be attractive to smallholders when short-term increases in income or welfare occur. Landscape models have shown the impacts of investing and implementing policy in 'business-as-usual' versus 'green' scenarios, such as allowing land swaps for permits granted within natural forest for oil palm expansion, so that plantations can expand only onto land that is already degraded, as well as tax concessions for plantations that expand only onto degraded land [261]. In a recent report, the International Union for the Conservation of Nature assessed climate-change mitigation activities across many regions of the world where REDD + policies likely would be implemented [262]. Examination of the social, economic and environmental trade-offs and potential synergies revealed that clear tenure and property rights, including rights of access, use and ownership, are essential for effective REDD + implementation To benefit local communities, including the most vulnerable, REDD + policies must enhance the ecosystem services upon which the rural poor are most dependent and leverage new financial resources to reward local communities for management. These opportunities can easily be lost if the vulnerable are explicitly excluded as beneficiaries (for example, because of unclear tenure) or high barriers to entry (for example, forest certification) [263].

Participatory, transparent, accountable governance can help achieve benefits of implementing REDD + policy by creating synergy between parties at multiple scales. A governance approach that facilitates harmonized goals and policies between civil society and engaged stakeholders focuses on the relationships among organizations rather than on new organizational structures and financing mechanisms. Public–private partnerships can improve the effectiveness of the biodiversity governance system and complement regional and multinational efforts [263]. In Cameroon, for example, nongovernmental organizations are implementing REDD + pilot projects and acting as bridges between the public and the state, both to create awareness among local communities and to voice concerns about social safeguards [264]. Such partnerships have helped government institutions organize international biodiversity governance around an ecosystem approach, largely by changing the scale and nature of the dialogue through a community of practice with institutions outside the immediate REDD + network [257].

Although REDD + will benefit from institutional interactions that build trust and reach eventual consensus on forming, coordinating and integrating policies that support livelihoods and resilience while sequestering C in forests, the definition of appropriate ecosystems for payments still is a major issue. As pointed out by Visseren-Hamakers and Verkooijen [257], it remains to be seen whether CSA, with its integrated planning of land, agriculture, forests, fisheries and water, will be included in policymaking steps towards broadening of the REDD + agenda.

Rural migration due to climate change

A worldwide transition toward urbanization is occurring, partly in response to climate change, although rural out-migration due to climate shocks, such as hurricanes, is better documented than gradual changes, such as lower rainfall in arid areas [43]. Migration within countries is complex, having both positive and negative impacts on adaptation and household resilience. Climate shocks and disasters can propel people living under vulnerable conditions into poverty traps that force migration out of rural areas [265], where men most often migrate, leaving the women and children with increased household and farming burdens [45]. Migration can be a beneficial strategy that spreads risks through resource diversification, such as remittances that bring money back to the household [266]. Livelihood and food security, as well as culture, affect who migrates, when, for what reasons and to which destinations [267] (Figure 9). Despite the material benefits that can result from mobility and migration, displacement of people from places that they value reduces culturally based activities, such as preplanning for specific climate-change events [42]. Migration can lead to inhabiting vulnerable urban locations, such as flood-prone areas [268], and increase inequities due to poverty and lack of social networks. Opportunities exist to improve structural and institutional frameworks to reduce migration from rural areas, including greater diversification of rural livelihood systems [149,269]; opportunities for public health, social equity and environmental welfare [270]; and connection of urban populations with local or regional food sources to support rural incomes [3,11,28].

Land scarcity and degradation are conducive to out-migration. In Guatemala, people from households affected by flooding or soil degradation were found to be more likely to leave settled rural areas for the forest frontier to engage in clearing of forests for agriculture [271]. Surprisingly, on the basis of employing a remote-sensing approach across Central and South America over a 10-year period, rural–urban migration was not observed to strongly affect the recovery of forest vegetation [272]. The researchers in that study found that a significant increase in woody vegetation occurred in only about half of the municipalities that lost population. Thus, depopulation does not necessarily imply land-use change. In their analysis of annual satellite land cover maps, they found that 180,000 km^2 of forest was lost

between 2001 and 2010, with the majority of deforestation occurring in South America (92%), particularly in Argentina, Brazil, Bolivia and Paraguay. Much of this land is in soybean production and cattle-grazing to meet the increasing global demand for meat. DeFries *et al.* [273] recently demonstrated that increases in rates of deforestation are closely linked to increases in urban populations and their demand for agricultural products rather than changes in rural populations. In Central America, temporary international migration of members of smallholder households has been indirectly associated with a lack of reforestation; remittances are spent on owning more land, and less household labour favours a transition to cattle production. This is relatively safe and risk-averse compared to row crop production, but it increases forest loss and land degradation and thus decreases the mitigation and adaptation potential [274].

Rural–rural migration offers a livelihood adaptation strategy for rural people facing stresses and shocks due to climate change, but it can also increase migrants' vulnerability. In Vietnam, migrants to the fertile Central Highlands aim to increase their economic livelihoods by producing coffee destined for international markets. Instead of settling permanently, many circulate between their new and origin communities because their social networks that remain at home allow them to avoid some of the risks of permanent relocation [275]. For example, family members in the community of origin may look after the migrants' children, take care of land and assets and provide access to loans. The lack of formal credit institutions at the new destination means that the community of origin may provide continual financial support instead of successful migrants' sending remittances home. Such social networks expose remaining household members to risk if ventures fail because of economic, social and environmental conditions. Both the migrants and origin households may then require loans to take further livelihood risks. In these cases, migration may drive both households into further poverty. Reforming Vietnam's household registration system to allow migrants access to banking, lending and other public services at their new locations could reduce the risks of such outcomes [275].

In the project 'Where the Rain Falls: Climate Change, Food and Livelihood Security, and Migration', researchers have examined rainfall, food security and human migration in eight countries in Asia, Africa and Latin America [267], mainly in agricultural areas. Four distinct household migration profiles were identified, varying along a spectrum from resilience, where migration is one of a variety of adaptation measures that progressively reduce climate sensitivity, to vulnerability, where migration either is difficult or exacerbates sensitivity to climatic stressors. Although national and regional contexts affect migration, household characteristics were discovered to be most

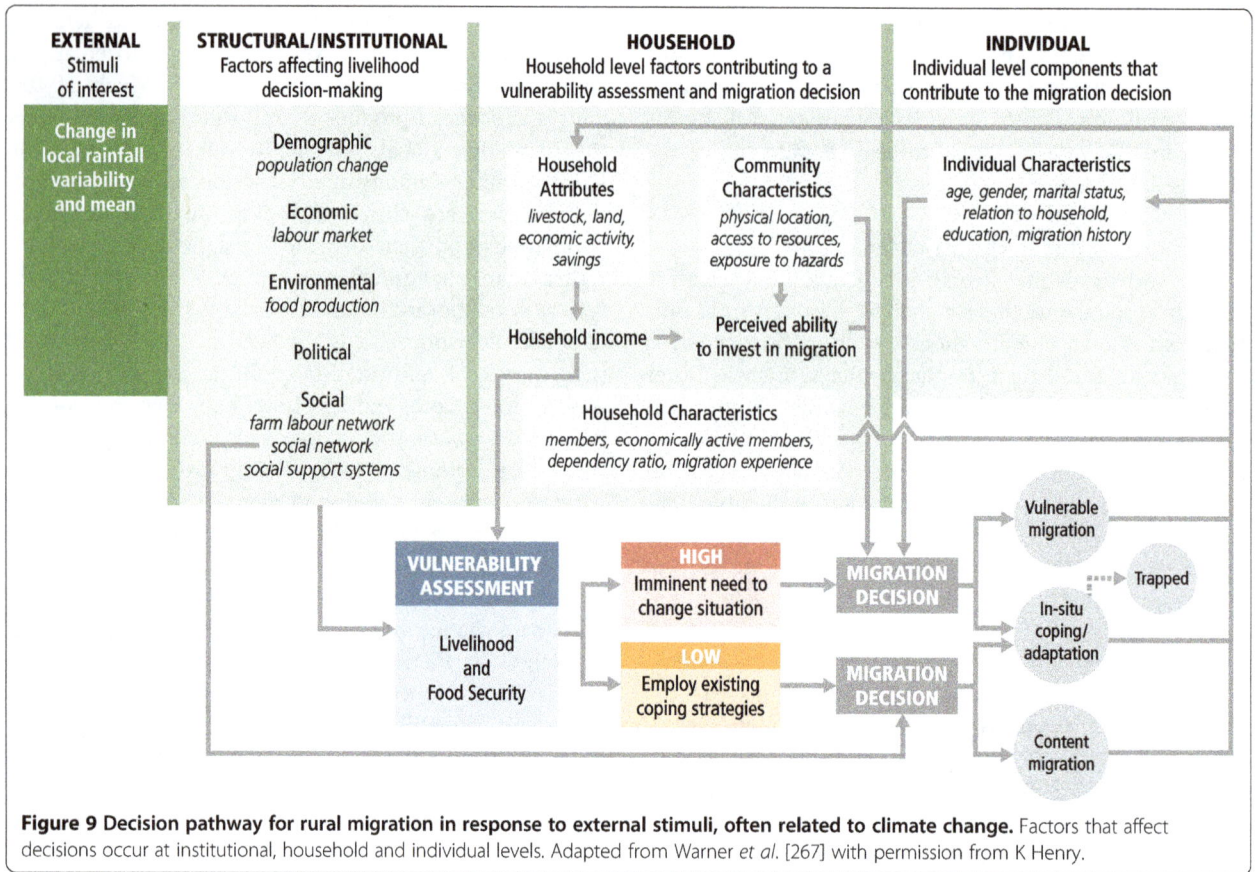

Figure 9 Decision pathway for rural migration in response to external stimuli, often related to climate change. Factors that affect decisions occur at institutional, household and individual levels. Adapted from Warner *et al*. [267] with permission from K Henry.

important for migration-related decisions and outcomes. For example, migration was generally erosive for the poor and those with small land holdings. Household size and composition, land ownership, asset base, degree of livelihood diversity and education levels were associated with migration strategies that increased resilience, such as non-agricultural jobs or diversified livelihoods [267]. One of the 'Where the Rain Falls' project case studies is the Mantaro Basin of Peru, where pressures to migrate stem from lower precipitation that reduced farmer and herder incomes [276]. Two livelihood and migration profiles in the Mantaro Basin were identified in response to climatic vulnerability. Lowland farmers who often commuted on a daily basis for casual urban employment used their proximity to the city to diversify their livelihoods. In contrast, herders farther from the city were forced to migrate for longer periods or permanently, in the absence of other options, and therefore were generally more vulnerable.

The act of migration has a risk dimension, whether it is a positive form of adaptation or part of erosive coping strategies. Understanding the cultural dimensions of risk-taking under climate uncertainty is crucial for determining migration decisions, especially as the necessity for climate-driven planned resettlement becomes more urgent [42]. Although outmigrants are mainly men, the outcomes of climate-change–induced migration are likely to be highly gendered because women are disproportionately affected. Women tend to be poorer and less educated and to have lower health status and limited direct access to, or ownership of, natural resources [45]. It will become more feasible to identify risk-prone agricultural areas and circumstances if models of biophysical aspects of climate change and land use also take into consideration factors that influence migration decisions, such as landlessness, land tenure and distribution issues, as well as the role of social networks that facilitate resilience and adaptation in rural areas as well as escape from poverty traps [167]. Climate-induced outmigration from rural areas involves mitigation and adaptation issues related to urban and peri-urban outcomes, such as increased GHG emissions due to urban sprawl on land that once supported food production [11]. Interdisciplinary work is needed to understand effective strategies for developing and preserving smallholder agriculture near cities, expanding urban and periurban agriculture, managing urban growth for farmland preservation, connecting agricultural producers with local urban markets, ensuring availability of agricultural labour and enabling diversified rural livelihood systems. Such strategies will have combined benefits for climate change mitigation and adaptation.

Metrics for vulnerability assessment, food security and ecosystem services in agricultural landscapes

Science-based actions within CSA require integrated data sets and sound metrics for testing hypotheses about feedback regarding climate, weather data products and agricultural productivity, such as the nonlinearity of temperature effects on crop yield [277], and the assessment of trade-offs and synergies that arise from different agricultural intensification strategies. Approaches range from the development of broad indicators for identifying differences in climate vulnerability over large spatial scales down to the use of finely disaggregated spatial metrics [278]. New and innovative research and policy designs, as well as cooperative arrangements among and between government agencies, research institutions and civil society, have the potential to implement monitoring and assessment systems for decision-making. Examples presented here demonstrate how biometeorological, economic and sociological indicators can be used in vulnerability assessments and show nuances that must be addressed with respect to scale.

Novel outcomes, such as nonlinear effects of climate change on agricultural productivity (for example, US maize), are emerging based on the use of large-scale data sets, indicating that environmental change may drive agricultural productivity in unexpected ways [277]. For example, Lobell et al. [5] examined harvest and daily weather data derived from more than 20,000 historical maize trials conducted by the International Maize and Wheat Improvement Center and private seed companies in Sub-Saharan Africa from 1999 to 2007. 'Optimal management' and 'drought stress' were the two most common scenarios under which maize was grown. Final yield was reduced to the following different extents due to warmer temperatures: by 1% under optimal rain-fed conditions and 1.7% under drought conditions for every degree day spent above 30°C. Lobell et al. [5] suggested that a 1°C warming would lead to negative yield where maize is presently grown under optimal management (roughly 65% of the area) in Sub-Saharan Africa, whereas all areas in this region would show decreased yield of as much as 20% under drought stress. Similarly, in the United States, which generates 40% of global maize production, predicted increases in interannual weather variability (temperature and precipitation) could result in an 18% decrease in maize yields by 2030 to 2050 in comparison to the period from 1980 to 2000, along with increasing volatility in annual yields [279]. Expansion of cropland in other regions and retention of speculative inventories (that is, holding volumes for higher price earnings) may offset the volatility. Here metrics of climate and indicators of crop productivity and other agronomic factors predicted crop response to climate warming and drought over a widespread region, setting the stage for more research on

how adaptation measures, such as improving soil moisture and breeding for drought and heat tolerance, could be used to reduce vulnerability in the future [5].

Metrics that incorporate human ecology are integral to enabling the CSA strategy. Vital Signs [280] is a monitoring programme for changes in human well-being, agriculture and ecosystem services and is designed to provide metrics in rapidly expanding and intensifying agricultural landscapes in Africa, leading to integrated approaches that support food security (Figure 10). A primary goal of Vital Signs is characterizing the uncertainty and quantifying the sampling intensity needed to achieve different levels of accuracy and statistical power to detect change. Information gathered in the initial phase will be further evaluated for its overall utility and delivery cost. Measurements collected by Vital Signs participants are based on hierarchical spatial scales to provide integrated information that can inform structural relationships and counterfactuals involved in decision-making from the global to household scale. The global perspective facilitates comparisons between different regions ($250,000 \text{ km}^2 \cdot \text{region}^{-1}$), whereas regional measurements deliver information at the scale on which agricultural investments are made. Information collected at the landscape scale (10 to 20 units per region) measures the relationships between agricultural intensification, water availability, soil health and other ecosystem services, together with human well-being. Plot-level (1 ha) data reflect agricultural production, including seed selection, fertilizer type and application rate, as well as crop yield response. At the household level, surveys are employed to collect information on health, nutritional status, income and assets. Stakeholder planning meetings and participatory research established both at the onset and throughout the project are integral to garnering active engagement in Vital Signs.

Prioritizing allocation of resources and focusing policies on vulnerable regions requires metrics to assess susceptibility to a lack of food security due to climate change [281]. Biophysical climate indicators derived from global climate-change models and food insecurity indicators (that is, availability, access and utilization) can serve as such metrics. As an example of this approach, Ericksen et al. evaluated hotspots of vulnerability using the overlap among indicators of global climate (for example, rainfall variability, number of reliable growing degree days, and change in mean annual temperature) and food security indicators across the global tropics [281]. The latter were composed of availability (for example, crop yield and mean food production indices), access (for example, GDP per capita, transport time to markets, and monthly staple food prices) and utilization (for example, malnutrition prevalence and proportion of the population using unimproved water source). Future

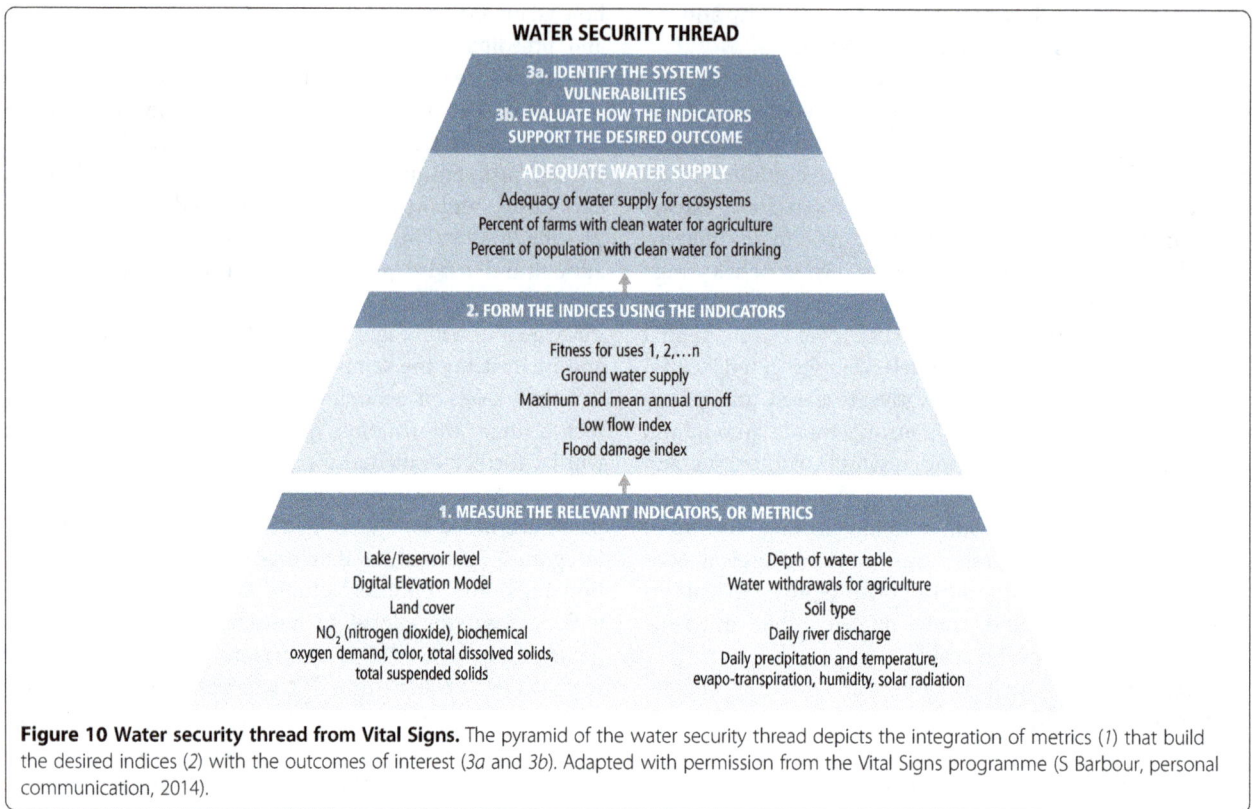

Figure 10 Water security thread from Vital Signs. The pyramid of the water security thread depicts the integration of metrics (*1*) that build the desired indices (*2*) with the outcomes of interest (*3a* and *3b*). Adapted with permission from the Vital Signs programme (S Barbour, personal communication, 2014).

vulnerability was depicted by existing resource pressure (for example, annual population growth and agricultural area per capita). The resulting index of vulnerability reflected three central components: exposure of populations to the impacts of climate change, sensitivity of food systems to these impacts and coping capacity of populations to address these impacts. With this vulnerability index, it was possible to rank the most highly exposed regions, leading to the emergence of southern Africa as a highly exposed region, as well as areas within Brazil, Mexico, Pakistan, India and Afghanistan. This approach is limited by the following factors: The data represent only current food security levels; data are gathered only at the national level, which masks variability within regions and among households; and other data are needed on climate-change exposure and on food security variables other than crop yields and utilization, such as food distribution and equity.

Systems delivering real-time indicators and metrics that are tied closely to management decisions and current conditions allow science and policymaking entities to progress from using lagging indicators to finding leading indicators that can be used to identify when and where thresholds of climate-change responses will occur [112]. Indicators and metrics are often used to support public goods and services, so better standards and codified practices that support shared vocabulary and ontology will reduce the costs and streamline efforts for curating and disseminating such information. Research designed to develop metrics that inform global to local social networks for data collection, sharing and integration can also be leveraged for extension efforts. The identification of efficient and location- and situation-specific sets of indicators will complement efforts to construct human capital, social awareness and consensus regarding specific issues, leading to action strategies and policy guidelines across various temporal and spatial scales.

Theme 3
Integrative and transformative institutional and policy issues: bridging across scales
Figures 11 and 12 provide an overview of some of the main points covered in each session of the 2013 Global Science Conference on Climate-Smart Agriculture. The relative emphasis on mitigation of GHG emissions versus adaptive capacity to climate change (or both) varied depending on the session topic. CSA strives for food security, adaptation, mitigation and resilience, but not all of these are achieved in the same context. The session topics often invoked multiple scientific disciplines to inform further action and problem-solving strategies in support of CSA goals in the context of the session topic, but further integration across these topics and disciplines is necessary. Scientific uncertainties are inherent in climate science,

SESSION CONTENT FROM THEME 1 (2.0):
FARM AND FOOD SYSTEM ISSUES: SUSTAINABLE INTENSIFICATION, AGRO-ECOSYSTEM MANAGEMENT, AND FOOD SYSTEMS

Mitigation and/or adaptation	Interdisciplinary science	Scientific uncertainties	Social controversies	Stakeholder programs mentioned	Social-ecological pathways
2.1 CROP PHYSIOLOGY & GENETICS					
Mainly adaptation	Molecular biology, ecophysiology, agronomy	Plant response to high [CO2] + high temperatures	Genetic engineering	Regional crop breeding networks	Participatory plant breeding
2.2 LIVESTOCK MANAGEMENT & ANIMAL HEALTH					
Both	Animal science, range science, veterinary medicine, economics	Feasibility of scaling up of combined mitigation and adaptation strategies	Emphasis on low GHG emissions under poverty	Global programs for REDD+; for managing diseases	Management decisions based on livelihood criteria
2.3 NITROGEN MANAGEMENT					
Both	Agronomy, plant nutrition biogeochemistry	Decision-support tools effective across agroecosystems	Emphasis on low GHG emissions under poverty	Agricultural extension programs	Knowledge systems to enhance adoption of practices
2.4 FARMER DECISION MAKING					
Both	Social & political sciences, agronomy	Overcoming barriers to adoption of new practices	Dealing with perceptions of low climate risk	Social networks in farming communities	Short- & long-term planning for uncertainty
2.5 CLIMATE RISK MANAGEMENT					
Adaptation	Economics, remote-sensing, agronomy	Design of safety nets for uncertain climate shocks	Insurance reduces incentives for adaptation	Index-based livestock insurance programs	Bundled strategies for household welfare
2.6 ENERGY & BIOFUELS					
Mainly mitigation	Biogeochemistry, life-cycle analysis, agronomy, economics, geography	Projections of costs & benefits from biofuels	Food price increase attributed to biofuels	Energy directives and action plans	Biofuel feedstock value chains that support food security

Figure 11 Conference session 2.0 content within theme 1. Farm and food system issues: sustainable intensification, agroecosystem management and food systems[a]. [a]GHG, Greenhouse gas; REDD+, Reducing Emissions from Deforestation and Forest Degradation.

given the difficulty of forecasting climate and its interactions with other aspects of human-induced environmental change. The examples that are mentioned here require intensified scientific activity, formation of knowledge networks, and involvement of many relevant stakeholders to obtain better information to support decision-making (see also [135-137,152]). Also, there are clear social controversies challenging CSA, often derived from assumptions and questions of equity and legitimacy, such as who will implement a response to climate change and how this will occur. To obtain buy-in from vulnerable populations and countries, such issues must gain the forefront in discussions of CSA science and policy among the diverse set of stakeholders described in the Introduction section above. Many of the stakeholder-driven programmes mentioned in the conference sessions exist at regional and global levels, as climate science is often funded for large-scale initiatives. As stated previously, this article and the conference presentations do not emphasize the local knowledge-to-action processes

that are essential for transformations towards climate preparedness. Nonetheless, some of the possible pathways towards such socioecological approaches to fostering greater participation and advancement of CSA objectives are shown for each of the session topics. Clearly, science must play an active and central role in developing the information base that will support food security, adaptation and mitigation in CSA and new types of inclusive, participatory decision-making as well as knowledge exchange processes [135,152].

Inter- and transdisciplinary scientific approaches are principal both to our understanding of how socioecological systems support the adaptive management and governance that are essential to long-term human provisioning of food and to the establishment of science–policymaking dialogues to plan for the future [47,282-284]. These actions are keys to assessing trade-offs of mitigation in context-specific situations, such that resource-poor farmers are supported rather than undermined by CSA. To realize the CSA objectives of increased food security,

SESSION CONTENT FROM THEME 2 (3.0):
LANDSCAPE AND REGIONAL ISSUES: LAND USE, ECOSYSTEM SERVICES AND REGIONAL RESILIENCE

Mitigation and/or adaptation	Interdisciplinary science	Scientific uncertainties	Social controversies	Stakeholder programs mentioned	Social-ecological pathways
3.1 CLIMATE CHANGE & FOOD SECURITY					
Adaptation	Agroecology, soil science, land use & geography, political science	Ability of models to simulate multiple types of processes	'Mitigate for 2 °C but adapt for 4 °C'	Policy makers	Strategize with spatial/temporal modeling & stakeholder dialog
3.2 SOIL CARBON					
Both	Soil science, soil microbiology, biogeochemistry, agroecology	Accurate models of C, GHG emissions & tradeoffs	Support to create multifunctional landscapes	Science-policy interface of UN conventions	Participatory research across landscapes
3.3 WATER MANAGEMENT					
Mainly adaptation	Hydrology, land use & geography, atmospheric science, economics	Forecasting future water availability and demand	Governance, e.g. failing water rights, access & storage	Regional civil society & policy makers	Multi-stakeholder dialog for vulnerability and adaptation scenarios
3.4 MANAGING FOREST BIODIVERSITY					
Both	Forestry, agroecology, land use & geography, economics, political science	Appropriate methods & ecosystems for payments & rewards	Excluding the vulnerable as beneficiaries of REDD+	Multi-scale & multi-institutional planning for REDD+	Public-private partnerships for socially 'green' & 'rights-based' approaches
3.5 RURAL MIGRATION					
Adaptation	Social and political sciences, land use & geography, economics	Quantifying impact of climate shocks on out-migration	Disproportionate adverse effects on women and children	Stakeholder participation not clearly identified	Social networks that provide income, ties to the land & resilience
3.6 METRICS FOR VULNERABILITY					
Both	Ecology, social sciences, remote-sensing, land use & geography, economics	Biophysical & social data at scales that detect climate effects	Choice of indicators to direct policy decisions	Government and policy makers	Stakeholder involvement in data collection & for use of metrics

Figure 12 Conference session 3.0 content from theme 2. Landscape and regional issues: land use, ecosystem services and regional resilience[a].
[a]GHG, Greenhouse gas; REDD +, Reducing Emissions from Deforestation and Forest Degradation.

resilience, mitigation and adaptation, scientific research supports awareness, analytical capacity and the evidence base to understand the impacts of climate change on agricultural growth strategies and food security, and identify climate smart options suitable to the local context [15]. How does research better inform the institutional, financial and knowledge-sharing arrangements to create a sense of possibility for transformative processes that reduce vulnerability and increases climate preparedness? Truly transformative solutions tap into a sense of possibility for positive action, and, as in business value propositions, there is a promise of goods and services to be delivered and experienced [47]. Yet, a 'doomsday' attitude has permeated much of the agricultural science regarding climate change, emphasizing harsh potential impacts under business-as-usual scenarios (Figure 2). Although it is effective in stimulating awareness and action in some sectors, CSA research is potentially more conducive to achieving food security, adaptation, mitigation and resilience. Examples include models that go beyond impacts to include adaptation and transformation at either the farm or landscape scale (for example, see [211]), capacity approaches to examine multifunctional solutions within the socioecological system and direct evidence for situations, options and scenarios which increase human behaviours that build natural capital and resilience. Action-oriented research can also show how public-private partnerships can be used successfully to develop technologies, policies and approaches that may lead to sustainable food production and consumption patterns in a changing climate.

Uncertainty is one of the most difficult obstacles to determining priorities for CSA research. Not only is future climate uncertain, but so also is the existence and operation of the institutions that are and will be involved in adaptation, mitigation and resilience. Uncertainty can breed scepticism about the urgency to plan for climate change, especially in agricultural communities and industries that already deal with large annual variability in production and prices. Thus, uncertainty is a barrier to mitigation and adaptation among some of the stakeholders whose investment, engagement and broad agricultural knowledge are critical for designing better research on coping strategies. CSA recognizes that the unfolding of decision-making processes, their translation into action and the formation of adaptive capacity depend on the socioecological contexts in which farmers are embedded (for example, the vital role of social networks in rural communities) (Figure 1). The ways of addressing uncertainty are likely to differ greatly among communities and socioecological systems, and research is needed to understand how to approach uncertainty in different contexts.

Although poverty can sometimes drive collective action, such as for improved food security in Kenya and Uganda through risk-sharing and pooling of labour and other limited assets [16], the least food-secure may be less likely to adopt new CSA practices because innovation implies additional costs before benefits can be realized [285]. Research on adoption of new farming technology and practices is needed to understand how upfront costs, lost income, worries about personal health and additional risks assumed during the conversion period can present formidable barriers to farmers [149], even if the new practices leverage ecological processes to improve sustainability and production [14]. For instance, diffusion of new germplasm with specialized traits (for example, drought tolerance) to targeted end users may suffer slow adoption even though new regional and local cultivars will likely be adapted to the range of conditions and management practices employed during climate change. To illustrate this point, modelled diffusion of a drought-tolerant variety among vulnerable (highly risk averse) farmers took four times longer than it did among those less vulnerable (less risk averse), underscoring the need for consideration of how seed prices affect the access of vulnerable farmers to new crop varieties [286]. Synthesis of information on how CSA practices have been facilitated by specific policy interventions, leading to broad community support, also is needed. A better understanding of how social benefits such as access to food and healthcare, rights to land and water, markets, and financing situations facilitate adoption of new farming practices or technologies will inform governance decisions [14].

Collective action for climate preparedness and problem-solving has already been effective in some situations. Safety nets for the poorest and most vulnerable households usually occur in the form of humanitarian relief and food aid, cash payments, agricultural inputs and public works [14], often after a critical event has occurred. Instead, communities can collectively plan safety-net strategies and resource transfers that are predictable and flexible enough to be scaled up and then scaled down when the crisis subsides [13,14]. CSA research on learning, knowledge-sharing and social network analysis can help build awareness, early-warning indicators and criteria for benefit transfers for disaster responses and also effectively combine local collective action with national and/or international aid. Enhancing human and social capital, such as for childhood nutrition, entrepreneurship by women, and synergies between fuel use and C sequestration in trees, also rehabilitates household and community assets. Proactive planning will be more effective than reactive responses to a disastrous climate event, and research can help increase understanding of how adaptation policies must be designed accordingly [204].

Furthermore, collective action at the institutional scale is essential to avoiding conflicts that result from climate change. For example, institutional transboundary water agreements are associated with lower risk of conflict during water scarcity, but even one weak link in the communication, coordination and cooperation between coriparian nations will reduce their adaptive capacity to respond to new changes in hydrology, thus increasing the potential for risk and disputes [206]. So far, climate change has rarely been incorporated into such agreements. Collective action at the institutional scale could also address changing migration patterns of rural–urban connections that are likely due to extreme climate events and climate change and which will have potentially large ramifications on food production and food security, land tenure and cultural integrity [42] (Figure 10). At this point, research is needed to better understand how climate affects the dynamics of the rural labour force and thus on the stability of local food production for rural communities and nearby cities [267,287].

To realize CSA, research on targeted financing is essential, especially in support of the most vulnerable. Upfront investment to plan and start implementation strategies is required, as is research to develop monitoring systems designed to track climate-related human responses by utilizing consistent metrics that demonstrate private benefits along with public goods (for example, GHG mitigation). Already existent funds, such as the Adaptation Fund established under the Kyoto Protocol [288], and the International Fund for Agricultural Development's Adaptation for Smallholder Agriculture Program [289], can improve smallholders' access to climate-smart assessments, technologies and institutions related to sustainable management of forests, providing up to 16 million additional jobs globally and increasing household income in rural areas as a result of restoring degraded forest [290]. Larger-scale investments, such as financing infrastructure for water resources and carbon capture, can potentially be provided by the Green Climate Fund [291], and private finance may also play a role. As climate finance develops, research shares a role in prioritizing investments and effective financing solutions and in monitoring outcomes.

Investment in research on food systems that are resilient to climate shocks may be more likely to occur if CSA expands beyond the agricultural sector. As examples, CSA research could more explicitly involve issues related to: (1) local, national and regional food trade, including governance and regulations, food safety, roads and infrastructure, and value chain coordination; (2) flexibility in financial arrangements, insurance and planning to cope with, and be responsive to, variability in climate and markets; and (3) integration of the interdisciplinary research to form a more holistic and service-oriented approach based on science to inform policy. For research to be utilized most effectively in policies related to CSA, pathways for communication of the latest scientific progress and research results must be established within relevant time frames. Communication must span sectors and scales in which policymakers and other stakeholders operate, crossing boundaries between scientists and local, regional and global actors such as nongovernmental organizations, governmental agencies, corporations and broad social and media networks [290].

CSA strategies support the realization of a broader green economy concept that acknowledges 'the sum total of all ecosystem services and how they collectively provide the complete life support system we need' [292], p. 9. In practice, market prices, costs, and benefits for the ecosystem services related to carbon sequestration, clean water production, flood protection and grass forage have been quantified. In Cameroon, for instance, the value (in $US\$\cdot ha^{-1}\cdot yr^{-1}$) attributed to the forest's contribution to climate and flood control is 1.3- to 2.6-fold greater than that of the timber, fuel wood and nontimber products. Coordinated action resulting from CSA and green economy research not only realizes the improvement of livelihoods and food security through mitigation and adaptation to climate change but also creates cobenefits for ecosystem services and sustainable use of natural capital and enables evaluation of a broader set of trade-offs associated with a certain course of action.

Conclusions

Disciplinary, interdisciplinary and transdisciplinary scientific approaches play a fundamental and profound role in developing understanding of the processes underlying CSA and serve as partners in enumerating priorities for CSA. They form a crucial element in the knowledge base needed to implement CSA actions and manifest future transformative changes in agriculture in a changing climate. Global science conferences on CSA have already been influential in assembling scientists and other stakeholders to share knowledge [17,49]. A third conference in Montpellier, France, is planned for 2015 with the following agenda items: discussion key scenarios in agriculture and food systems, identifying priorities for early

action and designing a roadmap for moving forward with an action plan. These objectives set the stage for a much stronger emphasis on knowledge-to-action frameworks, capacity-building and the changes in human behaviour and social infrastructure that are necessary for adaptation and resilience [133,152,293]. The momentum that has already built among the science community for CSA forms the foundation for critical engagement by more researchers in fundamental and applied studies. To this end, establishing a more formal governance mechanism to embed science in the information base for the CSA Alliance, would be a vital step in developing priorities, scientific engagement and funding to support the knowledge needed for policymaking decisions.

Competing interests
The authors declare that they have no competing interests.

Authors' contributions
KLS, AKH and LEJ cowrote the manuscript. AJB, MRC, AC, CJC, JLH, KH, JWH, WRH, LSJ, BMJ, EK, RL, LL, MNL, SM, RP, MPR, SSS, WMS, MS, PT, SJV, SMW and EW contributed content. All authors read and approved the final manuscript.

Acknowledgements
We acknowledge Jill E Walker for graphic design, Ria D'Aversa for maintaining the bibliographic content and Kayla Burns for copyediting. We also acknowledge all conference participants who shared their knowledge and ideas at the Global Science Conference for Climate-Smart Agriculture held in Davis, CA, USA, in March 2013. We thank the session leaders, who also are authors on this manuscript, for composing summaries of the information presented in individual sessions at the conference and synthesizing discussions by participants. These summaries formed the basis of this paper. This article was funded by a Programmatic Initiative from the College of Agriculture and Environmental Science at the University of California, Davis (to LEJ).

Author details
[1]Crops Pathology and Genetics Research Unit, Agricultural Research Service, United States Department of Agriculture (ARS/USDA), c/o Department of Viticulture and Enology, RMI North, Rm. 1151, 595 Hilgard Lane, Davis, CA 95616, USA. [2]Department of Land, Air and Water Resources, University of California at Davis, One Shields Avenue, Davis, CA 95616, USA. [3]Department of Plant Sciences, University of California at Davis, One Shields Avenue, Davis, CA 95616, USA. [4]Department of Agricultural and Resource Economics, University of California at Davis, One Shields Avenue, Davis, CA 95616, USA. [5]Climate Smart Agriculture Project, Food and Agriculture Organization of the U.N., Viale delle Terme di Caracalla, 00100 Rome, Italy. [6]eWater, University of Canberra Innovation Centre, Building 22, University Drive South, Bruce ACT 2617, Australia. [7]National Laboratory for Agriculture and the Environment, ARS/USDA, Ames, IA, USA. [8]Where the Rain Falls, CARE France, 71 rue Archereau, Paris 75019, France. [9]School of Global Environmental Sustainability, Colorado State University, 108 Johnson Hall, Fort Collins, CO 80523, USA. [10]Department of Biological and Agricultural Engineering, University of California at Davis, One Shields Avenue, Davis, CA 95616, USA. [11]Department of Animal Science, University of California at Davis, One Shields Avenue, Davis, CA 95616, USA. [12]Environmental Sciences, Wageningen University, P.O. Box 47, 6700AA Wageningen, the Netherlands. [13]Agricultural and Development Economic Analysis Division, Food and Agriculture Organization of the U.N., Viale delle Terme di Caracalla, 00100 Rome, Italy. [14]Department of Environmental Science and Policy, Center for Environmental Policy and Behavior, University of California at Davis, One Shields Avenue, Davis, CA 95616, USA. [15]Environment and Production Technology Division, International Food Policy Research Institute (IFPRI), 2033 K St., NW, Washington DC 20006-1002, USA. [16]World Agroforestry Center (ICRAF), P.O. Box 30677, 00100 Nairobi, Kenya. [17]Plant, International Maize and Wheat Improvement Center, Consultative Group on International Agricultural Research (CGIAR) Apdo, Postal 6-641, 06600 Mexico, D.F., Mexico. [18]Food- and Water-borne Disease Research Program, College of Veterinary

Medicine, Washington State University, PO Box 646610, Pullman, WA 99164-6610, USA. [19]Department of Environmental Science and Policy, University of California at Davis, One Shields Avenue, Davis, CA 95616, USA. [20]Plant Sciences, Wageningen University, P.O. Box 563, 6700AN Wageningen, the Netherlands. [21]Department of Landscape Architecture, University of California at Davis, One Shields Avenue, Davis, CA 95616, USA. [22]Climate Change, Agriculture and Food Security, Consultative Group on International Agricultural Research (CGIAR), Department of Plant and Environmental Sciences, University of Copenhagen, Rolighedsvej 21, DK-1958 Frederiksberg C, Denmark. [23]Gund Institute for Ecological Economics and Rubenstein School of Environment and Natural Resources, University of Vermont, 617 Main Street, Burlington, Vermont 05405, USA. [24]Department of Agriculture and Resource Economics, University of California at Davis, One Shields Avenue, Davis, CA 95616, USA.

References

1. Morton JF: **The impact of climate change on smallholder and subsistence agriculture.** *Proc Natl Acad Sci U S A* 2007, 104:19680–19685.
2. Reidsma P, Ewert F, Lansink AO, Leemans R: **Adaptation to climate change and climate variability in European agriculture: the importance of farm level responses.** *Eur J Agron* 2010, 32:91–102.
3. Vermeulen SJ, Campbell BM, Ingram JSI: **Climate change and food systems.** *Annu Rev Environ Resour* 2012, 35:195–222.
4. Intergovernmental Panel on Climate Change, Solomon S, Qin D, Manning M, Chen Z, Marquis M, Averyt KB Tignor MMB, Miller HL Jr: *Climate Change 2007: The Physical Science Basis (Contribution of Working Group I to the Fourth Assessment Report of the Intergovernmental Panel on Climate Change (IPCC)).* Cambridge, UK: Cambridge University Press; 2007. [http://www.ipcc.ch/pdf/assessment-report/ar4/wg1/ar4_wg1_full_report.pdf] (accessed 25 July 2014).
5. Lobell DB, Schlenker W, Costa-Roberts J: **Climate trends and global crop production since 1980.** *Science* 2011, 333:616–620.
6. Cline WR: *Global Warming and Agriculture: Impact Estimates by Country.* Washington DC: Center for Global Development; 2007.
7. Jarvis A, Lau C, Cook S, Wollenberg E, Hansen J, Bonilla O, Challinor A: **An integrated adaptation and mitigation framework for developing agricultural research: synergies and trade-offs.** *Exp Agric* 2011, 47:185–203.
8. Knox J, Hess T, Daccache A, Wheeler T: **Climate change impacts on crop productivity in Africa and South Asia.** *Environ Res Lett* 2012, 7:034032.
9. Rosenzweig C, Elliott J, Deryng D, Ruane AC, Müller C, Arneth A, Boote KJ, Folberth C, Glotter M, Khabarov N, Neumann K, Piontek F, Pugh TAM, Schmid E, Stehfest E, Yang H, Jones JW: **Assessing agricultural risks of climate change in the 21st century in a global gridded crop model intercomparison.** *Proc Natl Acad Sci U S A* 2014, 111:3268–3273.
10. Myers SS, Zanobetti A, Kloog I, Bloom AJ, Carlisle EA, Dietterich LH, Fitzgerald G, Hasegawa T, Holbrook NM, Huybers P, Leakey ADB, Nelson R, Ottman MJ, Raboy V, Sakai H, Sartor KA, Schwartz J, Seneweera S, Tausz M, Usui Y: **Rising CO_2 threatens food quality.** *Nature* 2014, 510:139–142.
11. Wheeler T, von Braun J: **Climate change impacts on global food security.** *Science* 2013, 341:508–513.
12. World Bank: *Climate-Smart Agriculture: Increased Productivity and Food Security, Enhancing Resilience and Reduced Carbon Emissions for Sustainable Development, Opportunities and Challenges for a Converging Agenda: Country Examples.* Washington, DC: World Bank; 2011.
13. World Bank: *World Development Report 2014: Risk and Opportunity: Managing Risk for Development.* Washington DC: World Bank; 2013. [http://siteresources.worldbank.org/EXTNWDR2013/Resources/8258024-1352909193861/8936935-1356011448215/8986901-1380046989056/WDR-2014_Complete_Report.pdf] (accessed 25 July 2014).
14. Food and Agriculture Organization of the United Nations (FAO): *Climate-Smart Agriculture Sourcebook.* Rome: FAO; 2013.
15. Climate Smart Agriculture: *The 2nd Global Science Conference on Climate-Smart Agriculture.* 2014. [http://climatesmart.ucdavis.edu/docs/CSANoteShort.pdf] (accessed 5 August 2014).
16. Neufeldt H, Jahn M, Campbell BM, Beddington JR, DeClerck F, De Pinto A, Gulledge J, Hellin J, Herrero M, Jarvis A, LeZaks D, Meinke H, Rosenstock T, Scholes M, Scholes R, Vermeulen S, Wollenberg E, Zougmoré R: **Beyond climate-smart agriculture: toward safe operating spaces for global food systems.** *Agric Food Secur* 2013, 2:12.
17. *Wageningen Statement: Climate-Smart Agriculture – Science for Action. The Global Science Conference on Climate-Smart Agriculture (GSCSA).* 2011. [http://ccafs.cgiar.org/sites/default/files/assets/docs/the_wageningen_statement__26-10_14_17_final.pdf] (accessed 25 July 2014).
18. Carpenter S, Walker B, Anderies JM, Abel N: **From metaphor to measurement: resilience of what to what?** *Ecosystems* 2001, 4:765–781.
19. Holling CS: **Understanding the complexity of economic, ecological, and social systems.** *Ecosystems* 2001, 4:390–405.
20. Walker B, Holling CS, Carpenter SR, Kinzig A: **Resilience, adaptability, and transformability in social-ecological systems.** *Ecol Soc* 2004, 9:5.
21. Obrist B: **Multi-layered social resilience: a new approach in mitigation research.** *Progr Dev Stud* 2010, 10:283–293.
22. Cumming GS: *Spatial Resilience in Social-Ecological Systems.* New York: Springer Science; 2011.
23. Cabell JF, Oelofse M: **An indicator framework for assessing agroecosystem resilience.** *Ecol Soc* 2012, 17:18.
24. Easterling DR: **Climate extremes: observations, modeling, and impacts.** *Science* 2000, 289:2068–2074.
25. Battisti DS, Naylor RL: **Historical warnings of future food insecurity with unprecedented seasonal heat.** *Science* 2009, 323:240–244.
26. Bloom AJ, Burger M, Rubio Asensio JS, Cousins AB: **Carbon dioxide enrichment inhibits nitrate assimilation in wheat and *Arabidopsis*.** *Science* 2010, 328:899–903.
27. Yin X: **Improving ecophysiological simulation models to predict the impact of elevated atmospheric CO_2 concentration on crop productivity.** *Ann Bot* 2013, 112:465–75.
28. Vermeulen SJ, Aggarwal PK, Ainslie A, Angelone C, Campbell BM, Challinor AJ, Hansen JW, Ingram JSI, Jarvis A, Kristjanson P, Lau C, Nelson GC, Thornton PK, Wollenberg E: **Options for support to agriculture and food security under climate change.** *Environ Sci Policy* 2012, 15:136–144.
29. Gutierrez AP, Ponti L, d'Oultremont T, Ellis CK: **Climate change effects on poikilotherm tritrophic interactions.** *Clim Change* 2007, 87(1 Suppl):S167–S192.
30. Antón J, Cattaneo A, Kimura S, Lankoski J: **Agricultural risk management policies under climate uncertainty.** *Glob Environ Change* 2013, 23:1726–1736.
31. Jackson LE, Pulleman MM, Brussaard L, Bawa KS, Brown GG, Cardoso IM, de Ruiter PC, García-Barrios L, Hollander AD, Lavelle P, Ouédraogo E, Pascual U, Setty S, Smukler SM, Tscharntke T, van Noordwijk M: **Social-ecological and regional adaptation of agrobiodiversity management across a global set of research regions.** *Glob Environ Change* 2012, 22:623–639.
32. Stringer LC, Dougill AJ, Thomas AD, Spracklen DV, Chesterman S, Speranza CI, Rueff H, Riddell M, Williams M, Beedy T, Abson DJ, Klintenberg P, Syampungani S, Powell P, Palmer AR, Seely MK, Mkwambisi DD, Falcao M, Sitoe A, Ross S, Kopolo G: **Challenges and opportunities in linking carbon sequestration, livelihoods and ecosystem service provision in drylands.** *Environ Sci Policy* 2012, 19–20:121–135.
33. Minang PA, Duguma LA, Bernard F, Mertz O, van Noordwijk M: **Prospects for agroforestry in REDD + landscapes in Africa.** *Curr Opin Environ Sustain* 2014, 6:78–82.
34. Elliott J, Deryng D, Müller C, Frieler K, Konzmann M, Gerten D, Glotter M, Flörke M, Wada Y, Best N, Eisner S, Fekete BM, Folberth C, Foster I, Gosling SN, Haddeland I, Khabarov N, Ludwig F, Masaki Y, Olin S, Rosenzweig C, Ruane AC, Satoh Y, Schmid E, Stacke T, Tang Q, Wisser D: **Constraints and potentials of future irrigation water availability on agricultural production under climate change.** *Proc Natl Acad Sci U S A* 2013, 111:3239–3244.
35. Meza FJ, Wilks DS, Gurovich L, Bambach N: **Impacts of climate change on irrigated agriculture in the Maipo Basin, Chile: reliability of water rights and changes in the demand for irrigation.** *J Water Resour Plan Manag* 2012, 138:421–430.
36. Nelson GC, Rosegrant MW, Palazzo A, Gray I, Ingersoll C, Robertson R, Tokgoz S, Zhu T, Sulser TB, Ringler C, Msangi S, You L: *Food Security, Farming, And Climate Change to 2050: Scenarios, Results, Policy Options.* Washington DC: International Food Policy Research Institute; 2010. [http://www.ifpri.org/sites/default/files/publications/rr172.pdf] (accessed 25 July 2014).
37. Thornton PK, Gerber PJ: **Climate change and the growth of the livestock sector in developing countries.** *Mitig Adapt Strateg Glob Change* 2010, 15:169–184.
38. Valin BH, Havlík P, Mosnier A, Herrero M, Schmid E, Obersteiner M: **Agricultural productivity and greenhouse gas emissions: trade-offs or synergies between mitigation and food security?** *Environ Res Lett* 2013, 8:035019.
39. Kraxner F, Nordström EM, Havlík P, Gusti M, Mosnier A, Frank S, Valin H, Fritz S, Fuss S, Kindermann G, McCallum I, Khabarov N, Böttcher H, See L, Aoki K,

Schmid E, Máthé L, Obersteiner M: **Global bioenergy scenarios–future forest development, land-use implications, and trade-offs.** *Biomass Bioenergy* 2013, **57**:86–96.

40. Mbow C, Van Noordwijk M, Luedeling E, Neufeldt H, Minang PA, Kowero G: **Agroforestry solutions to address food security and climate change challenges in Africa.** *Curr Opin Environ Sustain* 2014, **6**:61–67.

41. Adger WN: **Social and ecological resilience: are they related?** *Prog Hum Geogr* 2000, **24**:347–364.

42. Adger WN, Barnett J, Brown K, Marshall N, O'Brien K: **Cultural dimensions of climate change impacts and adaptation.** *Nat Clim Chang* 2012, **3**:112–117.

43. Gray CL, Mueller V: **Natural disasters and population mobility in Bangladesh.** *Proc Natl Acad Sci U S A* 2012, **109**:6000–6005.

44. Ifejika Speranza C: **Drought coping and adaptation strategies: Understanding adaptations to climate change in agro-pastoral livestock production in Makueni District.** *Kenya. Eur J Dev Res* 2010, **22**:623–642.

45. Chindarkar N: **Gender and climate change-induced migration: proposing a framework for analysis.** *Environ Res Lett* 2012, **7**:025601.

46. Villamor GB, van Noordwijk M, Djanibekov U, Chiong-Javier ME, Catacutan D: **Gender differences in land-use decisions: shaping multifunctional landscapes?** *Curr Opin Environ Sustain* 2014, **6**:128–133.

47. Vermeulen SJ, Challinor AJ, Thornton PK, Campbell BM, Eriyagama N, Vervoort JM, Kinyangi J, Jarvis A, Läderach P, Ramirez-Villegas J, Nicklin KJ, Hawkins E, Smith DR: **Addressing uncertainty in adaptation planning for agriculture.** *Proc Natl Acad Sci U S A* 2013, **110**:8357–8362.

48. Reed MS, Evely AC, Cundill G, Fazey I, Glass J, Laing A, Newig J, Parrish B, Prell C, Raymond C, Stringer LC: **What is social learning?** *Ecol Soc* 2010, **15**:r1.

49. Speelman EN, Groot JCJ, García-Barrios LE, Kok K, van Keulen H, Tittonell P: **From coping to adaptation to economic and institutional change–trajectories of change in land-use management and social organization in a Biosphere Reserve community, Mexico.** *Land Use Policy* 2014, **41**:31–44.

50. *The Davis Statement–Climate-Smart Agriculture Global Research Agenda: Science for Action.* Climate-Smart Agriculture: Global Science Conference, 19–23 March 2013, University of California, Davis, USA. [http://climatesmart. ucdavis.edu/docs/Davis_Statement_CSA.pdf] (accessed 26 July 2014).

51. de Graaf MA, van Groenigen KJ, Six J, Hungate B, van Kessel C: **Interactions between plant growth and soil nutrient cycling under elevated CO_2: a meta-analysis.** *Glob Chang Biol* 2006, **12**:2077–2091.

52. Thornton PK, Jones PG, Alagarswamy G, Andresen J, Herrero M: **Adapting to climate change: agricultural system and household impacts in East Africa.** *Agric Syst* 2010, **103**:73–82.

53. Haden VR, Niles MT, Lubell M, Perlman J, Jackson LE: **Global and local concerns: what attitudes and beliefs motivate farmers to mitigate and adapt to climate change?** *PLoS One* 2012, **7**:e52882.

54. Stavi I, Lal R: **Agroforestry and biochar to offset climate change: a review.** *Agron Sustain Dev* 2012, **33**:81–96.

55. Weber CL, Matthews HS: **Food-miles and the relative climate impacts of food choices in the United States.** *Environ Sci Technol* 2008, **42**:3508–3513.

56. Garnett T, Appleby MC, Balmford A, Bateman IJ, Benton TG, Bloomer P, Burlingame B, Dawkins M, Dolan L, Fraser D, Herrero M, Hoffmann I, Smith P, Thornton PK, Toulmin C, Vermeulen SJ, Godfray HCJ: **Sustainable intensification in agriculture: premises and policies.** *Science* 2013, **341**:33–34.

57. Godfray HC, Garnett T: **Food security and sustainable intensification.** *Philos Trans R Soc Lond B Biol Sci* 2014, **369**:20120273.

58. Tittonell P, Giller KE: **When yield gaps are poverty traps: the paradigm of ecological intensification in African smallholder agriculture.** *Field Crops Res* 2013, **143**:76–90.

59. George T: **Why crop yields in developing countries have not kept pace with advances in agronomy.** *Glob Food Sec* 2014, **3**:49–58.

60. Newton AC, Johnson SN, Gregory PJ: **Implications of climate change for diseases, crop yields and food security.** *Euphytica* 2011, **179**:3–18.

61. Mittler R, Blumwald E: **Genetic engineering for modern agriculture: challenges and perspectives.** *Annu Rev Plant Biol* 2010, **61**:443–462.

62. Peleg Z, Reguera M, Tumimbang E, Walia H, Blumwald E: **Cytokinin-mediated source/sink modifications improve drought tolerance and increase grain yield in rice under water-stress.** *Plant Biotechnol J* 2011, **9**:747–758.

63. Reguera M, Peleg Z, Blumwald E: **Targeting metabolic pathways for genetic engineering abiotic stress-tolerance in crops.** *Biochim Biophys Acta* 2012, **1819**:186–194.

64. Cossani CM, Reynolds MP: **Physiological traits for improving heat tolerance in wheat.** *Plant Physiol* 2012, **160**:1710–1718.

65. Reynolds M, Foulkes J, Furbank R, Griffiths S, King J, Murchie E, Parry M, Slafer G: **Achieving yield gains in wheat.** *Plant Cell Environ* 2012, **35**:1799–1823.

66. Pask AJD, Reynolds MP: **Breeding for yield potential has increased deep soil water extraction capacity in irrigated wheat.** *Crop Sci* 2013, **53**:2090–2104.

67. Pask A, Pietragalla J, Mullan D, Reynolds M: *Physiological Breeding II: A Field Guide To Wheat Phenotyping.* Mexico City: International Maize and Wheat Improvement Center (CIMMYT); 2011. [http://repository.cimmyt.org/xmlui/ bitstream/handle/10883/1288/96144.pdf] (accessed 26 July 2014).

68. Araus JL, Cairns JE: **Field high-throughput phenotyping: the new crop breeding frontier.** *Trends Plant Sci* 2014, **19**:52–61.

69. Peleg Z, Blumwald E: **Hormone balance and abiotic stress tolerance in crop plants.** *Curr Opin Plant Biol* 2011, **14**:290–295.

70. Reynolds M, Dreccer F, Trethowan R: **Drought-adaptive traits derived from wheat wild relatives and landraces.** *J Exp Bot* 2007, **58**:177–186.

71. Sadok W, Naudin P, Boussuge B, Muller B, Welcker C, Tardieu F: **Leaf growth rate per unit thermal time follows QTL-dependent daily patterns in hundreds of maize lines under naturally fluctuating conditions.** *Plant Cell Environ* 2007, **30**:135 146.

72. Wang X, Taub DR: **Interactive effects of elevated carbon dioxide and environmental stresses on root mass fraction in plants: a meta-analytical synthesis using pairwise techniques.** *Oecologia* 2010, **163**:1–11.

73. Roy SJ, Tucker EJ, Tester M: **Genetic analysis of abiotic stress tolerance in crops.** *Curr Opin Plant Biol* 2011, **14**:232–239.

74. Reynolds M, Manes Y, Izanloo A, Langridge P: **Phenotyping approaches for physiological breeding and gene discovery in wheat.** *Ann Appl Biol* 2009, **155**:309–320.

75. Pinto RS, Reynolds MP, Mathews KL, McIntyre CL, Olivares-Villegas JJ, Chapman SC: **Heat and drought adaptive QTL in a wheat population designed to minimize confounding agronomic effects.** *Theor Appl Genet* 2010, **121**:1001–1021.

76. Rebetzke GJ, Chenu K, Biddulph B, Moeller C, Deery DM, Rattey AR, Bennett D, Barrett-Leonard G, Mayer JE: **A multisite managed environmental facility for targeted trait and germplasm phenotyping.** *Funct Plant Biol* 2013, **40**:1–13.

77. Olivares-Villegas JJ, Reynolds MP, McDonald GK: **Drought-adaptive attributes in the Seri/Babax hexaploid wheat population.** *Funct Plant Biol* 2007, **34**:189–203.

78. Bouteillé M, Rolland G, Balsera C, Loudet O, Muller B: **Disentangling the intertwined genetic bases of root and shoot growth in Arabidopsis.** *PLoS One* 2012, **7**:e32319.

79. Tester M, Langridge P: **Breeding technologies to increase crop production in a changing world.** *Science* 2010, **327**:818–822.

80. Langridge P, Fleury D: **Making the most of "omics" for crop breeding.** *Trends Biotechnol* 2011, **29**:33–40.

81. Heslot N, Akdemir D, Sorrells ME, Jannink JL: **Integrating environmental covariates and crop modeling into the genomic selection framework to predict genotype by environment interactions.** *Theor Appl Genet* 2014, **127**:463–480.

82. Philippot L, Hallin S: **Towards food, feed and energy crops mitigating climate change.** *Trends Plant Sci* 2011, **16**:476–80.

83. Reynolds MP, Hellin J, Govaerts B, Kosina P, Sonder K, Hobbs P, Braun B: **Global crop improvement networks to bridge technology gaps.** *J Exp Bot* 2012, **63**:1–12.

84. Ceccarelli S, Galie A, Grando S: **Participatory breeding for climate change-related traits.** In *Genomics and Breeding for Climate-Resilient Crops: Concepts and Strategies, Volume 1.* 8th edition. Edited by Kole C. Berlin: Springer Science & Business; 2013:331–376.

85. Opio C, Gerber P, Mottet A, Falcucci A, Tempio G, MacLeod M, Vellinga T, Henderson B, Steinfeld H: *Greenhouse Gas Emissions From Ruminant Supply Chains: A Global Life Cycle Assessment.* Rome: Food and Agriculture Organization of the United Nations (FAO); 2013. [http://www.fao.org/ docrep/018/i3461e/i3461e.pdf] (accessed 26 July 2014).

86. Thornton PK, van de Steeg J, Notenbaert A, Herrero M: **The impacts of climate change on livestock and livestock systems in developing countries: a review of what we know and what we need to know.** *Agric Syst* 2009, **101**:113–127.

87. Sutherst RW: **Implications of global change and climate variability for vector-borne diseases: generic approaches to impact assessments.** *Int J Parasitol* 1998, **28**:935–945.

88. Hristov AN, Oh J, Lee C, Meinen R, Montes F, Ott T, Firkins J, Rotz A, Dell C, Adesogan A, Yang W, Tricarico J, Kebreab E, Waghorn G, Dijkstra J, Oosting S: In *Mitigation of Greenhouse Gas Emissions in Livestock Production: A Review of Technical Options for Non-CO_2 Emissions*, FAO Animal Production and Health Paper No 177. Edited by Gerber PJ, Henderson B, Makkar HPS. Rome: Food and Agriculture Organization of the United Nations; 2013. [http://www.fao.org/docrep/018/i3288e/i3288e.pdf] (accessed 26 July 2014).

89. Niles MT: **Achieving social sustainability in animal agriculture: challenges and opportunities to reconcile multiple sustainability goals.** In *Sustainable Animal Agriculture*. Edited by Kebreab E. Wallingford, UK: CABI; 2013:193–211.

90. Herrero M, Thornton PK, Bernués A, Baltenweck I, Vervoort J, van de Steeg J, Makokha S, van Wijk MT, Karanja S, Rufino MC, Staal SJ: **Exploring future changes in smallholder farming systems by linking socio-economic scenarios with regional and household models.** *Glob Environ Change* 2014, **24**:165–182.

91. Thornton PK, Herrero M: **Potential for reduced methane and carbon dioxide emissions from livestock and pasture management in the tropics.** *Proc Natl Acad Sci U S A* 2010, **107**:19667–19672.

92. Herrero M, Thornton PK: **Livestock and global change: emerging issues for sustainable food systems.** *Proc Natl Acad Sci U S A* 2013, **110**:20878–20881.

93. Doré T, Makowski D, Malézieux E, Munier-Jolain N, Tchamitchian M, Tittonell P: **Facing up to the paradigm of ecological intensification in agronomy: revisiting methods, concepts and knowledge.** *Eur J Agron* 2011, **34**:197–210.

94. Silvestri S, Osano P, de Leeuw J, Herrero M, Ericksen P, Hariuki J, Njuki J, Notenbaert A, Bedelian C: *Greening Livestock: Assessing the Potential of Payment for Environmental Services in Livestock-Inclusive Agricultural Production Systems in Developing Countries*, ILRI Position Paper. Nairobi: International Livestock Research Institute (ILRI); 2012.

95. Rufino MC, Thornton PK, Ng'ang'a SK, Mutie I, Jones PG, van Wijk MT, Herrero M: **Transitions in agro-pastoralist systems of East Africa: impacts on food security and poverty.** *Agric Ecosyst Environ* 2013, **179**:215–230.

96. Thornton PK, Boone RB, Galvin KA, Burnsilver SB, Waithaka MM, Kuyiah J, Karanja S, González-Estrada E, Boone B: **Coping strategies in livestock-dependent households in East and Southern Africa: a synthesis of four case studies.** *Hum Ecol* 2007, **35**:461–476.

97. Alvarez S, Rufino MC, Vayssières J, Salgado P, Tittonell P, Tillard E, Bocquier F: **Whole-farm nitrogen cycling and intensification of crop-livestock systems in the highlands of Madagascar: an application of network analysis.** *Agric Syst* 2014, **126**:25–37.

98. Bengoumi M, Vias G, Faye B: **Camel milk production and transformation in Sub-Saharan Africa**. In *Desertification Combat and Food Safety: the Added Value of Camel Producers*. Edited by Faye B, Esenov P. Amsterdam: IOS Press; 2005:200–208.

99. Jones PG, Thornton PK: **Croppers to livestock keepers: livelihood transitions to 2050 in Africa due to climate change.** *Environ Sci Policy* 2009, **12**:427–437.

100. Bustamante MMC, Nobre CA, Smeraldi R, Aguiar APD, Barioni LG, Ferreira LG, Longo K, May P, Pinto AS, Ometto JPHB: **Estimating greenhouse gas emissions from cattle raising in Brazil.** *Clim Change* 2012, **115**:559–577.

101. Maia SMF, Ogle SM, Cerri CEP, Cerri CC: **Effect of grassland management on soil carbon sequestration in Rondonia and Mato Grosso states, Brazil.** *Geoderma* 2009, **149**:84–91.

102. Newbold CJ, López S, Nelson N, Ouda JO, Wallace RJ, Moss AR: **Propionate precursors and other metabolic intermediates as possible alternative electron acceptors to methanogenesis in ruminal fermentation *in vitro*.** *Br J Nutr* 2007, **94**:27–35.

103. Norad: the Norwegian Agency for Development Cooperation. [http://www.norad.no/en/front-page;jsessionid=AA171500C5301B88715DA5E9EE23C307] (accessed 26 July 2014).

104. Morse SS, Mazet JAK, Woolhouse M, Parrish CR, Carroll D, Karesh WB, Zambrana-Torrelio C, Lipkin WI, Daszak P: **Prediction and prevention of the next pandemic zoonosis.** *Lancet* 2012, **380**:1956–1965.

105. US Agency for International Development (USAID): *Pandemic Influenza and Other Emerging Threats*. Washington, DC: USAID; 2013. [http://www.usaid.gov/news-information/fact-sheets/emerging-pandemic-threats-program] (accessed 26 July 2014).

106. Foley P, Crosson P, Lovett DK, Boland TM, O'Mara FP, Kenny D: **Whole-farm systems modelling of greenhouse gas emissions from pastoral suckler beef cow production systems.** *Agric Ecosyst Environ* 2011, **142**:222–230.

107. Rotz C, Montes F, Chianese DS: **The carbon footprint of dairy production systems through partial life cycle assessment.** *J Dairy Sci* 2010, **93**:1266–1282.

108. Del Prado A, Misselbrook T, Chadwick D, Hopkins A, Dewhurst RJ, Davison P, Butler A, Schröder J, Scholefield D: **SIMS$_{DAIRY}$: a modelling framework to identify sustainable dairy farms in the UK framework description and test for organic systems and N fertiliser optimisation.** *Sci Total Environ* 2011, **409**:3993–4009.

109. Reisinger A, Havlik P, Riahi K, Vliet O, Obersteiner M, Herrero M: **Implications of alternative metrics for global mitigation costs and greenhouse gas emissions from agriculture.** *Clim Change* 2012, **117**:677–690.

110. Havlík P, Valin H, Herrero M, Obersteiner M, Schmid E, Rufino MC, Mosnier A, Thornton PK, Böttcher H, Conant RT, Frank S, Fritz S, Fuss S, Kraxner F, Notenbaert A: **Climate change mitigation through livestock system transitions.** *Proc Natl Acad Sci U S A* 2014, **111**:3709–3714.

111. *Feed the Future Innovation Lab Progress Report: Alignment with Feed the Future*, Collaborative Research on Adapting Livestock Systems to Climate Change. Fort Collins, CO: Colorado State University, US Agency for International Development; 2013. [http://lcccrsp.org/wp-content/uploads/2013/06/SMALLAnnualReport_Indicators2.pdf] (accessed 26 July 2014).

112. Rockström J, Steffen W, Noone K, Persson A, Chapin FS, Lambin EF, Lenton TM, Scheffer M, Folke C, Schellnhuber HJ, Nykvist B, de Wit CA, Hughes T, van der Leeuw S, Rodhe H, Sorlin S, Snyder PK, Costanza R, Svedin U, Falkenmark M, Karlberg L, Corell RW, Fabry VJ, Hansen J, Walker B, Liverman D, Richardson K, Crutzen P, Foley JA: **A safe operating space for humanity.** *Nature* 2009, **461**:472–475.

113. Sutton MA, Oenema O, Erisman JW, Leip A, van Grinsven H, Winiwarter W: **Too much of a good thing.** *Nature* 2011, **472**:159–161.

114. Smith P, Martino D, Cai Z, Gwary D, Janzen H, Kumar P, McCarl B, Ogle S, O'Mara F: **Agriculture.** In *Climate Change 2007: Mitigation*, Contribution of Working Group III to the Fourth Assessment Report of the Intergovernmental Panel on Climate Change. Edited by Metz B, Davidson O, Bosch P, Dave R, Meyer L. Cambridge, UK: Cambridge University Press; 2007:497–540. [https://www.ipcc.ch/pdf/assessment-report/ar4/wg3/ar4_wg3_full_report.pdf] (26 July 2014).

115. Halvorson AD, Del Grosso SJ, Alluvione F: **Tillage and inorganic nitrogen source effects on nitrous oxide emissions from irrigated cropping systems.** *Soil Sci Soc Am J* 2010, **74**:436–445.

116. Akiyama H, Yan X, Yagi K: **Evaluation of effectiveness of enhanced-efficiency fertilizers as mitigation options for N_2O and NO emissions from agricultural soils: meta-analysis.** *Glob Chang Biol* 2009, **16**:1837–1846.

117. Parkin TB, Hatfield JL: **Enhanced efficiency fertilizers: effect on nitrous oxide emissions in Iowa.** *Agron J* 2014, **106**:694–702.

118. Hatfield JL, Parkin TB: **Enhanced efficiency fertilizers: effect on agronomic performance of corn in Iowa.** *Agron J* 2014, **106**:771–780.

119. Kallenbach CM, Rolston DE, Horwath WR: **Cover cropping affects soil N_2O and CO_2 emissions differently depending on type of irrigation.** *Agric Ecosyst Environ* 2010, **137**:251–260.

120. Singh A, Kaur J: **Impact of conservation tillage on soil properties in rice-wheat cropping system.** *Agric Sci Res J* 2012, **2**:30–41.

121. Mandal B, Majumder B, Adhya TK, Bandyopadhyay PK, Gangopadhyay A, Sarkar D, Kundu MC, Choudhury SG, Hazra GC, Kundu S, Samantaray RN, Misra AK: **Potential of double-cropped rice ecology to conserve organic carbon under subtropical climate.** *Glob Chang Biol* 2008, **14**:2139–2151.

122. Smith P, Haberl H, Popp A, Erb KH, Lauk C, Harper R, Tubiello FN, de Siqueira Pinto A, Jafari M, Sohi S, Masera O, Böttcher H, Berndes G, Bustamante M, Ahammad H, Clark H, Dong H, Elsiddig EA, Mbow C, Ravindranath NH, Rice CW, Robledo Abad C, Romanovskaya A, Sperling F, Herrero M, House JI, Rose S: **How much land-based greenhouse gas mitigation can be achieved without compromising food security and environmental goals?** *Glob Chang Biol* 2013, **19**:2285–302.

123. Joseph S, Khoi DD, Hien NV, Anh ML, Nguyen HH, Hung TM, Yen NT, Thomsen MF, Lehmann J, Chia CH: *North Vietnam Villages Lead the Way in the Use of Biochar: Building on an Indigenous Knowledge Base*. International Biochar Initiative. [http://www.biochar-international.org/sites/default/files/CARE_Vietnam_1.31.2012.pdf] (accessed 26 July 2014).

124. Hatfield JL, Parkin TB, Sauer TJ, Prueger JH: **Mitigation opportunities from land management practices in a warming world: increasing potential sinks.** In *Managing Agricultural Greenhouse Gases: Coordinated Agricultural Research through GRACEnet to Address Our Changing Climate*. Edited by

Liebig MA, Franzluebbers AT, Follett RF. Waltham, MA: Academic Press; 2012:487–504.

125. Jensen ES, Peoples MB, Boddey RM, Gresshoff PM, Hauggaard-Nielsen H, Alves B, Morrison MJ: **Legumes for mitigation of climate change and the provision of feedstock for biofuels and biorefineries: a review.** *Agron Sustain Dev* 2012, **32:**329–364.

126. Kim DG: **Estimation of net gain of soil carbon in a nitrogen-fixing tree and crop intercropping system in Sub-Saharan Africa: results from re-examining a study.** *Agrofor Syst* 2012, **86:**175–184.

127. Corre-Hellou G, Dibet A, Hauggaard-Nielsen H, Crozat Y, Gooding M, Ambus P, Dahlmann C, von Fragstein P, Pristeri A, Monti M, Jensen ES: **The competitive ability of pea-barley intercrops against weeds and the interactions with crop productivity and soil N availability.** *Field Crops Res* 2011, **122:**264–272.

128. Delgado JA, Kowalski K, Tebbe C: **The first Nitrogen Index app for mobile devices: using portable technology for smart agricultural management.** *Comput Electron Agric* 2013, **91:**121–123.

129. Fischer RA, Edmeades GO: **Breeding and cereal yield progress.** *Crop Sci* 2010, **50**(Suppl 1):S85–S98.

130. Grassini P, Eskridge KM, Cassman KG: **Distinguishing between yield advances and yield plateaus in historical crop production trends.** *Nat Commun* 2013, **4:**2918.

131. van Ittersuma MK, Cassman KG, Grassini P, Wolfa J, Tittonell P, Hochman Z: **Yield gap analysis with local to global relevance—a review.** *Field Crops Res* 2013, **143:**4–17.

132. van Warta JK, Kersebaum C, Peng S, Milner M, Cassman KG: **Estimating crop yield potential at regional to national scales.** *Field Crops Res* 2012, **143:**34–43.

133. Reed MS, Podesta G, Fazey I, Geeson N, Hessel R, Hubacek K, Letson D, Nainggolan D, Prell C, Rickenbach MG, Ritsema C, Schwilch G, Stringer LC, Thomas AD: **Combining analytical frameworks to assess livelihood vulnerability to climate change and analyse adaptation options.** *Ecol Econ* 2013, **94:**66–77.

134. Rogers EM: *Diffusion of Innovations.* 5th edition. New York: Free Press; 2003.

135. Kristjanson P, Reid RS, Dickson N, Clark WC, Romney D, Puskur R, MacMillan S, Grace D: **Linking international agricultural research knowledge with action for sustainable development.** *Proc Natl Acad Sci U S A* 2009, **106:**5047–5052.

136. Spielman DJ, Ekboir J, Davis K: **The art and science of innovation systems inquiry: applications to Sub-Saharan African agriculture.** *Technol Soc* 2009, **31:**399–405.

137. Vervoort JM, Thornton PK, Kristjanson P, Forch W, Ericksen PJ, Kok K, Ingram JSI, Herrero M, Palazzo A, Helfgott AES, Wilkinson A, Havlík P, Mason-D'Croz D, Jost C: **Challenges to scenario-guided adaptive action on food security under climate change.** *Glob Environ Change* in press. doi:10.1016/j.gloenvcha.2014.03.001.

138. Arbuckle JG, Prokopy LS, Haigh T, Hobbs J, Knoot T, Knutson C, Loy A, Mase AS, McGuire J, Morton LW, Tyndall J, Widhalm M: **Climate change beliefs, concerns, and attitudes toward adaptation and mitigation among farmers in the Midwestern United States.** *Clim Change* 2013, **117:**943–950.

139. Prokopy LS, Haigh T, Mase AS, Angel J, Hart C, Knutson C, Lemos MC, Lo YJ, McGuire J, Morton LW, Perron J, Todey D, Widhalm M: **Agricultural advisors: a receptive audience for weather and climate information?** *Weather Clim Soc* 2013, **5:**162–167.

140. Baumgart-Getz A, Prokopy LS, Floress K: **Why farmers adopt best management practice in the United States: a meta-analysis of the adoption literature.** *J Environ Manage* 2012, **96:**17–25.

141. Food and Agriculture Organization of the United Nations (FAO): *Investing in Sustainable Agricultural Intensification: The Role of Conservation Agriculture. A Framework for Action.* Rome: FAO; 2008. [http://www.fao.org/ag/ca/doc/proposed_framework.pdf] (accessed 26 July 2014).

142. Shaxson F, Kassam A, Friedrich T, Adekunle A: *Conservation Agriculture: Looking beneath the Surface.* Nakuru, Kenya: Food and Agriculture Organization of the United Nations (FAO). Conservation Agriculture Workshop; 2008. [http://www.fao.org/ag/ca/CA-Publications/Nanyuki%202008.pdf] (accessed 4 August 2014).

143. Ndlovu PV, Mazvimavi K, An H, Murendo C: **Productivity and efficiency analysis of maize under conservation agriculture in Zimbabwe.** *Agric Syst* 2014, **124:**21–31.

144. Giller KE, Witter E, Corbeels M, Tittonell P: **Conservation agriculture and smallholder farming in Africa: the heretics' view.** *Field Crops Res* 2009, **114:**23–34.

145. Knowler D, Bradshaw B: **Farmers' adoption of conservation agriculture: a review and synthesis of recent research.** *Food Policy* 2007, **32:**25–48.

146. Giller KE, Tittonell P, Rufino MC, van Wijk MT, Zingore S, Mapfumo P, Adjei-Nsiah S, Herrero M, Chikowo R, Corbeels M, Rowe EC, Baijukya F, Mwijage A, Smith J, Yeboah E, van der Burg WJ, Sanogo OM, Misiko M, de Ridder N, Karanja S, Kaizzi C, K'ungu J, Mwale M, Nwaga D, Pacini C, Vanlauwe B: **Communicating complexity: integrated assessment of trade-offs concerning soil fertility management within African farming systems to support innovation and development.** *Agric Syst* 2011, **104:**191–203.

147. Arslan A, McCarthy N, Lipper L, Asfaw S, Cattaneo A: *Adoption and Intensity of Adoption of Conservation Farming Practices in Zambia. ESA Working Paper No. 13-01.* Rome: Food and Agriculture Organization of the United Nations; 2013. [http://www.fao.org/3/a-aq288e.pdf] (accessed 26 July 2014).

148. Harvey CA, Chacón M, Donatti CI, Garen E, Hannah L, Andrade A, Bede L, Brown D, Calle A, Chará J, Clement C, Gray E, Hoang MH, Minang P, Rodríguez AM, Seeberg-Elverfeldt C, Semroc B, Shames S, Smukler S, Somarriba E, Torquebiau E, van Etten J, Wollenberg E: **Climate-smart landscapes: opportunities and challenges for integrating adaptation and mitigation in tropical agriculture.** *Conserv Lett* 2013, **7:**77–90.

149. Jerneck A, Olsson L: **Food first! Theorising assets and actors in agroforestry: risk evaders, opportunity seekers and "the food imperative" in Sub-Saharan Africa.** *Int J Agric Sustain* 2013, **12:**1–22.

150. Jerneck A, Olsson L: **More than trees! Understanding the agroforestry adoption gap in subsistence agriculture: Insights from narrative walks in Kenya.** *J Rural Stud* 2013, **32:**114–125.

151. Adger N, Barnett J, Dabelko G: **Climate and war: a call for more research.** *Nature* 2013, **498:**171.

152. Cash DW, Borck JC, Patt AG: **Countering the loading-dock approach to linking science and decision making: comparative analysis of El Niño/Southern Oscillation (ENSO) forecasting systems.** *Sci Technol Human Values* 2006, **31:**465–494.

153. Hayward T: **Climate change and ethics.** *Nat Clim Chang* 2012, **2:**843–848.

154. Markowitz EM, Shariff AF: **Climate change and moral judgment.** *Nat Clim Chang* 2012, **2:**243–247.

155. Barnett BJ, Barrett CB, Skees JR: **Poverty traps and index-based risk transfer products.** *World Dev* 2008, **36:**1766–1785.

156. Lybbert TJ, McPeak J: **Risk and intertemporal substitution: livestock portfolios and off-take among Kenyan pastoralists.** *J Dev Econ* 2012, **97:**415–426.

157. Carter MR: **Designed for development impact: next generation index insurance for smallholder farmers.** In *Protecting the Poor: A Microinsurance Compendium, Volume II: Part IV. General Insurance.* Edited by Churchill C, Matul M. Geneva: International Labour Office; 2012:238–257. ISBN 978-92-2-125744-8 [http://www.ilo.org/wcmsp5/groups/public/@dgreports/@dcomm/@publ/documents/publication/wcms_175786.pdf] (accessed 26 July 2014).

158. McDowell JZ, Hess JJ: **Accessing adaptation: multiple stressors on livelihoods in the Bolivian highlands under a changing climate.** *Glob Environ Change* 2012, **22:**342–352.

159. Adger WN: **Vulnerability.** *Glob Environ Change* 2006, **16:**268–281.

160. Valdivia C, Seth A, Gilles JL, García M, Jiménez E, Cusicanqui J, Navia F, Yucra E: **Adapting to climate change in Andean ecosystems: landscapes, capitals, and perceptions shaping rural livelihood strategies and linking knowledge systems.** *Ann Assoc Am Geogr* 2010, **100:**818–834.

161. Johnson L: **Index insurance and the articulation of risk-bearing subjects.** *Environ Plan A* 2013, **45:**2663–2681.

162. Patt A, Suraez P, Hess U: **How do small-holders understand insurance, and how much do they want it? Evidence from Africa.** *Glob Environ Change* 2010, **20:**153–161.

163. Traerup S: **Informal networks and resilience to climate change impacts: a collective approach to index insurance.** *Glob Environ Change* 2012, **22:**255–267.

164. Chantarat S, Mude AG, Barrett CB, Carter MR: **Designing index-based livestock insurance for managing asset risk in Northern Kenya.** *J Risk Insur* 2013, **80:**205–237.

165. BASIS, United States Agency for International Development, Food and Agriculture Organization of the United Nations, Micro-Insurance Innovation Facility of the International Labour Organization, Oxfam America: *Feed the Future BASIS Assets and Market Innovation Lab: Index Insurance Innovation Initiative (I4).* Davis, CA: BASIS Assets and Market Access CRSP. [http://basis.ucdavis.edu/projects/i4-index-info/] (accessed 26 July 2014).

166. International Livestock Research Institute (ILRI): **Index Based Livestock Insurance.** [http://livestockinsurance.wordpress.com/] (accessed 26 July 2014).

167. Chantarat S, Barrett CB: **Social network capital, economic mobility and poverty traps.** *J Econ Inequal* 2011, **10**:299–342.

168. Karlan D, Osei RD, Osei-Akoto I, Udry C: *Agricultural Decisions after Relaxing Credit and Risk Constraints (NBER Working Paper No 18463).* Cambridge, MA: National Bureau of Economic Research; 2012:1–65.

169. McIntosh C, Sarris A, Papadopoulos F: **Productivity, credit, risk, and the demand for weather index insurance in smallholder agriculture in Ethiopia.** *Agric Econ* 2013, **44**:399–417.

170. Carter MR, Lybbert TJ: **Consumption versus asset smoothing: testing the implications of poverty trap theory in Burkina Faso.** *J Dev Econ* 2012, **99**:255–264.

171. Antón J, Kimura S, Lankoski J, Cattaneo A: **A comparative study of risk management in agriculture under climate change.** *OECD Food, Agriculture and Fisheries Papers* 2012, **58**:1–89.

172. Klein D, Luderer G, Kriegler E, Strefler J, Bauer N, Leimbach M, Popp A, Dietrich JP, Humpenöder F, Lotze-Campen H, Edenhofer O: **The value of bioenergy in low stabilization scenarios: an assessment using REMIND-MAgPIE.** *Clim Change* 2014, **123**:705.

173. Lotze-Campen H, von Lampe M, Kyle P, Fujimori S, Havlik P, van Meijl H, Hasegawa T, Popp A, Schmitz C, Tabeau A, Valin H, Willenbockel D, Wise M: **Impacts of increased bioenergy demand on global food markets: an AgMIP economic model intercomparison.** *Agric Econ* 2014, **45**:103–116.

174. US Department of Agriculture (USDA): *World Agricultural Supply and Demand Estimates (WASDE-459).* Washington, DC: USDA; 2008. [http://usda.mannlib.cornell.edu/usda/waob/wasde//2000s/2008/wasde-06-10-2008.pdf] (accessed 26 July 2014).

175. Mitchell D: *A Note on Rising Food Prices (Policy Research Working Paper 4682).* Washington, DC: World Bank Development Prospects Group; 2008. [http://www-wds.worldbank.org/external/default/WDSContentServer/WDSP/IB/2008/07/28/000020439_20080728103002/Rendered/PDF/WP4682.pdf] (accessed 26 July 2014).

176. Zhang W, Yu EA, Rozelle S, Yang J, Msangi S: **The impact of biofuel growth on agriculture: Why is the range of estimates so wide?** *Food Policy* 2013, **38**:227–239.

177. Schmitz C, van Meijl H, Kyle P, Nelson GC, Fujimori S, Gurgel A, Havlik P, Heyhoe E, D'Croz DM, Popp A, Sands R, Tabeau A, van der Mensbrugghe D, von Lampe M, Wise M, Blanc E, Hasegawa T, Kavallari A, Valin H: **Land-use change trajectories up to 2050: insights from a global agro-economic model comparison.** *Agric Econ* 2014, **45**:69–84.

178. Fujimori S, Matsuoka Y: **Development of method for estimation of world industrial energy consumption and its application.** *Energy Econ* 2011, **33**:461–473.

179. van Meijl H, van Rheenen T, Tabeau A, Eickhout B: **The impact of different policy environments on agricultural land use in Europe.** *Agric Ecosyst Environ* 2006, **114**:21–38.

180. Clarke L, Edmonds J, Krey V, Richels R, Rose S, Tavoni M: **International climate policy architectures: overview of the EMF 22 International Scenarios.** *Energy Econ* 2009, **31**(Suppl 2):S64–S81.

181. Edmonds JA, Reilly JM: *Global Energy: Assessing the Future.* New York: Oxford University Press; 1985.

182. Havlik P, Valin H, Mosnier A, Obersteiner M, Baker JS, Herrero M, Rufino MC, Schmid E: **Crop productivity and the global livestock sector: implications for land use change and greenhouse gas emissions.** *Am J Agric Econ* 2012, **95**:442–448.

183. Lotze-Campen H: **Improved data for integrated modeling of global environmental change.** *Environ Res Lett* 2011, **6**:041002.

184. Popp A, Dietrich JP, Lotze-Campen H, Klein D, Bauer N, Krause M, Beringer T, Gerten D, Edenhofer O: **The economic potential of bioenergy for climate change mitigation with special attention given to implications for the land system.** *Environ Res Lett* 2011, **6**:034017.

185. Bondeau A, Smith P, Zaehle SON, Schaphoff S, Lucht W, Cramer W, Gerten D, Lotze-Campen H, Muller C, Reichstein M: **Modelling the role of agriculture for the 20th century global terrestrial carbon balance.** *Glob Chang Biol* 2007, **13**:679–706.

186. Rost S, Gerten D, Bondeau A, Lucht W, Rohwer J, Schaphoff S: **Agricultural green and blue water consumption and its influence on the global water system.** *Water Resour Res* 2008, **44**, W09405.

187. Leimbach M, Bauer N, Baumstark L, Edenhofer O: **Mitigation costs in a globalized world: climate policy analysis with REMIND-R.** *Environ Model Assess* 2009, **15**:155–173.

188. Popp A, Krause M, Dietrich JP, Lotze-Campen H, Leimbach M, Beringer T, Bauer N: **Additional CO_2 emissions from land use change—forest conservation as a precondition for sustainable production of second generation bioenergy.** *Ecol Econ* 2012, **74**:64–70.

189. Hoefnagels R, Junginger M, Resch G, Matzenberger J, Panzer C: *Development of a Tool to Model European Biomass Trade,* Report for IEA Bioenergy Task 40. Utrecht, the Netherlands: Science Technology and Society, Copernicus Institute, University of Utrecht; 2011. [http://www.bioenergytrade.org/downloads/development-of-a-tool-to-model-european-biomas.pdf] (accessed 26 July 2014).

190. Sperling D, Yeh S: **Toward a global low carbon fuel standard.** *Transp Policy* 2010, **17**:47–49.

191. Tokgoz S, Zhang W, Msangi S, Bhandary P: **Biofuels and the future of food: competition and complementarities.** *Agriculture* 2012, **2**:414–435.

192. Lotze-Campen H, Popp A, Beringer T, Müller C, Bondeau A, Rost S, Lucht W: **Scenarios of global bioenergy production: the trade-offs between agricultural expansion, intensification and trade.** *Ecol Modell* 2010, **221**:2188–2196.

193. van Dam J, Junginger M, Faaij APC: **From the global efforts on certification of bioenergy towards an integrated approach based on sustainable land use planning.** *Renew Sustain Energy Rev* 2010, **14**:2445–2472.

194. Msangi S, Evans M: **Biofuels and developing economies: is the timing right?** *Agric Econ* 2013, **44**:501–510.

195. Beddington JR, Asaduzzaman M, Clark M: **The role for scientists in tackling food insecurity and climate change.** *Agric Food Secur* 2012, **1**:10.

196. Verburg PH, Ellis EC, Letourneau A: **A global assessment of market accessibility and market influence for environmental change studies.** *Environ Res Lett* 2011, **6**:034019.

197. Lugato E, Panagos P, Bampa F, Jones A, Montanarella L: **A new baseline of organic carbon stock in European agricultural soils using a modelling approach.** *Glob Chang Biol* 2014, **20**:313–26.

198. Smith P, Davies CA, Ogle S, Zanchi G, Bellarby J, Bird N, Boddey RM, McNamara NP, Powlson D, Cowie A, van Noordwijk M, Davis SC, Richter DDB, Kryzanowski L, Wijk MT, Stuart J, Kirton A, Eggar D, Newton-Cross G, Adhya TK, Braimoh AK: **Towards an integrated global framework to assess the impacts of land use and management change on soil carbon: current capability and future vision.** *Glob Chang Biol* 2012, **18**:2089–2101.

199. Ewert F, van Ittersum MK, Heckelei T, Therond O, Bezlepkina I, Andersen E: **Scale changes and model linking methods for integrated assessment of agri-environmental systems.** *Agric Ecosyst Environ* 2011, **142**:6–17.

200. Therond HB, Oomen R, Russell G, Ewert F: **Using a cropping system model at regional scale: low-data approaches for crop management information and model calibration.** *Agric Ecosyst Environ* 2011, **142**:85–94.

201. Grace PR, Antle J, Aggarwal PK, Ogle S, Paustian K, Basso B: **Soil carbon sequestration and associated economic costs for farming systems of the Indo-Gangetic Plain: a meta-analysis.** *Agric Ecosyst Environ* 2012, **146**:137–146.

202. de Groot RS, Alkemade R, Braat L, Hein L, Willemen L: **Challenges in integrating the concept of ecosystem services and values in landscape planning, management and decision making.** *Ecol Complex* 2010, **7**:260–272.

203. Olsson L, Jerneck A: **Farmers fighting climate change-from victims to agents in subsistence livelihoods.** *Wiley Interdiscip Rev Clim Chang* 2010, **1**:363–373.

204. You L, Wood S, Wood-Sichra U: **Generating plausible crop distribution maps for Sub-Saharan Africa using a spatially disaggregated data fusion and optimization approach.** *Agric Syst* 2009, **99**:126–140.

205. Milman A, Bunclark L, Conway D, Adger WN: **Assessment of institutional capacity to adapt to climate change in transboundary river basins.** *Clim Chang* 2013, **121**:755–770.

206. Rivington M, Matthews KB, Buchan K, Miller DG, Bellocchi G, Russell G: **Climate change impacts and adaptation scope for agriculture indicated by agro-meteorological metrics.** *Agric Syst* 2013, **114**:15–31.

207. Trnka M, Brázdil R, Olesen JE, Eitzinger J, Zahradníček P, Kocmánková E, Dobrovolný P, Štěpánek P, Možný M, Bartošová L, Hlavinka P, Semerádová D, Valášek H, Havlíček M, Horáková V, Fischer M, Žalud Z: **Could the changes in regional crop yields be a pointer of climatic change?** *Agric For Meteorol* 2012, **166–167**:62–71.

208. Del Prado A, Crosson P, Olesen JE, Rotz CA: Whole-farm models to quantify greenhouse gas emissions and their potential use for linking climate change mitigation and adaptation in temperate grassland ruminant-based farming systems. *Animal* 2013, **7**:373–385.

209. Gourdji SM, Sibley AM, Lobell DB: Global crop exposure to critical high temperatures in the reproductive period: historical trends and future projections. *Environ Res Lett* 2013, **8**:024041.

210. Smith P, Olesen JE: Synergies between the mitigation of, and adaptation to, climate change in agriculture. *J Agric Sci* 2010, **148**:543–552.

211. Acosta L, Klein RJT, Reidsma P, Metzger MJ, Rounsevell MDA, Leemans R, Schröter D: A spatially explicit scenario-driven model of adaptive capacity to global change in Europe. *Global Environ Change* 2013, **23**:1211–1224.

212. Schröter D, Cramer W, Leemans R, Prentice IC, Araújo MB, Arnell NW, Bondeau A, Bugmann H, Carter TR, Gracia CA, de la Vega-Leinert AC, Erhard M, Ewert F, Glendining M, House JI, Kankaanpää S, Klein RJT, Lavorel S, Lindner M, Metzger MJ, Meyer J, Mitchell TD, Reginster I, Rounsevell M, Sabaté S, Sitch S, Smith B, Smith J, Smith P, Sykes MT, *et al*: Ecosystem service supply and vulnerability to global change in Europe. *Science* 2005, **310**:1333–1337.

213. Sustainable farm Management Aimed at Reducing Threats to SOILs under climate change: SmartSOIL. [http://smartsoil.eu] (accessed 26 July 2014).

214. Ingram J, Mills J, Frelih-Larsen A, Davis M: *Uptake of Soil Management Practices and Experiences with Decisions Support Tools: Analysis of the Consultation with the Farming Community (Project No 289684)*. SmartSOIL; 2012. [http://smartsoil.eu/fileadmin/www.smartsoil.eu/WP5/D5_1_Final.pdf] (accessed 26 July 2014).

215. White JW, Hoogenboom G, Kimball BA, Wall GW: Methodologies for simulating impacts of climate change on crop production. *Field Crops Res* 2011, **124**:357–368.

216. Hansen J, Sato M, Ruedy R: Perception of climate change. *Proc Natl Acad Sci U S A* 2012, **109**:415–423.

217. Moriondo M, Giannakopoulos C, Bindi M: Climate change impact assessment: the role of climate extremes in crop yield simulation. *Clim Change* 2010, **104**:679–701.

218. Lobell DB: Errors in climate datasets and their effects on statistical crop models. *Agr Forest Meteorol* 2013, **170**:58–66.

219. Challinor A, Martre P, Asseng S, Thornton P, Ewert F: Making the most of climate impacts ensembles. *Nat Clim Change* 2014, **4**:77–80.

220. Peters GP, Andrew RM, Boden T, Canadell JG, Ciais P, Le Quéré C, Marland G, Raupach MR, Wilson C: The challenge to keep global warming below 2°C. *Nat Clim Chang* 2012, **3**:4–6.

221. Jordan A, Rayner T, Schroeder H, Adger N, Anderson K, Bows A, Le Quéré C, Joshi M, Mander S, Vaughan N, Whitmarsh L: Going beyond two degrees? The risks and opportunities of alternative options. *Clim Policy* 2013, **13**:751–769.

222. Petherick A: Seeking a fair and sustainable future. *Nat Clim Change* 2014, **4**:81–83.

223. Lal R, Lorenz K, Hüttl RF, Schneider BU, von Braun J: *Ecosystem Services and Carbon Sequestration in the Biosphere*. New York: Springer Science; 2013.

224. Qadir M, Noble AD, Chartres C: Adapting to climate change by improving water productivity of soils in dry areas. *Land Degrad Develop* 2013, **21**:12–21.

225. Cochard R: Natural hazards mitigation services of carbon-rich ecosystems. In *Ecosystem Services and Carbon Sequestration in the Biosphere*. Edited by Lal R, Lorenz K, Hüttl RF, Schneider BU, von Braun J. New York: Springer Science; 2013:221–293.

226. Colomb V, Touchemoulin O, Bockel L, Chotte JL, Martin S, Tinlot M, Bernoux M: Selection of appropriate calculators for landscape-scale greenhouse gas assessment for agriculture and forestry. *Environ Res Lett* 2013, **8**:015029.

227. Henry M, Tittonell P, Manlay RJ, Bernoux M, Albrecht A, Vanlauwe B: Agriculture, ecosystems and environment biodiversity, carbon stocks and sequestration potential in aboveground biomass in smallholder farming systems of western Kenya. *Agric Ecosyst Environ* 2009, **129**:238–252.

228. Tittonell PA: *Msimu wa Kupanda, targeting resources within diverse, heterogeneous and dynamic farming systems of East Africa*. PhD thesis. Wageningen University, Production Ecology and Resource Conservation; 2007. [http://edepot.wur.nl/121949] (accessed 26 July 2014).

229. Claessens L, Antle JM, Stoorvogel JJ, Valdivia RO, Thornton PK, Herrero M: A method for evaluating climate change adaptation strategies for small-scale farmers using survey, experimental and modeled data. *Agric Syst* 2012, **111**:85–95.

230. Maskell LC, Crowe A, Dunbar MJ, Emmett B, Henrys P, Keith AM, Norton LR, Scholefield P, Clark DB, Simpson IC, Smart SM: Exploring the ecological constraints to multiple ecosystem service delivery and biodiversity. *J Appl Ecol* 2013, **50**:561–571.

231. Setälä H, Bardgett RD, Birkhofer K, Brady M, Byrne L, de Ruiter PC, de Vries FT, Gardi C, Hedlund K, Hemerik L, Hotes S, Liiri M, Mortimer SR, Pavao-Zuckerman M, Pouyat R, Tsiafouli M, Putten WH: Urban and agricultural soils: conflicts and trade-offs in the optimization of ecosystem services. *Urban Ecosyst* 2013, **17**:239–253.

232. Williams A, Hedlund K: Indicators and trade-offs of ecosystem services in agricultural soils along a landscape heterogeneity gradient. *Appl Soil Ecol* 2014, **77**:1–8.

233. Wada Y, Wisser D, Eisner S, Flörke M, Gerten D, Haddeland I, Hanasaki N, Masaki Y, Portmann FT, Stacke T, Tessler Z, Schewe J: Multimodel projections and uncertainties of irrigation water demand under climate change. *Geophys Res Lett* 2013, **40**:4626–4632.

234. Sood A, Muthuwatta L, McCartney M: A SWAT evaluation of the effect of climate change on the hydrology of the Volta River basin. *Water Int* 2013, **38**:297–311.

235. Taylor RG, Scanlon B, Doll P, Rodell M, van Beek R, Wada Y, Longuevergne L, Leblanc M, Famiglietii JS, Edmunds M, Konikow L, Green TR, Chen J, Taniguchi M, Bierkens MFP, MacDonald A, Fan Y, Maxwell RM, Yechieli Y, Gurdak JL, Allen DM, Shamsudduha M, Hiscock K, Yeh PJF, Holman I, Treidel H: Ground water and climate change. *Nat Clim Chan* 2013, **3**:321–330.

236. Dye PJ: Climate, forest, and streamflow relationships in South African afforested catchments. *Commonw Forest Rev* 1996, **75**:31–38.

237. Khan S, Hanjra MA: Footprints of water and energy inputs in food production–global perspectives. *Food Policy* 2009, **34**:130–140.

238. Duncan MJ: Hydrological impacts of converting pasture and gorse to pine plantation, and forest harvesting, Nelson, New Zealand. *J Hydrol* 1995, **34**:15–41.

239. Bosch JM, Hewlett JD: A review of catchment experiments to determine the effect of vegetation changes on water yield and evapotranspiration. *J Hydrol* 1982, **55**:3–23.

240. Scanlon BR, Jolly I, Sophocleous M, Zhang L: Global impacts of conversions from natural to agricultural ecosystems on water resources: quantity versus quality. *Water Resour Res* 2007, **43**, W034037.

241. Jobbágy EG, Jackson RB: Patterns and mechanisms of soil acidification in the conversion of grasslands to forests. *Biogeochemistry* 2003, **64**:205–229.

242. Jobbágy EG, Jackson RB: Groundwater use and salinization with grassland afforestation. *Glob Chan Biol* 2004, **10**:1299–1312.

243. Farley KA, Kelly EF, Hofstede RGM: Soil organic carbon and water retention following conversion of grasslands to pine plantations in the Ecuadorian Andes. *Ecosystems* 2004, **7**:729–739.

244. Schoups G, Hopmans JW, Young CA, Vrugt JA, Wallender WW, Tanji KK, Panday S: Sustainability of irrigated agriculture in the San Joaquin Valley, California. *Proc Natl Acad Sci U S A* 2005, **102**:15352–15356.

245. Scott CA, Varady RG, Meza F, Montaña E, De Raga B, Luckman B, Martius C: Science-policy dialogues for water security. *Sci Policy Sustain Dev* 2012, **54**:37–41.

246. Kam SP, Badjeck M, Teh L, Teh L, Tran N: *Autonomous Adaptation to Climate Change by Shrimp and Catfish Farmers in Vietnam's Mekong River Delta*, Working Paper 2012-24. Penang, Malaysia: Worldfish. [http://www.worldfishcenter.org/resource_centre/WF_3395.pdf] (accessed 26 July 2014).

247. Duarte CM, Losada IJ, Hendriks IE, Mazarrasa I, Marbà N: The role of coastal plant communities for climate change mitigation and adaptation. *Nat Clim Chang* 2013, **3**:961–968.

248. van Noordwijk M, Namirembe S, Catacutan D, Williamson D, Gebrekirstos A: Pricing rainbow, green, blue and grey water: tree cover and geopolitics of climatic teleconnections. *Curr Opin Environ Sustain* 2014, **6**:41–47.

249. Spracklen DV, Arnold SR, Taylor CM: Observations of increased tropical rainfall preceded by air passage over forests. *Nature* 2012, **489**:282–285.

250. McElrone AJ, Shapland TM, Calderon A, Fitzmaurice L, Paw UKT, Snyder RL: Surface renewal: an advanced micrometeorological method for measuring and processing field-scale energy flux density data. *J Vis Exp* 2013, **82**:e50666.

251. Shapland TM, Snyder RL, Paw UKT, McElrone AJ: Thermocouple frequency response compensation leads to convergence of the surface renewal alpha calibration. *Agric For Meteorol* 2014, 189–190:36–47.

252. Clausen J, Jägerskog A: *Feeding a Thirsty World: Challenges and Opportunities for a Water and Food Secure Future*, SIWI Report No 31. Stockholm: Stockholm International Water Institute (SIWI); 2012. [http://www.worldwaterweek.org/documents/Resources/Reports/Feeding_a_thirsty_world_2012worldwaterweek_report_31.pdf] (accessed 26 July 2014).

253. Mukherji A, Facon T, de Fraiture C, Molden D, Chartres C: Growing more food with less water: how can revitalizing Asia's irrigation help? *Water Policy* 2012, 14:430–446.

254. Mbow C, Smith P, Skole D, Duguma L, Bustamante M: Achieving mitigation and adaptation to climate change through sustainable agroforestry practices in Africa. *Curr Opin Environ Sustain* 2014, 6:8–14.

255. Mbow C, van Noordwijk M, Prabhu R, Simons T: Knowledge gaps and research needs concerning agroforestry's contribution to sustainable development goals in Africa. *Curr Opin Environ Sustain* 2014, 6:162–170.

256. van Noordwijk M, Hoang MH, Neufeldt H, Öborn I, Yatich T: *How Trees and People Can Co-adapt to Climate Change: Reducing Vulnerability in Multifunctional Agroforestry Landscapes*. Nairobi: World Agroforestry Centre; 2011. [http://www.asb.cgiar.org/PDFwebdocs/PDFwebdocs/How%20trees%20and%20people%20can%20co-adapt%20to%20climate%20change.pdf] (accessed 26 July 2014).

257. Visseren-Hamakers I, Verkooijen P: The practice of interaction management: enhancing synergies among multilateral REDD+ institutions. In *Forest and Nature Governance: A Practice Based Approach (World Forests Series Vol 14)*. Edited by Arts B, Behagel J, van Bommel S, de Koning J, Turnhout E. New York: Springer Science; 2013:133–149.

258. Ordonez JC, Luedeling E, Kindt R, Tata HL, Harja D, Jamnadass R, van Noordwijk M: Constraints and opportunities for tree diversity management along the forest transition curve to achieve multifunctional agriculture. *Curr Opin Environ Sustain* 2014, 6:54–60.

259. Geldenhuys CJ, Ham C, Ham H: Sustainable Forest Management in Africa: Some Solutions to Natural Forest Management Problems in Africa. In *Proceedings of the Sustainable Forest Management in Africa Symposium. 3–7 November 2008, Stellenbosch, South Africa*. Edited by Geldenhuys CJ, Ham C, Ham H. Matieland, South Africa: Department of Forest and Wood Science, Stellenbosch University; 2011:1–538. [http://scholar.sun.ac.za/handle/10019.1/17385] (26 July 2014).

260. Thorlakson T, Neufeldt H: Reducing subsistence farmers' vulnerability to climate change: evaluating the potential contributions of agroforestry in western Kenya. *Agric Food Secur* 2012, 1:15.

261. van Paddenburg A, Bassi AM, Buter E, Cosslett CE, Dean A: *Heart of Borneo: Investing in Nature for a Green Economy*. WWF Heart of Borneo (HoB) Global Initiative; 2012. [http://awsassets.wwf.org.au/downloads/fl013_heart_of_borneo_investing_in_nature_for_a_green_economy_jan12.pdf] (accessed 26 July 2014).

262. Parrotta JA, Wildburger C, Mansourian S: *Understanding Relationships between Biodiversity, Carbon, Forests and People: The Key to Achieving REDD+ Objectives*, A Global Assessment Report Prepared by the Global Forest Expert Panel on Biodiversity, Forest Management, and REDD+ (IUFRO World Series Vol 31). Vienna: International Union of Forestry Research Organizations (IUFRO); 2012. [http://theredddesk.org/resources/understanding-relationships-between-biodiversity-carbon-forests-and-people-key-achieving] (accessed 26 July 2014).

263. Visseren-Hamakers IJ, Gupta A, Herold M, Peña-Claros M, Vijge MJ: Will REDD + work? The need for interdisciplinary research to address key challenges. *Curr Opin Environ Sustain* 2012, 4:590–596.

264. Somorin OA, Visseren-Hamakers IJ, Arts B, Sonwa DJ, Tiani A-M: REDD + policy strategy in Cameroon: actors, institutions and governance. *Environ Sci Policy* 2014, 35:87–97.

265. Wrathall DJ: Migration amidst social-ecological regime shift: the search for stability in Garífuna villages of Northern Honduras. *Hum Ecol* 2012, 40:583–596.

266. Gibson MA, Gurmu E: Rural to urban migration is an unforeseen impact of development intervention in Ethiopia. *PLoS One* 2012, 7:e48708.

267. Warner K, Afifi T, Henry K, Rawe T, Smith C, de Sherbinin A: *Where the Rain Falls: Climate Change, Food and Livelihood Security, and Migration*, Global Policy Report of Where the Rain Falls Project. Bonn: CARE France and United Nations University Institute for Environment and Human Security (UNU-EHS); 2012. [http://www.ehs.unu.edu/file/get/10569.pdf] (accessed 26 July 2014).

268. Jha AK, Bloch R, Lamond J: *Cities and Flooding: A Guide to Integrated Urban Flood Risk Management for the 21st Century*. Washington, DC: World Bank; 2012. [https://www.gfdrr.org/sites/gfdrr.org/files/publication/World_Bank_Cities_and_Flooding_Guidebook.pdf] (accessed 26 July 2014).

269. Laube W, Schraven B, Awo M: Smallholder adaptation to climate change: dynamics and limits in Northern Ghana. *Clim Change* 2011, 111:753–774.

270. Jankowska MM, Lopez-Carr D, Funk C, Husak GJ, Chafe ZA: Climate change and human health: Spatial modeling of water availability, malnutrition, and livelihoods in Mali, Africa. *Appl Geogr* 2012, 33:4–15.

271. López-Carr D: Agro-ecological drivers of rural out-migration to the Maya Biosphere Reserve, Guatemala. *Environ Res Lett* 2012, 7:045603.

272. Aide TM, Clark ML, Grau R, López-Carr D, Levy MA, Redo D, Bonilla-Moheno M, Riner G, Andrade-Núñez MJ, Muñiz M: Deforestation and reforestation of Latin America and the Caribbean (2001–2010). *Biotropica* 2013, 45:262–271.

273. DeFries RS, Rudel T, Uriarte M, Hansen M: Deforestation driven by urban population growth and agricultural trade in the twenty-first century. *Nat Geosci* 2010, 3:178–181.

274. Davis J, Lopez-Carr D: Migration, remittances and smallholder decision-making: implications for land use and livelihood change in Central America. *Land Use Policy* 2014, 38:319–329.

275. Winkels A: Migration, social networks and risk: the case of rural-to-rural migration in Vietnam. *J Vietnam Stud* 2013, 7:92–121.

276. Ho R, Milan A: *Rainfall, Food Security and Human Mobility. Case Study: Peru, "Where the Rain Falls" Project. Results from Huancayo Province, Junín Region (Report no 5)*. Bonn: United Nations University Institute for Environment and Human Security (UNU-EHS); 2012. [http://wheretherainfalls.org/wp-content/uploads/2012/11/Peru_CSR.pdf] (accessed 26 July 2014).

277. Schlenker W, Roberts MJ: Nonlinear temperature effects indicate severe damages to U.S. crop yields under climate change. *Proc Natl Acad Sci U S A* 2009, 106:15594–15598.

278. Auffhammer M, Hsiang SM, Schlenker W, Sobel A: Using weather data and climate model output in economic analyses of climate change. *Rev Environ Econ Policy* 2013, 7:181–198.

279. Urban D, Roberts MJ, Schlenker W, Lobell DB: Projected temperature changes indicate significant increase in interannual variability of U.S. maize yields. *Clim Change* 2012, 112:525–533.

280. Vital Signs. [http://vitalsigns.org/] (accessed 26 July 2014).

281. Ericksen P, Thornton P, Notenbaert A, Cramer L, Jones P, Herrero M: *Mapping Hotspots of Climate Change and Food Insecurity in the Global Tropics. (CCAFS Report no 5)*. Copenhagen: CGIAR Research Program on Climate Change, Agriculture and Food Security (CCAFS); 2011. [http://ccafs.cgiar.org/publications/mapping-hotspots-climate-change-and-food-insecurity-global-tropics#.U_f7CflSa3I] (accessed 26 July 2014).

282. Holm P, Goodsite ME, Cloetingh S, Agnoletti M, Moldan B, Lang DJ, Leemans R, Moeller JR, Buendía MP, Pohl W, Scholz RW, Sors A, Vanheusden B, Yusoff K, Zondervan R: Collaboration between the natural, social and human sciences in global change research. *Environ Sci Policy* 2013, 28:25–35.

283. Leemans R, Solecki W: Redefining environmental sustainability. *Curr Opin Environ Sustain* 2013, 5:3–4.

284. Andersson E, Gabrielsson S: 'Because of poverty we had to come together': collective action for improved food security in rural Kenya and Uganda. *Int J Agr Sustain* 2012, 10:245–262.

285. Kristjanson P, Neufeldt H, Gassner A, Mango J, Kyazze FB, Desta S, Sayula G, Thiede B, Förch W, Thornton PK, Coe R: Are food insecure smallholder households making changes in their farming practices? Evidence from East Africa. *Food Secur* 2012, 4:381–397.

286. Lybbert TJ, Bell A: Why drought tolerance is not the new *Bt*. *Nat Biotechnol* 2010, 28:553–554.

287. Shah AK, Mullainthan S, Shafir E: Some consequences of having too little. *Science* 2012, 338:682–685.

288. Adaptation Fund. [https://www.adaptation-fund.org/] (accessed 26 July 2014).

289. International Fund for Agricultural Development (IFAD): *Adaptation for Smallholder Agriculture Programme*. Rome: [http://www.ifad.org/climate/asap/index.htm] (accessed 26 July 2014).

290. UNEP International Resource Panel Working Group on Reducing Emissions from Deforestation and Forest Degradation (REDD+) and a Green Economy: *Building natural capital: how REDD+ can support a green economy*. Nairobi: United Nations Environment Programme (UNEP);

2014. [http://www.un-redd.org/portals/15/documents/IRPBuildingNatural CapitalthroughREDDMarch2014finallowres_EN.pdf] (accessed 26 July 2014).

291. **Green Climate Fund.** [http://www.gcfund.org/home.html] (accessed 26 July 2014).

292. United Nations Environment Programme (UNEP): *The Role of Ecosystems in Developing a Sustainable 'Green Economy',* UNEP Policy Series: Ecosystems Management Policy Brief 2-2010. Nairobi: UNEP; 2010. [http://www.unep. org/ecosystemmanagement/Portals/7/Documents/policy%20series%202% 20-%20small.pdf] (26 July 2014).

293. Bernard F, van Noordwijk M, Luedeling E, Villamor GB, Sileshi GW, Namirembe S: **Social actors and unsustainability of agriculture.** *Curr Opin Environ Sustain* 2014, **6:**155–161.

3

Overcoming barriers to trust in agricultural biotechnology projects

Obidimma C Ezezika[1,2,3*], Abdallah S Daar[1,4,5]

Abstract

Background: Nigeria, Africa's most populous country, has been the world's largest cowpea importer since 2004. The country is currently in the early phases of confined field trials for two genetically modified crops: *Bacillus thuringiensis* (Bt) cowpea and nutritionally enhanced cassava ("BioCassava Plus"). Using the bio-safety guidelines process as a backdrop, we evaluate the role of trust in the operation of the Cowpea Productivity Improvement Project, which is an international agricultural biotechnology public-private partnership (PPP) aimed at providing pest-resistant cowpea varieties to Nigerian farmers.

Methods: We reviewed the published literature and collected data through direct observations and semi-structured, face-to-face interviews. Data were analyzed based on emergent themes to create a comprehensive narrative on how trust is understood and built among the partners and with the community.

Results: Our findings highlight the importance of respecting mandates and eliminating conflicts of interest; holding community engagement initiatives early on; having on-going internal discussion and planning; and serving a locally-defined need. These four lessons could prove helpful to other agricultural biotechnology initiatives in which partners may face similar trust-related challenges.

Conclusions: Overcoming challenges to building trust requires concerted effort throughout all stages of project implementation. Currently, plans are being made to backcross the cowpea strain into a local variety in Nigeria. The development and adoption of the Bt cowpea seed hinges on the adoption of a National Biosafety Law in Nigeria. For countries that have decided to adopt biotech crops, the Nigerian cowpea experiment can be used as a model for other West African nations, and is actually applied as such in Ghana and Burkina Faso, interested in developing a Bt cowpea.

Background

Nigeria and the cowpea legume

Cowpea (known as black-eyed peas in North America) is the most important indigenous African legume because of its ability to grow in drought-prone areas and improve soil fertility. It is the favored crop of small-scale, low-income farmers in Africa [1-3]. It is also a high-protein, low-cost staple food and an important cash crop for farmers [1]. In 2009, the International Institute for Tropical Agriculture (IITA) reported that approximately 7.56 million tonnes of cowpea were produced globally. Nigeria is the largest producer and consumer of cowpea in the world, accounting for approximately five million of the 12.76 million total hectares of land devoted to cowpea growth [4,2,5]. The enormous demand for cowpeas in Nigeria has made it the world's largest cowpea importer since 2004 [6].

However, cowpea is affected by a number of environmental stressors. The cowpea pod borer (*Maruca vitrata*) is particularly detrimental. Reports have shown that severe infestations of the *Maruca* can cause crop yield losses as high as 70-80% [7]. Insecticides against the *Maruca* exist but have not been widely adopted by farmers due to prohibitive costs and significant health hazards [7]. Overall, insecticides have proven ineffective in combating the *Maruca*.

* Correspondence: obidimma.ezezika@srcglobal.org
[1]Sandra Rotman Centre, University Health Network and University of Toronto, Toronto, Ontario, Canada

The purpose of the Cowpea Productivity Improvement Project

The Cowpea Productivity Improvement Project is an agricultural biotechnology (agbiotech) public-private partnership (PPP) that emerged from the recognition of the damage the *Maruca* pest was doing to cowpea varieties in Africa. This project is the result of a network of individuals and organizations, listed in Table 1, who committed to working together to advance the conventional cowpea seed. This partnership brings together a variety of public, private, research, funding and advocacy organizations (see Additional file 1 for a complete description of these organizations) that fall into three broad categories: project development and technical expertise, funding, and regulatory. These categories denote the primary roles that each group has taken, but are not mutually exclusive.

The *Maruca*-resistant cowpea was developed by an international agbiotech PPP coordinated by the African Agricultural Technology Foundation (AATF), a not-for-profit organization that facilitates and promotes PPPs for the benefit of resource-poor smallholder farmers in sub-Saharan Africa. The Monsanto Company, a multinational agricultural biotechnology company, donated the Bt gene to AATF on a humanitarian basis under a royalty-free license [8]. The Institute for Agricultural Research (IAR) in Zaria, Nigeria was the body responsible for the *cry1Ab* gene introgression into local cowpea varieties. The *cry1Ab* gene is a gene derived from the soil bacterium, *Bacillus thuringiensis*, which produces specialized proteins to kill specific insects. In this case, the gene is intended to kill the *Maruca* pod borer [8]. Field testing was then carried out in specific locations in Nigeria.

The main objective of the project is to "enable smallholder farmers in sub-Saharan Africa to have access to farmer-preferred, elite cowpea varieties with resistance to insect pests, especially the pod borer *Maruca vitrata*" [7].

Table 1 Organizations involved in the Cowpea Productivity Improvement Project

Partner	Responsibility	Sector	Contribution
Commonwealth Scientific and Industrial Research Organisation (CSIRO)	Core Partner: Researching transformation methods	Public	Project Development and Technical
International institute of Tropical Agriculture (IITA)	Core Partner: Working with CSIRO on transformation	Public	Project Development and Technical
Monsanto Company	Core Partner: Provides insecticide gene, *Cry1ab*	Private	Project Development and Technical
The Kirkhouse Trust	Core Partner: Provides funding for research	Public	Funding
Institute of Agricultural Research (IAR), Zaria, Nigeria National Agricultural Research Systems in target countries of West Africa	Core Partner: Institutional biosafety committee; submitted application for confined field trials (CFT) Responsible for gene introgression into local cowpea varieties	Public	Project Development and Technical
Institut de l'Environnement et Recherches Agricoles (INERA), Burkina Faso	Core Partner: Helped identify field testing sites, CFT applications	Public	Project Development and Technical
National Biotechnology Development Agency (NABDA), Nigeria	Core Partner: Technical implementation, and regulatory affairs.	Public	Technology development
African Agricultural Technology Foundation (AATF)	Core Partner: Technology acquisition, seed issues	Public	Project Development and Technical
Program for Biosafety Systems (PBS)	Regulatory support to NBC/Nigeria and CFT implementation at IAR. Technical training.	Public	Regulatory support
United States Agency for International Development (USAID)	Funder to CSIRO's efforts, AATF and PBS	Public	Funder
Agricultural Research Council of Nigeria (ARCN)	Supervises agricultural research	Public	Project Development and Technical
National Agricultural Research Systems (NARS) in target countries of West Africa	Gene introgression into local cowpea varieties	Public	Project Development and Technical
Rockefeller Foundation	Contributed funds to CSIRO efforts	Public	Funder
National Biosafety Committee (NBC)	Serve as advisory body to the Competent National Authority for Biosafety in Nigeria (Ministry of Environment)	Public	Regulatory authority

The achievement of this entails three specific objectives: 1) obtain the *cry1Ab* gene from Monsanto, 2) facilitate regulatory compliance for the development of a *Maruca*-resistant cowpea variety, and 3) provide product stewardship for the altered variety [7]. The expected outcomes of the project are increased food production, increased trade between regions in Africa, and elimination of the use of pesticides that are harmful to both humans and the environment [3].

Brief literature review of the history of the Cowpea Productivity Improvement Project

The Cowpea Productivity Improvement Project is rooted in a longstanding international effort by a group of dedicated scientists to apply the power of research to pressing, humanitarian issues. While the Cowpea Productivity Improvement Project technically began as a project in 2003, its roots reach back to the late 1980s and early 1990s at Purdue University. During this period, scientists from the United States, Italy and Japan, among others, were investigating the potential of genetically modifying the cowpea for specific resistances [9].

In 2001, key international cowpea stakeholders convened for a conference held in Dakar, Senegal and sponsored by a number of institutes including the Food and Agriculture Organization (FAO), IITA, Purdue University and the Rockefeller Foundation. The goal of the meeting was to investigate ways in which the availability of cowpea could be increased [10]. This diverse group of individuals formed the core of the Network for the Genetic Improvement of Cowpea for Africa (NGICA), one of the key partners in the project [10]. In 2002, NGICA convened a workshop in Capri, Italy for those actively engaged in the funded research projects on genetic transformation of cowpea. This workshop sought to encourage discussion between researchers in order to strengthen the NGICA network [11]. Despite NGICA's efforts to encourage scientific research, NGICA remained more of a consortium than a formal organization with an office and staff. Members included researchers from all over Africa as well as the United States and Australia, making the operation of the organization virtual.

NGICA, knowing that a formal organization was necessary to push the project forward, contacted the newly formed AATF as a potential partner. It was believed that AATF had the organizational capacity to coordinate partners and negotiate agreements that NGICA did not have. AATF took on a strong leadership role in the cowpea project and coordinated a second meeting, which was held at their headquarters in Nairobi, Kenya in 2003. The purpose of this meeting was to develop a plan for an AATF/NGICA cowpea technology transfer project. Five constraint areas were identified and a task force was developed to assess the progress of each sector and develop

recommendations for the future. The areas include: seed production and access; field production; storage and utilization; marketing; and intellectual property [12]. The findings that resulted from these task forces determined the agenda for the next planned meeting.

In 2004, a Cowpea Stakeholders Workshop was held to develop a strategic plan for the improvement of cowpea [13]. This workshop integrated knowledge from different stakeholder groups and established technical milestones. Intellectual property rights issues were addressed and a plan to acquire the *cry1ab* gene from the Monsanto Company was developed [13,14]. The major outputs of this included the development of a project framework for the Cowpea Productivity Improvement Project including technical plans, a funding breakdown, and a management system structure.

Between 2005 and 2007, countries interested in the Cowpea Productivity Improvement Project went into negotiation to formally define partner roles and develop formal partnership agreements. During the same period, the Commonwealth Scientific and Industrial Research Organisation (CSIRO) in Australia, funded by the Rockefeller Foundation, had begun an experiment to see if genes could be inserted into cowpea varieties. In 2005, CSIRO scientists successfully inserted transgenes into cowpeas that were transmitted to their progeny [15]. In 2007, the AATF had successfully negotiated access to Monsanto's *cry1ab* gene and the gene had been incorporated into the cowpea at CSIRO in Australia. Testing of this variety against pests had also begun in Australia. The variety was being tested against the *Helicoverpa armigera* pest, as the *Maruca* is not present in Australia [16].

In 2007, a meeting was held in Accra, Ghana to develop the application for genetically modified (GM) cowpea confined field trials (CFTs) in Nigeria. By this point, an initial procedure of transforming the Bt gene into a Nigerian wild-type variety had occurred in Australia. However, the challenge of bringing the seed to Nigeria for CFTs remained [16]. Nigeria at this point had ratified the Cartagena Protocol in 2000 and had developed National Biosafety Guidelines through the United Nations Environment Programme-Global Environment Facility (UNEP-GEF) in 2005. These interim guidelines allow for the testing of GM crops for research purposes, but do not allow for commercialization. An application for CFTs was developed after this meeting and submitted to the National Biosafety Committee of Nigeria for review [16].

In 2008, despite the need to evaluate the crops in the field, the hope of experimental trials of Bt cowpea in Nigeria seemed slim. In order to keep the project moving, the partners collectively decided to initiate the first CFTs of the Bt cowpea in August 2008 in Puerto Rico [16]. The trials in Puerto Rico were important to evaluate the efficacy of the Bt events against the pod borer in the field.

Trust in agbiotech initiatives

Nigeria is in the midst of developing its regulatory capacity and is in the final stages of passing a formal biosafety law for the safe development and use of GM crops in the country [17,18]. Such biosafety capacity provides an important backdrop for the development of biotech development initiatives such as the Cowpea Productivity Improvement Project, which is the focus of this study.

While the goal of the Cowpea Productivity Improvement Project is to provide pest-resistant cowpea varieties to Nigerian farmers, a great obstacle to achieving success with such initiatives is the distrust that may exist among government, regulatory authorities, and the private sector in Nigeria. Trust is important in PPPs, in which groups must collaborate in order to complete complex, long-term tasks [19-21]. In the context of agbiotech projects, these tasks could fall under processes such as product development, regulatory approvals, communication with the public, engagement with farmers, seed multiplication, and breeding, to name a few. Yet, one barrier to establishing trust in past partnerships in Nigeria has been the awareness of corruption at the government level, such as those between civil society organizations and the government [22]. In addition, a history of corruption in Nigerian corporate governance has posed a challenge to attracting foreign investment in the past [23].

Furthermore, establishing and maintaining trust is critical in the agbiotech field as the introduction of transgenic crops can often be contentious and hindered by public distrust [24,25]. Particularly, a frequently noted source of distrust in GM crops is a lack of information and awareness across various levels of the public sector. The lack of knowledge about GM crops among both scientists and community members in Nigeria has led to the perception that agricultural biotechnology is risky. Among community members surveyed before participating in informational workshops, the greatest concerns about genetic modification revolved around food safety, ethical issues, and cost [26]. In another survey conducted in southwestern Nigeria, half of the respondents perceived GM crops to be of high risk [27]. For these reasons, the issue of trust in the Bt cowpea project in Nigeria is a critical area worthy of being studied.

This study constitutes one in a series of eight case studies investigating the role of trust in agbiotech PPPs and the adoption of GM crops in sub-Saharan Africa. There are three specific goals of this study: 1) describe effective trust-building practices in agbiotech projects, 2) describe the challenges to trust-building in successful and unsuccessful agbiotech PPPs, and 3) determine what makes these practices effective or ineffective through the identification of barriers to trust and trust-enhancing practices. To our knowledge, it is the first globally-focused case study series of its kind to investigate the role of trust in the successes and failures in PPPs. This study provides insight for potential funders of agbiotech projects as well as for researchers, farmers and others interested in such projects.

Methods

Data was collected by reviewing academic articles, news articles and publicly available project documents on the cowpea project. We also conducted semi-structured, face-to-face interviews with eleven individuals. These interviewees came from a variety of organizations (listed in Table 1) associated with the project.

Interviewees were identified first by making a list of key individuals associated with the project based on the stakeholders identified within the research protocol. This list was then populated further through snowball sampling by engaging with partners involved in the project and stakeholder informants who were familiar with the Cowpea Productivity Improvement Project through the Sandra Rotman Centre's Social Audit Project [28]. Potential interviewees were sent an invitation, which included an explanation of the case study series, to participate in the interview. Those who consented to participate were informed that the interview would be recorded, transcribed verbatim, and then analyzed. The interviews took place primarily in Nigeria, with some taking place in the United States and one in Tanzania (in accordance to the location in which each stakeholder was based). The interview guide included questions on the interviewees' background, their understanding of the project, and their interpretation of the word *trust*. The interview explored perceptions of trust within the partnership and the public, apparent challenges to trust, and observed trust-building practices. Interviewees were also asked to provide suggestions on how to improve agbiotech PPPs (see Additional file 2 for sample questions from the interview guide).

The interviews were transcribed and the analysis was performed by reading through the transcripts several times, identifying trends and organizing them into major themes. The results were critically analyzed and compiled to create a comprehensive narrative on how trust is understood and built within the project and between the project and the community that it aims to serve.

We received Research Ethics Board (REB) approval for conducting the case study. Participants provided signed consent to participate in the study and to have the interviews recorded.

Results and discussion

When developing the interview guide for this study, we did not provide a definition of trust but instead asked the interviewees for their own definition and understanding of trust. While the interviewees linked trust to a number of factors, the strongest link was made to the character of the individual or institution. The most frequent descriptions of trust that our interviewees provided were associated with fostering transparency, honesty, accountability, and dependability. However, the diversity of actors involved in the Cowpea Productivity Improvement Project renders trust to exist on different planes, and the level of trust in each of these planes was not identical. For example, we observed a high level of trust among the core partners and a low level of trust between the project partners and the public.

Based on the results of this study, we have derived four key lessons, from which partners in other agbiotech PPPs can learn and use as a guide for building and fostering trust.

1. Build up, not over: acknowledge existing mandates and institutions

As the promoter of biotechnology development for Nigeria, it was particularly important for the National Biotechnology Development Agency (NABDA) to recognize each research institution's mandate and existing capabilities when contributing to biotechnology efforts. A government interviewee stated that the government wants to support projects at institutes whose mandates contain no conflicts of interest. This approach not only reduces the redundancies caused by duplication of labor, but also mitigates the potential tensions that could arise among partners.

Some interviewees were concerned that the project might encourage mandate-stealing or result in institutions encroaching on other institutions' area of expertise. At the time of the study, the cowpea project had not encountered any major conflicts in this regard.

Ambiguities in the mandates of partners and organizations involved in the project were, however, a source of distrust in the project. Interviewees stressed that conflicts over ownership and project direction can arise when various stakeholders have diverging understandings about their role in the project. An interviewee cited AATF as an example of an organization that was successful in clarifying its mandate. Their bringing transparency to their plans suggested that both sides of the partnership would be closely involved throughout the entire process.

2. Educate the public now: information dissemination in the development stages of the project

Although the commercialization of a Bt cowpea crop is many years away, educating the public was seen as vital to facilitating adoption of the crop, since public trust in biotechnology had already been low. Interviewees mentioned that there were high levels of skepticism within the community due to a lack of familiarity with the genetic transformation process or incorrect knowledge about GM crops. Before introducing a finished GM crop to the Nigerian market, farmers, journalists, politicians and society at large must first be educated about biotechnology in a language that is understandable. One communications officer noted that people will be more inclined to trust if they understand how a technology was arrived at. To this effect, information-sharing workshops were hosted for farmers, media, and community stakeholders in order to educate the public about the project and GM crops. These efforts were improved by using the appropriate language for different stakeholder groups and making the information as accessible as possible.

Perceptions of GM crops

A government representative noted that most anti-GM groups counter claims supporting GM without empirical evidence. An interviewee saw some anti-GM groups as purposely wanting to undermine good relations [trust] between project partners and community members. Moreover, communicating accurate information about GM crops to the public was seen as a challenge by some interviewees because of a lack of interest and engagement by scientists and their institutions, as well as the perceived salience of negative gossip among Nigerians regarding GM crops.

Debunking misperceptions

An interviewee from one of the partner organizations noted that communication with farmers in the beginning will encourage uptake of the product. One approach adopted with regard to this was the organizing of annual visits to the cowpea project lab so farmers and end consumers can see the products being developed. "Farm walks" were also hosted for media representatives and community members to allow them to explore the fields and see the Bt cowpea crops first-hand.

Early information dissemination in the form of flyers, radio broadcasts and educational workshops for community members led to the creation of GM 'ambassadors' - community members who understood the benefits of biotechnology and were willing to discuss it informally to others in the community. An interviewee recounted that the Institute for Agricultural Research in Zaria – where the project is being hosted – had many of their students attend this workshop. The students, who did not have much knowledge about biotechnology issues prior to the workshop, became ambassadors that day.

In addition, workshops were held to encourage media, both pro- and anti-GM, to promote accurate information and demystify the concept of GM. A very good example of this as witnessed by one of the authors (OE) was the Open Forum Agricultural Biotechnology (OFAB), which is an organization of public forums on biotechnology that are held on a monthly basis.

Overall, trust between the partners and the community was brought about by engagement and open communication, and reflected in the farmers' and the public's receptiveness toward the anticipated product and the increased awareness around agricultural biotechnology.

3. Team-building as trust-building: establish a core team that works and is results-oriented

Constructing a core team with defined roles, commitment to the partnership, common goals, and the ability to meet deadlines, manage funds, and communicate effectively is essential to building trust.

Keeping commitments is critical

Failing to consistently meet deadlines and follow through on commitments was identified as a challenge to building trust within the partnerships. This, according to one government official, was a major challenge to building trust.

A common complaint among interviewees was that some partners were inconsistently meeting deadlines, poorly managing funds, or distracted by other projects. These issues were listed as significant challenges to building trust. Despite these shortcomings, however, one interviewee described the need to keep all partners, particularly public partners, included in order to prevent violating public trust in the project. Effective team-building during the beginning stages of the project is an important practice that can obviate these issues. This would entail establishing and clearly articulating partner roles and the ownership that each partner has over the project when designing the project.

Effective team building

When team-building is done correctly, the resulting trust among partners can lead to the establishment and continuation of a highly functional partnership [20]. In fact, one of the most prevalent indicators of trust within the relationship was the existence of the partnership itself. An interviewee from the Nigerian government stated that a major indicator of trust is if the partnership is still intact. An officer from the AATF echoed this statement and said that trust helps the project move at the pace expected by partners at any point in time. The trust established by effective team-building can have many positive outcomes, including the pursuit of common goals and open communication among partners [29].

Communication and free expression among partners is also an essential outcome of trust [19]. One researcher stated that communication is absolutely critical for trust building. Another interviewee stated that the ability of partners to express themselves freely – along with any problems or concerns of theirs – is an indication that trust exists in the partnership.

Regular internal communication between partners was also a key driver of project development. Meetings were held monthly and allowed partners to conceptualize project direction, allocate responsibility and establish milestones and objectives. Project partners are also kept updated through the use of bulletins circulated regularly

among stakeholders. In addition, high-level expert consultations were held in order to anticipate future challenges. Ongoing daily interactions between partners were also important in ensuring that the team remains intact.

Similarly, critical scientific discussion of the project was a practice that fostered trust within the scientific community. Merits and plans of the cowpea project were debated in a research program review committee, in which the scientists sit together and criticize different projects.

Effective trust-building and team-building enabled two important milestones to be reached: the continuation of funding for the project and the effective delivery of an end product. On the issue of funding and donor support, an interviewee from one of the partner organizations stated that trust motivated the management of his institution to support and continue funding the project. The delivery, eventually, of an end product to the farmers through an efficient and accountable system was viewed as the most obvious result of trust within a partnership. As stated by another interviewee, trust enables a more efficient way of delivering.

4. Partnering for a purpose: work with local organizations to serve local needs

The impetus for the cowpea project emerged from a shared recognition of the severity of the pest problem burdening cowpea growers in Nigeria and the inadequacy of regular pest control methods to effectively address it. The need for the project, then, was *locally* defined by those working with the farmers and addressed by different African agricultural associations. However, the idea that a foreign company is driving the project was deemed a huge challenge that may promote the perception that the private sector has come to exploit poor farmers.

Responding to a locally-defined need was linked to the success and longevity of the project. In addition, one interviewee noted the importance of local capacity-building in research development and infrastructure, which had been integrated into the project. By addressing both the *Maruca* pest issue and capacity building, the project served two important locally-defined needs.

Negative perceptions of Monsanto and private sector involvement in product development

Preconceived notions about the private sector varied substantially. Some interviewees saw the Monsanto Company as a positive contributor while others saw it as negative. For example, many interviewees remarked on the minimal role that the private sector had played. An AATF representative stated that, aside from donating the gene, the Monsanto Company does not seem to have anything to do with the actual project. The limited role of Monsanto in the partnership was believed by some to reduce the high level of public suspicion and skepticism of private sector involvement in the project. One interviewee said that Monsanto's limited role was important in preventing damage to the trust that had been built within the partnership.

There were still, however, negative perceptions of foreign product development, since the Bt gene was donated by Monsanto and transformation research completed by CSIRO in Australia. An AATF representative stated that he had received negative comments, such as the following: *Why are these genes not being used here to develop our own varieties, to have our scientists empowered to use these technologies and to learn. So why is it being developed outside and then we just test and then disseminate... [W]hy not advance our own laboratories?* African leadership in the project, such as that of the AATF, was therefore seen as an important factor that built trust between the project and the public.

Despite such perceptions, private sector involvement is regarded by the stakeholders we interviewed as an important component of a PPP's delivery of a successful product. One interviewee, contemplating the nature of the cowpea project as a PPP, expressed concern about the ability of public sector institutions to successfully steward a product such as a GM crop over the long-term and maintain the quality of the seeds and therefore believes that the project must involve more private sector actors.

Conclusions

As the Cowpea Productivity Improvement Project in Nigeria demonstrates, trust is critical in building successful agbiotech PPPs. Overcoming challenges to trust-building requires concerted effort throughout all stages of project implementation. From this case study, four key lessons were drawn on building and maintaining trust among the partners in a project and between the project and the community. First, effective team-building is essential to building trust and ensuring that each partner can and is willing to fulfill their contributions to the project. Second, continuous communication and information sharing among partners and between the partners and the community is an important trust-building practice; this can be accomplished by designing novel and engaging activities that encourage awareness building, such as "farm walks" and the creation of GM 'ambassadors.' Third, building *up*, not *over*, existing institutions and mandates serves to encourage trust-building by preventing inter-institutional conflicts and mandate-stealing. Fourth, focusing on a locally-defined need and utilizing local organizations through partnership will foster trust and improve project success and longevity. These lessons on trust-building can and should be applied to future projects and agbiotech PPPs. Plans are currently being made to backcross the cowpea strain into a local variety in Nigeria. The further

development and adoption of the Bt cowpea seed hinges on, among other things, the adoption of a National Biosafety Law in Nigeria. If the current Biosafety Bill becomes law, some estimates expect that a Bt *Maruca*-resistant cowpea will be available by 2014 for farmers to use [16]. With such a law in place, it is likely that the partners will try to negotiate access to a second gene for the cowpea from Monsanto in order to make the cowpea variety sustainable on the seed market. The Nigerian cowpea experiment can be used as a model for other West African nations, and is actually applied as such in Ghana and Burkina Faso, interested in developing a Bt cowpea.

Acknowledgements

The authors are grateful to each of the participants who contributed substantial time and effort to this study. The authors also thank Kathryn Barber, Jessica Oh, Jocalyn Clark, Hassan Massum, and Jill Murray for comments on earlier drafts of the manuscript. Special thanks to Jill Murray for scheduling the interviews.

This project was funded by the Bill & Melinda Gates Foundation and supported by the Sandra Rotman Centre, an academic centre at the University Health Network and University of Toronto. The findings and conclusions contained within are those of the authors and do not necessarily reflect official positions or policies of the foundation.

This article has been published as part of *Agriculture & Food Security* Volume 1 Supplement 1, 2012: Fostering innovation through building trust: lessons from agricultural biotechnology partnerships in Africa. The full contents of the supplement are available online at http://www.agricultureandfoodsecurity.com/supplements/1/S1. Publication of this supplement was funded by the Sandra Rotman Centre at the University Health Network and the University of Toronto. The supplement was devised by the Sandra Rotman Centre.

Author details

[1]Sandra Rotman Centre, University Health Network and University of Toronto, Toronto, Ontario, Canada. [2]African Centre for Innovation and Leadership Development, Federal Capital Territory, Abuja, Nigeria. [3]Dalla Lana School of Public Health, University of Toronto, Toronto, Canada. [4]Grand Challenges Canada. [5]Dalla Lana School of Public Health and Department of Surgery, University of Toronto, Toronto, Canada.

Authors' contributions

Study conception and design: OCE and ASD. Data collection: OCE. Analysis and interpretation of data: OCE. Draft of the manuscript: OCE and ASD. Critical revision of the manuscript for important intellectual content: OCE and ASD. All authors read and approved the final manuscript.

Competing interests

The authors declare that they have no competing interests.

References

1. Kushwaha S, Musa AS, Lowenberg-DeBoer J, Fulton J: **Consumer Acceptance of GMO Cowpeas in sub-Saharan African.** *Proceedings of the American Agricultural Economics Association Annual Meeting: 3 August 2004: 1-4 August 2004; Denver, CO 2004.*

2. IITA. [http://old.iita.org/cms/details/cowpea_project_details.aspx?zoneid=63&articleid=269].

3. Thomson JA: **The role of biotechnology for agricultural sustainability in Africa.** *Phil Trans R Soc B* 2008, **363**:905-913.

4. Coulibaly O, Aitchedji C, Gbegbelegbe S, Mignouna H, Lowenberg-DeBoer J: **Baseline Study for Impact Assessment of High Quality Insect Resistant Cowpea in West Africa.** *African Agricultural Foundation* 2008, 1-51.

5. USAID: **The Study of the Cowpea Value Chain in Nigeria, from a Pro-Poor and Gender Perspective.** 2008, 1-56.

6. David MA: **GAIN Report: Nigeria Agricultural Biotechnology Annual Report.** United States Department of Agriculture Foreign Agricultural Service; 2009:**NI9008**:1-12.

7. African Agricultural Technology Foundation. [http://www.aatf-africa.org/userfiles/Cowpea-brief.pdf].

8. African Agricultural Technology Foundation: **Maruca-Resistant Cowpea: Frequently Asked Questions.** [http://www.aatf-africa.org/userfiles/CowpeaFAQ.pdf].

9. Ferry RL, Singh BB: **Cowpea genetics: a review of the recent literature.** In *Advances in Cowpea Research.* Ibadan, Nigeria: IITA:JIRCAS;Singh BB, Mohan Raj DR, Dashiell KE, Jackai LEN 1997:13-29.

10. Murdock LL: **Proceedings of the The Dakar Symposium/Workshop on the Genetic Improvement of Cowpea: 8-12 January 2001; Dakar, Senegal.** 2001.

11. NGICA: **Proceedings of the Workshop on the Genetic Transformation of Cowpea: October 31-November 2 2002; Capri, Italy.** 2002.

12. African Agricultural Technology Foundation: **Proceedings of the Constraints to Cowpea Production and Utilization in Sub-Saharan African Small Group Meeting: 10-11 July 2003; AATF Headquarters, ILRI, Nairobi, Kenya.** 2003.

13. **A plan to apply technology in the improvement of cowpea productivity and utilisation for the benefit of farmers and consumers in Africa.** In *Proceedings of the Small Group Meeting: 10-12 Feb 2004; Accra, Ghana* Majiwa P, Odera M, Muchiri N, Omanya G, Werehire P 2004.

14. Boadi RY, Bokanga M: **The African Agricultural Technology Foundation Approach to IP Management.** In *Intellectual Property Management in Health and Agricultural Innovation: A Handbook of Best Practices* Krattiger A, Mahoney RT, Nelson L 2007, 1765-1774.

15. Popelka JC, Gollasch S, Moore A, Molvig L, Higgins TJV: **Genetic transformation of cowpea (Vigna unguiculata L.) and stable transmission of the transgenes to progeny.** *Plant Cell Reports* 2006, **24(4)**:304-312.

16. African Agricultural Technology Foundation: **Proceedings of the Bt Cowpea Field Trial Planning Meeting Report: 17-18 December 2008; Donald Danforth Plant Science Center, St Louis, Missouri.** 2008.

17. Abutu A: **Nigeria: Experts Task President Jonathan on Biosafety Bill.** All Africa Press; 2012.

18. Okafor J: **Nigeria: Biosafety Bill Nears Final 'Push'.** All Africa Press; 2011.

19. Edelenbos J, Klijn E: **Trust in Complex Decision-Making Networks: A theoretical and Empirical Exploration.** *Administration & Society* 2007, **39(1)**:25-50.

20. Brewer B, Hayllar MR: **Building public trust through public-private partnerships.** *International Review of Administrative Sciences* 2005, **71(3)**:475-492.

21. White-Cooper S, Dawkins NU, Kamin SL, Anderson LA: **Community-institutional partnerships: understanding trust among partners.** *Health Educ Behav* 2009, **36(2)**:334-347.

22. Essia U, Yearoo A: **Strengthening civil society organizations/government partnership in Nigeria.** *International NGO Journal* 2009, **4(9)**:368-374.

23. Okike ENM: **Corporate Governance in Nigeria: The Status Quo.** *Corporate Governance* 2007, **15(2)**:173-193.

24. Friedberg SE, Horowitz L: **Converging Networks and Clashing Stories: South Africa's Agricultural Biotechnology Debate.** *Africa Today* 2004, **51(1)**:3-25.

25. Stone GD: **Both Sides Now. Fallacies in the Genetic-Modification Wars, Implications for Developing Countries and Anthropological Perspectives.** *Current Anthropology* 2002, **43(4)**:611-630.

26. Oladele OI, Adekoya AE: **Improving Technology Perception through Information and Education: A case of Biotechnology in Nigeria.** *Agriculturae Conspectus Scientificus* 2008, **73(4)**:239-243.

27. Alarima CI: **Knowledge and Perception of Genetically Modified Foods Among Agricultural Scientists in South-West Nigeria.** *OIDA International Journal of Sustainable Development* 2011, **2(6)**:77-88.

28. Ezezika OC, Thomas F, Lavery JV, Daar AS, Singer PA: **A Social Audit Model for Agro-biotechnology Initiatives in Developing Countries: Accounting for Ethical, Social, Cultural and Commercialization Issues.** *Journal of Technology Management and Innovation* 2009, **4(3)**:24-33.

29. Clark RD: **Components of selected public-private partnerships to build new schools in California.** University of La Verne; 2002.

4

Development of a core collection of *Triticum* and *Aegilops* species for improvement of wheat for activity against chronic diseases

Meenakshi Santra, Shawna B Matthews and Henry J Thompson[*]

Abstract

Background: The objective of this study was to develop a core collection of *Triticum* and *Aegilops* species as a resource for the identification and characterization of wheat lines with preventive activity against chronic diseases. Given that cancer is the leading cause of mortality in the world and shares risk factors with obesity, type-2 diabetes, and cardiovascular disease, and given that wheat has been reported to protect against these diseases, the core collection was developed based on cancer prevalence.

Methods: The Germplasm Resources Information Network (GRIN) database was used to identify *Triticum* and *Aegilops* species grown in regions of the world that vary in cancer prevalence based on the International Agency for Cancer Research GLOBOCAN world map of cancer statistics (2008). Cancer incidence data drove variety selection with secondary consideration of ploidy, center of origin, and climate.

Results: Analysis indicated that the geographic regions from which wheat is considered to have originated have a lower incidence of cancer than other geographic regions (*P* <0.01), so wheat lines from countries that comprise the 'Fertile Crescent' were highly represented in the core collection. A total of 188 lines were selected from 62,571 accessions maintained by GRIN. The accessions identified comprised two genera and 14 taxa of 10 species within 19 groups from 82 countries. The core collection is comprised of 153 spring, 25 winter, and five facultative selections of wheat.

Conclusions: A diverse core collection of wheat germplasm has been established from a range of regions worldwide. This core collection will be used to identify wheat lines with activity against chronic diseases using anticancer activity as a screening tool.

Keywords: *Aegilops*, Anticancer activity, Cancer, Core collection, *Triticum*, Wheat germplasm

Background

The consumption of whole grains has long been associated with a healthy lifestyle and chronic disease prevention; in particular, multiple studies have correlated whole-wheat consumption with protection against chronic diseases including cardiovascular disease, stroke, type 2 diabetes, and cancer at multiple sites [1-8]. However, these studies have failed to discriminate between the type of wheat that is consumed and the chronic disease protective effect observed. Specifically, the USDA has germplasm from 62,571 distinct wheat varieties; given the well-described differences in agronomic traits as a result of genetic polymorphisms within wheat species, as well as the recently characterized metabolite differences between and within wheat species and subspecies [9], further investigation of the chronic disease preventive capacity of individual wheat varieties is required.

In this paper, we propose that a neglected opportunity in the field of diet and chronic disease prevention is the use of staple food crops with defined bioactivity for daily consumption [10]. The rationale underlying this approach recognizes that societies have chosen their staple food crops, which are affordable and generally available to all individuals across socioeconomic strata, and that societies willingly consume these staples in large quantities on a

* Correspondence: henry.thompson@colostate.edu
Cancer Prevention Laboratory, Colorado State University, 1173 Campus Delivery, Fort Collins, CO 80523, USA

daily basis. These consumption patterns thus provide a stable flow of health beneficial phytochemicals in much the same way that an oral drug is taken to maintain plasma concentrations of the active ingredient in a beneficial range [11]. Further research on bioactivity of specific varieties of these staple food crops is critical, given that major chronic diseases, including obesity, type-2 diabetes, cardiovascular disease, stroke, and cancer, account for over 60% of deaths worldwide [12,13], are interrelated at the molecular and cellular levels and share many common risk factors [14-16], and, most importantly, are also considered preventable through lifestyle choices of which diet is considered to play a prominent role [17-19].

While concern exists that the genetic factors driving the occurrence and progression of cancer and other chronic diseases are so powerful that diet can have little impact, most evidence indicates that the key strategy to conquering chronic diseases like cancer is through prevention particularly when the prevention strategy is routinized from 'womb to tomb' (reviewed in [20]). However, in addition to the general presumption that all varieties of a particular staple food crop are created equal with respect to health benefits, one of the challenges of this approach is the assumption that the ingredients which a food is processed into, rather than the food itself, is the most critical factor accounting for health benefits [10]. The work reported in this paper was initiated to provide a resource for evaluating the first premise, that is, that all botanically defined lines of wheat (*Triticum* and *Aegilops* species) have equivalent chronic disease fighting activity with anticancer activity providing a focal point for analysis. Cancer was chosen because among these chronic diseases, the prevalence of cancer continues to increase globally and cancer is now the leading cause of chronic disease related mortality in the world [21]; furthermore previously published reports have described an inverse association between wheat consumption and cancer incidence.

Wheat is ranked second, after rice, among all members of the Poaceae family in terms of the amount consumed by the global population [22]. Wheat is used in the preparation of a wide variety of foods for everyday use, including bread, pasta/macaroni/noodles, bulgur, cookies, biscuits, cakes, cereals, pizza, vermicelli, couscous, pastry, and chapatti/flatbread [23,24]. It is also fermented to make beer and other alcoholic beverages. Wheat's role as a primary human dietary component is due to its large grain size, agronomic adaptability, ease of storage, and nutritional quality. While a limited number of wheat lines account for most of wheat products consumed globally due to the emergence of global industrial food systems, some ancient wheat lines- such as einkorn and emmer- are still consumed as cereal substitutes in Middle Eastern countries, where wheat is considered to have originated [25]. These grains are very small and difficult to harvest and clean. As such, they are often used in porridge or soup without grinding or processing. In the Arab world (including Iraq, Syria, and Tunisia), soft green (immature) wheat grains, mostly domesticated tetraploid emmer, are sundried and roasted to make a food called Freekeh. In addition, people in Arab countries routinely mix Freekeh made from domesticated landraces of wheat grains with meat and spices in their daily foods.

As noted above, wheat is consumed in large amounts worldwide, but the type of wheat and the manner in which it is consumed differ markedly depending on geographic region. Because of the novel events underlying the domestication of wheat, there are major genetic differences among the types of wheat commonly consumed. As a result of this inherent diversity, the Germplasm Resources Information Network (GRIN) has accumulated over 62,571 wheat-related accessions [26]. The general approach to working with such a large resource is to devise a strategy by which to pick a representative sample of lines from the total resource (collection) that is small enough to manage for use in research yet large enough to capture the diversity of the population for the trait(s) of interest. The resulting subsample of germplasm is referred to as a core collection [27]. Herein, a core collection of wheat lines for future use in chronic disease prevention research is described.

Methods
Source of plant materials
The *Triticum* and *Aegilops* collections at GRIN (USDA/ARS, Aberdeen, Idaho) include 59,564 and 2,650 accessions, respectively [26]. Only *Triticum* and *Aegilops* were selected for establishing this core collection with the reasoning that these two genera comprise the majority of wheat lines that have emerged due to domestication through natural selection and polyploidization. The accessions chosen for inclusion in the core collection are described in Table 1. All available information was obtained on selected accessions, including passport information, characterization, and evaluation.

Criteria of selection
Cancer statistics
The data used were based on GLOBOCAN 2008 cancer statistics [21]. GLOBOCAN's cancer statistics are based on the incidence of all cancers using the age-standardized rate (ASR). Our intent was to select wheat lines attributed to specific countries identified in the GLOBOCAN global map (Figure 1) that showed wide variations in cancer incidence rates, under the presumption that these wheat lines, and their close relatives, are likely to be consumed in greater amounts in those countries.

Table 1 A summary of the *Triticum* and *Aegilops* species used for core collection

Species	Genome	Total active accessions at GRIN	Total number in core collection	Base species (%)
Aegilops speltoides var. *speltoides*	BB/GG	9	3	33.33
Aegilops speltoides var. *ligustica*	BB/GG	11	1	9.10
Aegilops tauschii	DD	200	1	0.50
Triticum aestivum subsp. *aestivum*	BBA^{u}A^{u}DD	46,225	80	0.17
Triticum aestivum subsp. *compactum*	BBA^{u}A^{u}DD	118	1	0.85
Triticum aestivum subsp. *spelta*	BBA^{u}A^{u}DD	1,292	6	0.46
Triticum hybrid	BBAADD	219	1	0.46
Triticum monococcum subsp. *aegilopoides*	A^{m}A^m	826	4	0.48
Triticum timopheevii subsp. *armeniacum*	GGA^{u}A^u	249	1	0.40
Triticum turgidum subsp. *carthlicum*	BBA^{u}A^u	95	2	2.11
Triticum turgidum subsp. *dicoccon*	BBA^{u}A^u	622	10	1.61
Triticum turgidum subsp. *dicoccoides*	BBA^{u}A^u	777	2	0.26
Triticum turgidum subsp. *durum*	BBA^{u}A^u	8,526	63	0.74
Triticum turgidum subsp. *paleocolchicum*	BBA^{u}A^u	4	1	25
Triticum turgidum subsp. *turanicum*	BBA^{u}A^u	108	1	0.93
Triticum turgidum subsp. *turgidum*	BBA^{u}A^u	1,054	6	0.57
Triticum urartu	A^{u}A^u	245	3	1.22
Triticosecale sp.	BBA^{u}A^{u}RR	1,985	1	0.05
Triticum zhukovskyi	GGA^{u}A^{u}A^{m}A^m	6	1	16.67
Total		**62,571**	**188**	**0.30**

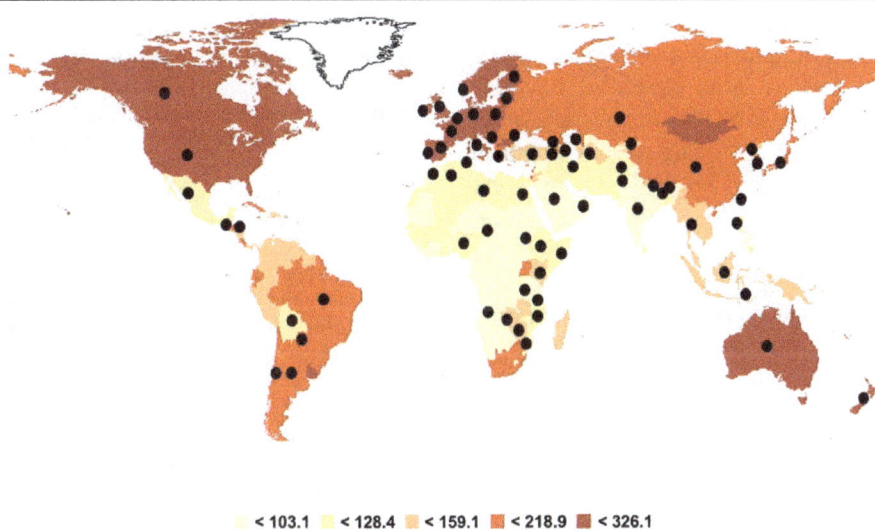

Source: GLOBOCAN cancer statistics, 2008. Estimated age-standardized incidence rate per 100,000 residents for all cancers, excluding non-melanoma skin cancer, both sexes and all ages.

Figure 1 A world map of cancer incidence displaying geographic distribution of core collection of wheat germplasm. Estimated age-standardized incidence rate (ASR) per 100,000 residents for all cancers, excluding non-melanoma skin cancer, both sexes and all ages based on GLOBOCAN Cancer statistics, 2008. Each black dot represented a wheat growing country of the world. Four colors ranging from very light yellow to dark brown described the ASR from <103.1 to >326.1 per 100,000 individuals.

Centers of origin

Archeological evidence indicates that Armenia, Iran, Iraq, Lebanon, Israel, Jordan, Syria, and Turkey were the centers of origin for wheat germplasm [28]. Cancer statistics also indicated that the occurrence of cancer is very low in these areas, supporting the possibility that the wheat species cultivated and consumed locally provide anticancer protection. Wheat lines from these countries were highly represented in the core collection.

Regression analysis

To determine whether the relationship between wheat consumption and cancer incidence was related to geographic origin of wheat, data were collected from the Food and Agriculture Organization of the United Nations (FAOSTAT) from 2007, operationally defined as kg wheat products consumed per capita per year, and from the GLOBOCAN resource from 2008, operationally defined as ASR of cancer incidence at all sites excluding non-melanoma skin cancer. Countries without data for both parameters were excluded from analyses, resulting in a total of 165 countries for the global analysis (Figure 2A) and a subset of the global analysis using 19 Near Eastern countries which are geographically proximate to the origin of wheat (Figure 2B). The countries included in the latter analysis were Armenia, Azerbaijan, Cyprus, Egypt, Georgia, Iran, Israel, Jordan, Kuwait, Lebanon, Pakistan, Saudi Arabia, the Syrian Arab Republic, Tajikistan, Turkey, Turkmenistan, United Arab Emirates, Uzbekistan, and Yemen. No wheat consumption data were available for Iraq through FAOSTAT. Log_{10}-transformed ASRs were regressed on wheat consumption data (10 kg/capita/year) in linear regression analysis using GraphPad Prism vs. 5.02 (GraphPad Software, San Diego, CA, USA). Fit parameters for each analysis, including slope, Y-intercept, R^2, and line equations are provided in the figure legend (Figure 2).

Other considerations

The species of *Triticum* and *Aegilops* include germplasm with three ploidy levels: diploid with genomes A^m, B, D, and G; tetraploid with BA^u and GA^u genomes; and hexaploid with BA^uD genomes [28]. Selections within each genome and ploidy level were represented in the core collection.

Results and discussion

To our knowledge, there are no published core collections of wheat that have been specifically developed to permit the investigation of wheat for human health benefits and particularly for reducing chronic disease risk using anticancer activity as a screening tool. Thus, the approach used was necessarily descriptive in nature. Rather than enforcing established criteria usually implemented for the

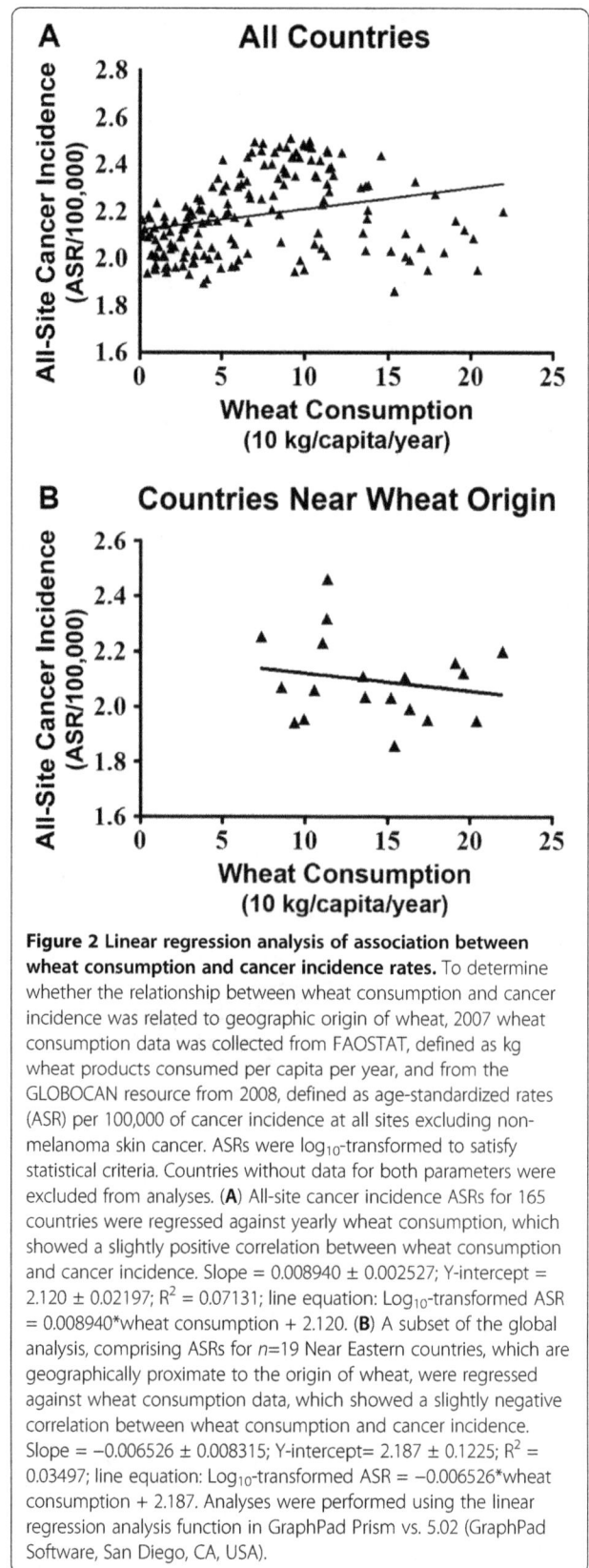

Figure 2 Linear regression analysis of association between wheat consumption and cancer incidence rates. To determine whether the relationship between wheat consumption and cancer incidence was related to geographic origin of wheat, 2007 wheat consumption data was collected from FAOSTAT, defined as kg wheat products consumed per capita per year, and from the GLOBOCAN resource from 2008, defined as age-standardized rates (ASR) per 100,000 of cancer incidence at all sites excluding non-melanoma skin cancer. ASRs were log_{10}-transformed to satisfy statistical criteria. Countries without data for both parameters were excluded from analyses. (**A**) All-site cancer incidence ASRs for 165 countries were regressed against yearly wheat consumption, which showed a slightly positive correlation between wheat consumption and cancer incidence. Slope = 0.008940 ± 0.002527; Y-intercept = 2.120 ± 0.02197; R^2 = 0.07131; line equation: Log_{10}-transformed ASR = 0.008940*wheat consumption + 2.120. (**B**) A subset of the global analysis, comprising ASRs for n=19 Near Eastern countries, which are geographically proximate to the origin of wheat, were regressed against wheat consumption data, which showed a slightly negative correlation between wheat consumption and cancer incidence. Slope = −0.006526 ± 0.008315; Y-intercept= 2.187 ± 0.1225; R^2 = 0.03497; line equation: Log_{10}-transformed ASR = −0.006526*wheat consumption + 2.187. Analyses were performed using the linear regression analysis function in GraphPad Prism vs. 5.02 (GraphPad Software, San Diego, CA, USA).

development of a core collection for agronomic traits such as disease or pest resistance, or post-harvest processing characteristics [27], cancer incidence data drove variety selection with secondary consideration of ploidy, center of origin, and climate.

Cancer statistics

A world map generated from the GLOBOCAN cancer database is shown (Figure 1) and was used to identify the countries from which germplasm was selected from the GRIN domain collection. Based on GLOBOCAN 2008 statistics, the lowest incidence rates of cancer occur in middle Africa, northern Africa, south central Asia, western Africa, eastern Africa, Central America, and western Asia. Interestingly, western Asia, or the Fertile Crescent region between the Tigris and Euphrates river basins, has been determined to be the geographic center of origin for wheat [28]. To explore the relationship between wheat consumption and cancer incidence as it relates to geographic origin of wheat, we used linear regression analysis. When cancer incidence rates for 165 countries were regressed on wheat consumption data in those countries, a slight positive correlation between these parameters was found (Figure 2A, slope = +0.0089 increase in log_{10}-transformed ASR per 10 kg/capita/year increase in wheat consumption). This translates to an increase of 1.02% in cancer incidence for each 10 kg/capita/year increase in wheat consumption. Conversely, when this analysis was confined to 19 countries near the geographic origin of wheat, a slight negative correlation (Figure 2B, slope = −0.0065 increase in log_{10}-transformed ASR per 10 kg/capita/year increase in wheat consumption). This translates to a reduction of 0.99% in cancer incidence for each 10 kg/capita/year increase in wheat consumption. Hence, the slightly positive association between global cancer incidence rates and wheat consumption is reversed when the analysis is restricted only to countries near the geographic origin of wheat (P <0.01). Many other factors are likely involved in the observed correlations between global wheat consumption and cancer incidence rates. Thus, it is important to underscore that there is no evidence of a causal link between these parameters but rather these analyses support the use of cancer incidence by geographic locale as an objective albeit arbitrary tool to guide wheat line selection for the core collection.

The core collection of wheat germplasm was selected from regions with lower incidence rates of cancer, as well as regions such as North America, Europe, and Oceania (Australia and New Zealand) with higher incidence rates of cancer; evaluation of germplasm from regions of both low and high cancer rates is critical for assessing differences in the type of wheat consumed which may impact cancer activity. In choosing multiple

selections from within a country, we were unable to follow the procedure of constant, proportional, and logarithmic selection, as all genotypes were not available in each country [27]. In addition, maintaining uniform diversity around the world was impossible as there are large differences in the total number of accessions in each country that was considered.

Characterization of selected lines

A total of 62,571 accessions at GRIN were designated as the source collection, which represented two genera and 14 taxa. From the source collection, 188 accessions were selected for the core collection, which is 0.3% of the source collection. The global distribution of the wheat lines in the core collection is shown (Figure 1, Table 1) and provides a summary of the *Triticum* and *Aegilops* species comprising the core collection. The core collection consisted of two genera and 14 taxa of 10 species and was comprised of 19 groups. These 188 accessions belonged to 82 different countries with three different climates (tropical, subtropical, and temperate). The plant introduction number, plant name, taxon, original source, selection criteria, growth habit, and probable market classes are shown (Table 2). Probable market classes were determined by visual observation of the germplasm using a grain color standard and may be changed during future evaluation.

Climate

There are several climates in which the domestication of wheat occurred: tropical, subtropical, and temperate, and three types of wheat resulted: winter, spring, and facultative. They differ in temperature response due to the presence and absence of dominant vernalization genes [29,30]. The three types of wheat are presented in the core collection.

Limitations

Cancer prevalence rates among countries are subject to a host of genetic and environmental determinants. Despite associations reported between wheat consumption and cancer risk, there is no direct causal evidence that a particular wheat variety reduces the cancer rate within a specific country [19]. Nonetheless, the overall cancer rate in a country provided an objective albeit arbitrary criterion for selecting wheat lines for inclusion in the core collection. The usefulness of this approach will be determined as screening for anticancer activity in laboratory model systems progresses. Another limitation is that many core collections of crop species are between 5% and 10% of the domain in size, and thus the core collection reported is relatively small in comparison (Table 1). However, there are examples of core collections <5% of the domain in size. For example, the international barley

Table 2 Detailed information for 188 *Triticum* and *Aegilops* germplasm collected from GRIN platform through USDA

No.	ID	Plant name	Taxon	Country	Selection criteria	Growth habit	Probable market class
1	PI 542274	84TK081-057	*A. speltoides var. speltoides*	Turkey	Center of origin	NA	NA
2	PI 560529	TU85-008-01	*A. speltoides var. speltoides*	Turkey	Center of origin	NA	NA
3	PI 449338	AE 081D	*A. speltoides var. speltoides*	Israel	Center of domestication	NA	NA
4	PI 560528	TU85-015-02-2	*A. speltoides var. ligustica*	Turkey	Center of origin	NA	NA
5	PI 603233	TA 1669	*A. tauschii*	Azerbaijan	D genome donor	NA	NA
6	Cltr 180	China	*T. aestivum subsp. aestivum*	China	Least breast cancer, highest producer of wheat	W	SRW
7	Cltr 14319	India Hybrid 65	*T. aestivum subsp. aestivum*	India	Least breast cancer, second highest producer of wheat	S	SWS
8	Cltr 15136	American 378	*T. aestivum subsp. aestivum*	Sudan	Northern Africa	S	SWS
9	PI 9871	Erivan	*T. aestivum subsp. aestivum*	Armenia	Center of origin	S	HRS
10	PI 9872	Galgalos	*T. aestivum subsp. aestivum*	Armenia	Center of origin	S	SWS
11	PI 52323	Little Joss	*T. aestivum subsp. aestivum*	UK	Northern Europe	W	SRW
12	PI 54431	Triminia	*T. aestivum subsp. aestivum*	Libya	Middle East	S	HRS
13	PI 62004	NA	*T. aestivum subsp. aestivum*	China	Least breast cancer	F	NA
14	PI 81791	Sapporo Haru Komugi Jugo	*T. aestivum subsp. aestivum*	Japan	Eastern Asia	S	SWS
15	PI 82469	Poubiru	*T. aestivum subsp. aestivum*	North Korea	Eastern Asia	S	HRS
16	PI 87117	Ejuiea	*T. aestivum subsp. aestivum*	South Korea	Eastern Asia	W	HRW
17	PI 94418	99	*T. aestivum subsp. aestivum*	Russia	Fourth highest producer of wheat	W	HRW
18	PI 116232	Solid Straw Tuscan	*T. aestivum subsp. aestivum*	NZ	High cancer incidence	W	SWW
19	PI 124818	Cross No. 7	*T. aestivum subsp. aestivum*	NZ	High cancer incidence	S	HRS
20	PI 139599	Egypt NA 101	*T. aestivum subsp. aestivum*	Egypt	Western Asia	S	SWS
21	PI 155315	Yemen	*T. aestivum subsp. aestivum*	Yemen	Western Asia	S	SRS
22	PI 165208	Sert Bolvadin	*T. aestivum subsp. aestivum*	Turkey	Center of origin	S	SWS
23	PI 178383	6256	*T. aestivum subsp. aestivum*	Turkey	Center of origin	W	SRW
24	PI 184994	Snogg II	*T. aestivum subsp. aestivum*	Norway	Northern Europe	S	HRS
25	PI 190451	Snogg	*T. aestivum subsp. aestivum*	Norway	Northern Europe	S	SRS
26	PI 191334	Marzuolo 87	*T. aestivum subsp. aestivum*	Italy	Southern Europe	W	HWW
27	PI 265482	Olympia	*T. aestivum subsp. aestivum*	Finland	Northern Europe	S	SWS
28	PI 266879	S995	*T. aestivum subsp. aestivum*	Iraq	Center of origin	S	HWS
29	PI 266880	S997	*T. aestivum subsp. aestivum*	Iraq	Center of origin	S	HRS
30	PI 274505	NA	*T. aestivum subsp. aestivum*	Thailand	South Eastern Asia	S	SWS
31	PI 283150	Horani Nawawi	*T. aestivum subsp. aestivum*	Jordan	Center of origin	S	SWS
32	PI 297005	AFRICA MAYO	*T. aestivum subsp. aestivum*	Kenya	Eastern Africa	S	HRS
33	PI 347003	White Shanazi	*T. aestivum subsp. aestivum*	Afghanistan	South Central Asia	S	SWS
34	PI 350308	DACIA	*T. aestivum subsp. aestivum*	Romania	Eastern Asia	W	HRW
35	PI 384399	Nigeria-2	*T. aestivum subsp. aestivum*	Nigeria	Western Africa	S	HWS
36	PI 406475	2	*T. aestivum subsp. aestivum*	Nepal	South Central Asia	S	HRS
37	PI 414975	No. 6	*T. aestivum subsp. aestivum*	Indonesia	South Central Asia	S	HRS
38	PI 480034	MG 31147	*T. aestivum subsp. aestivum*	Ethiopia	Eastern Africa	S	SWS
39	PI 487292	SY 270	*T. aestivum subsp. aestivum*	Jordan	Center of origin	S	HRS
40	PI 519554	PAKISTAN 20	*T. aestivum subsp. aestivum*	Kenya	Eastern Africa	S	SWS

Table 2 Detailed information for 188 *Triticum* and *Aegilops* germplasm collected from GRIN platform through USDA *(Continued)*

41	PI 532053	96	*T. aestivum subsp. aestivum*	Egypt	Northern Africa	S	SWS
42	PI 585019	15007	*T. aestivum subsp. aestivum*	Saudi Arabia	Western Asia	S	HRS
43	PI 585024	15063	*T. aestivum subsp. aestivum*	Saudi Arabia	Western Asia (a part of Core 95)	S	HRS
44	PI 603919	UCRBW98-2	*T. aestivum subsp. aestivum*	USA	DNA segment from Spelta and HWS Pavon (bread wheat), Northern America, a part of core 6	S	SWS
45	PI 648392	KUNDAN	*T. aestivum subsp. aestivum*	India	Least breast cancer, second highest producer of wheat	S	HWS
46	Cltr 14352	II-50-25	*T. aestivum subsp. aestivum*	Paraguay	South America	S	HRS
47	PI 10611	Talimka	*T. aestivum subsp. aestivum*	Turkmenistan	South Central Asia	S	HWS
48	PI 61693	294	*T. aestivum subsp. aestivum*	Malawi	Eastern Africa	S	HWS
49	PI 91235	Cagayan	*T. aestivum subsp. aestivum*	Philippines	South East Asia	S	HRS
50	PI 125088	Ile de France	*T. aestivum subsp. aestivum*	France	Western Europe, more cancer	S	HWS
51	PI 174657	Ile de France	*T. aestivum subsp. aestivum*	France	Western Europe, more cancer	S	SRS
52	PI 182665	9915	*T. aestivum subsp. aestivum*	Lebanon	Western Asia	S	HWS
53	PI 191701	B 256 F.S. 1354	*T. aestivum subsp. aestivum*	Mozambique	Eastern Africa	F	SWS
54	PI 191744	FL S Aurora	*T. aestivum subsp. aestivum*	Mozambique	Eastern Africa	S	SWS
55	PI 203081	Sabanero	*T. aestivum subsp. aestivum*	Tanzania	Eastern Africa	S	HRS
56	PI 231115	II-2734-2c(1–2)x2T	*T. aestivum subsp. aestivum*	Guatemala	Central America	S	HRS
57	PI 234233	Idaho 1877 NR AE	*T. aestivum subsp. aestivum*	Zambia	Eastern Africa	S	HRS
58	PI 278386	Morocco 58	*T. aestivum subsp. aestivum*	Morocco	Northern Africa	S	HRS
59	PI 278395	Poland 2	*T. aestivum subsp. aestivum*	Poland	Eastern Europe	W	HRW
60	PI 313098	Quern	*T. aestivum subsp. aestivum*	Ireland	Northern Europe	S	HRS
61	PI 344018	Mistura Pinto	*T. aestivum subsp. aestivum*	Angola	Middle Africa, least cancers	S	SWS
62	PI 344019	Saraiva Vieira	*T. aestivum subsp. aestivum*	Angola	Middle Africa, least cancers	S	SWS
63	PI 351474	Reval	*T. aestivum subsp. aestivum*	Estonia	Northern Europe	W	HRW
64	PI 351870	10180-54-29	*T. aestivum subsp. aestivum*	Burundi	Eastern Africa	S	SWS
65	PI 374248	BOL-17	*T. aestivum subsp. aestivum*	Chad	Middle Africa, least cancers	S	HRS
66	PI 374249	BOL-19	*T. aestivum subsp. aestivum*	Chad	Middle Africa, least cancers	S	HRS
67	PI 374254	34335	*T. aestivum subsp. aestivum*	Mali	Western Africa	S	HWS
68	PI 384399	Nigeria-2	*T. aestivum subsp. aestivum*	Nigeria	Western Africa	S	HWS
69	PI 410425	ILICHEVKA	*T. aestivum subsp. aestivum*	Kazakhstan	South Central Asia	W	SRW
70	PI 428690	LEUCURUM 3	*T. aestivum subsp. aestivum*	Uzbekistan	South Central Asia	S	SRS
71	PI 470905	MG 18060	*T. aestivum subsp. aestivum*	Algeria	Northern Africa	S	SWS
72	PI 480481	R-124	*T. aestivum subsp. aestivum*	Bolivia	South America	S	SRS
73	PI 481713	41	*T. aestivum subsp. aestivum*	Bhutan	South Central Asia	W	HRW
74	PI 481715	Ka	*T. aestivum subsp. aestivum*	Bhutan	South Central Asia	S	SWS
75	PI 486155	CHIWORE	*T. aestivum subsp. aestivum*	Zimbabwe	Eastern Africa, least breast cancer	S	HRS
76	PI 486156	GWEBI	*T. aestivum subsp. aestivum*	Zimbabwe	Eastern Africa, least breast cancer	S	SWS
77	PI 486157	RUSAPE	*T. aestivum subsp. aestivum*	Zimbabwe	Eastern Africa, least breast cancer	S	HRS
78	PI 490405	Koira alkuna	*T. aestivum subsp. aestivum*	Mali	Western Africa	S	HWS
79	PI 494926	ZFA 3145	*T. aestivum subsp. aestivum*	Zambia	Eastern Africa	S	HRS
80	PI 532301	Alas	*T. aestivum subsp. aestivum*	Oman	Western Asia	S	SRS
81	PI 573754	NSGC 531	*T. aestivum subsp. aestivum*	Hondurus	Central America	S	HRS

Table 2 Detailed information for 188 *Triticum* and *Aegilops* germplasm collected from GRIN platform through USDA *(Continued)*

82	PI 591964	EMBRAPA 16	*T. aestivum subsp. aestivum*	Brazil	South America	S	HRS
83	PI 639354	TJK03-128	*T. aestivum subsp. aestivum*	Tazikistan	South Central Asia	S	SWS
84	PI 648894	Dickson's No. 444	*T. aestivum subsp. aestivum*	Argentina	South America	S	HWS
85	Cltr 14108	Chinese Spring	*T. aestivum subsp. aestivum*	USA	Northern America	S	SRS
86	PI 434642	TINCURRIN	*T. aestivum subsp. compactum*	Australia	More cancer especially skin cancer	S	Club wheat
87	PI 190963	Spelta Hohenheim	*T. aestivum subsp. spelta*	Portugal	Southern Europe	S	Spelt
88	PI 348710	69Z6.894	*T. aestivum subsp. spelta*	Spain	Southern Europe	S	Spelt
89	PI 355625	Spelta 34	*T. aestivum subsp. spelta*	Belgium	Western Europe	S	Spelt
90	PI 591895	NA	*T. aestivum subsp. spelta*	Germany	Central Europe	W	Spelt
91	PI 367199	128	*T. aestivum subsp. spelta*	Afghanistan	Central Asia	W	Spelt
92	PI 538510	G2830	*T. monococcum subsp. aegilopoides*	Iraq	Center of origin, Western Asia	S	Wild einkorn
93	PI 167526	2485	*T. monococcum subsp. aegilopoides*	Turkey	Western Asia	S	Wild einkorn
94	PI 266844	87	*T. monococcum subsp. aegilopoides*	Uk	Northern Europe	S	Wild einkorn
95	PI 427990	G3114	*T. monococcum subsp. aegilopoides*	Lebanon	Center of domestication	W	Wild einkorn
96	PI 427304	G1764	*T. timopheevii subsp. armeniacum*	Armenia	Center of origin, Western Asia	W	Wild timopheevii
97	PI 538478	G2633	*T. timopheevii subsp. armeniacum*	Iraq	Center of domestication	W	Wild timopheevii
98	PI 70738	22	*T. turgidum subsp. carthlicum*	Iraq	Center of origin, Western Asia	S	Persian wheat
99	PI 532501	H83-1537	*T. turgidum subsp. carthlicum*	Soviet Union		S	Persian wheat
100	PI 471808	G-485-5 M	*T. turgidum subsp. dicoccoides*	Israel	Center of domestication	W	Wild emmer
101	PI 471778	G-40-1-2B-1 M	*T. turgidum subsp. dicoccoides*	Israel	Center of domestication	W	Wild emmer
102	PI 2789	Yaroslav Spring	*T. turgidum subsp. dicoccon*	Russia	Fourth highest producer of wheat	S	Cultivated emmer
103	PI 73388	2868	*T. turgidum subsp. dicoccon*	Armenia	Center of origin, Western Asia	S	Cultivated emmer
104	PI 79899	N-64	*T. turgidum subsp. dicoccon*	China	Least breast cancer, highest producer of wheat	S	Cultivated emmer
105	PI 190920	2323A	*T. turgidum subsp. dicoccon*	Portugal	Southern Europe	S	Cultivated emmer
106	PI 191390	Rufum	*T. turgidum subsp. dicoccon*	Ethiopia	Eastern Africa	S	Cultivated emmer
107	PI 308879	NA	*T. turgidum subsp. dicoccon*	Spain	Southern Europe	S	Cultivated emmer
108	PI 355483	T 563	*T. turgidum subsp. dicoccon*	Spain	Southern Europe	S	Cultivated emmer
109	PI 355485	T 567	*T. turgidum subsp. dicoccon*	Spain	Southern Europe	S	Cultivated emmer
110	PI 499973	KU 1533	*T. turgidum subsp. dicoccon*	Armenia	Center of origin, Western Asia	S	Cultivated emmer
111	PI 154582	NA	*T. turgidum subsp. dicoccon*	Taiwan	Eastern Asia	S	Cultivated emmer
112	Cltr 15185	Hudeiba 154	*T. turgidum subsp. durum*	Sudan	Northern Africa	S	Durum or macaroni
113	PI 9130	Saragolla	*T. turgidum subsp. durum*	Italy	Southern Europe	S	Durum or macaroni

Table 2 Detailed information for 188 *Triticum* and *Aegilops* germplasm collected from GRIN platform through USDA *(Continued)*

114	PI 54432	Tripshiro	*T. turgidum subsp. durum*	Libya	Northern Africa	S	Durum or macaroni
115	PI 67341	Huguenot	*T. turgidum subsp. durum*	Australia	More cancer especially skin cancer	S	Durum or macaroni
116	PI 78809	CI 10107	*T. turgidum subsp. durum*	Georgia	Western Asia	S	Durum or macaroni
117	PI 81792	Marching No. 8	*T. turgidum subsp. durum*	Japan	Eastern Asia	S	Durum or macaroni
118	PI 94701	390	*T. turgidum subsp. durum*	Ancient Palestine	Western Asia	S	Durum or macaroni
119	PI 133459	Durum H2	*T. turgidum subsp. durum*	Egypt	Center of origin, Northern Africa	S	Durum or macaroni
120	PI 153726	Sicilian	*T. turgidum subsp. durum*	N. Africa	Less cancers	S	Durum or macaroni
121	PI 157955	Francesone	*T. turgidum subsp. durum*	Italy	Southern Europe	S	Durum or macaroni
122	PI 174645	Huguenot	*T. turgidum subsp. durum*	Australia	More cancer especially skin cancer	S	Durum or macaroni
123	PI 182113	S-44	*T. turgidum subsp. durum*	Pakistan	South Central Asia	S	Durum or macaroni
124	PI 184532	Russia	*T. turgidum subsp. durum*	Russia	Europe	S	Durum or macaroni
125	PI 208908	Mendola	*T. turgidum subsp. durum*	Iraq	Center of origin	S	Durum or macaroni
126	PI 208910	Sin El-Jamil	*T. turgidum subsp. durum*	Iraq	Center of origin	S	Durum or macaroni
127	PI 210848	7979	*T. turgidum subsp. durum*	Iran	South Central Asia	S	Durum or macaroni
128	PI 221702	11	*T. turgidum subsp. durum*	Indonesia	Southern Eastern Asia	S	Durum or macaroni
129	PI 231380	Saragolla	*T. turgidum subsp. durum*	Italy	Oldest durum variety in Italy	S	Durum or macaroni
130	PI 261823	Namra	*T. turgidum subsp. durum*	Saudi Arabia	Western Asia	S	Durum or macaroni
131	PI 265017	796	*T. turgidum subsp. durum*	Serbia	Southern Europe	S	Durum or macaroni
132	PI 278223	Gartons Early Cone	*T. turgidum subsp. durum*	UK	Northern Europe	W	Durum or macaroni
133	PI 278258	Greece 1	*T. turgidum subsp. durum*	Greece	Southern Europe	S	Durum or macaroni
134	PI 278509	Valencia 6	*T. turgidum subsp. durum*	Spain	Southern Europe	S	Durum or macaroni
135	PI 278553	Tripolitco	*T. turgidum subsp. durum*	Cyprus	Western Asia	S	Durum or macaroni
136	PI 283853	China 34	*T. turgidum subsp. durum*	China	Least breast cancer, highest producer of wheat	S	Durum or macaroni
137	PI 306571	R.S.N.	*T. turgidum subsp. durum*	Italy	Southern Europe	S	Durum or macaroni
138	PI 325850	PW 3	*T. turgidum subsp. durum*	India	Least breast cancer, second highest producer of wheat	S	Durum or macaroni
139	PI 361149	Bijaga Yellow	*T. turgidum subsp. durum*	India	Least breast cancer, second highest producer of wheat	S	Durum or macaroni
140	PI 362046	C 1138/63	*T. turgidum subsp. durum*	Romania	Eastern Europe	W	Durum or macaroni

Table 2 Detailed information for 188 *Triticum* and *Aegilops* germplasm collected from GRIN platform through USDA
(Continued)

141	PI 422295	QUILAFEN	*T. turgidum subsp. durum*	Chile	South America	S	Durum or macaroni
142	PI 422297	SINCAPE 90	*T. turgidum subsp. durum*	Italy	Southern Europe	S	Durum or macaroni
143	PI 422312	MACS-45	*T. turgidum subsp. durum*	India	Least breast cancer, second highest producer of wheat	S	Durum or macaroni
144	PI 428458	Egypt Local No. 8	*T. turgidum subsp. durum*	Egypt	Center of origin, Northern Africa	S	Durum or macaroni
145	PI 428468	JORDAN 38	*T. turgidum subsp. durum*	Jordan	Center of origin, Western Asia	S	Durum or macaroni
146	PI 428469	JORDAN 40	*T. turgidum subsp. durum*	Jordan	Center of origin, Western Asia	S	Durum or macaroni
147	PI 462107	172	*T. turgidum subsp. durum*	Yemen	Western Asia	S	Durum or macaroni
148	PI 480347	MG 31577	*T. turgidum subsp. durum*	Ethiopia	Eastern Africa	S	Durum or macaroni
149	PI 496260	MEDORA	*T. turgidum subsp. durum*	Canada	North America	S	Durum or macaroni
150	PI 519864	DURUM VARIETY 24	*T. turgidum subsp. durum*	Mexico	North America	S	Durum or macaroni
151	PI 520393	TUNISIAN DURUM 1	*T. turgidum subsp. durum*	Tunisia	Northern Africa	S	Durum or macaroni
152	PI 520394	TUNISIAN DURUM 8	*T. turgidum subsp. durum*	Tunisia	Northern Africa	S	Durum or macaroni
153	PI 520414	ICD 7780-5AP -OSH-OAP	*T. turgidum subsp. durum*	Syria	Center of origin, Western Asia	S	Durum or macaroni
154	PI 520415	SYRIAN DURUM 27	*T. turgidum subsp. durum*	Syria	Center of origin, Western Asia	S	Durum or macaroni
155	PI 542464	SHORT SARAGOLLA	*T. turgidum subsp. durum*	USA	High cancer incidence	S	Durum or macaroni
156	PI 585025	15017	*T. turgidum subsp. durum*	Saudi Arabia	Western Asia	S	Durum or macaroni
157	Cltr 14374	497-360	*T. turgidum subsp. durum*	Lebanon	Western Asia	S	Durum or macaroni
158	Cltr 14802	ELS 6404-122	*T. turgidum subsp. durum*	Eritrea	Eastern Africa	S	Durum or macaroni
159	PI 5465	Candeal	*T. turgidum subsp. durum*	Argentina	South America	F	Durum or macaroni
160	PI 5639	Kubanka	*T. turgidum subsp. durum*	Kazakhstan	South Central Asia, largest consumer of wheat	S	Durum or macaroni
161	PI 35314	1809a	*T. turgidum subsp. durum*	Kyrgyzstan	South Central Asia	F	Durum or macaroni
162	PI 50929	933	*T. turgidum subsp. durum*	Kyrgyzstan	South Central Asia	S	Durum or macaroni
163	PI 61108	6951	*T. turgidum subsp. durum*	Turkmenistan	South Central Asia	S	Durum or macaroni
164	PI 89642	NA	*T. turgidum subsp. durum*	Hondurus	Central America	S	Durum or macaroni
165	PI 278384	Morocco C10895	*T. turgidum subsp. durum*	Morocco	Northern Africa	W	Durum or macaroni
166	PI 286066	NA	*T. turgidum subsp. durum*	Poland	Eastern Europe	S	Durum or macaroni
167	PI 384401	Wurno 2	*T. turgidum subsp. durum*	Nigeria	Western Africa	S	Durum or macaroni

Table 2 Detailed information for 188 *Triticum* and *Aegilops* germplasm collected from GRIN platform through USDA (*Continued*)

168	PI 519759	D 73121	*T. turgidum subsp. durum*	Algeria	Northern Africa	S	Durum or macaroni
169	PI 520164	ALGERIA LINE 47	*T. turgidum subsp. durum*	Algeria	Northern Africa	S	Durum or macaroni
170	PI 532289	Musane	*T. turgidum subsp. durum*	Oman	Western Asia	S	Durum or macaroni
171	PI 565208	Chaggo	*T. turgidum subsp. durum*	Bolivia	South America	S	Durum or macaroni
172	PI 592019	VATAN	*T. turgidum subsp. durum*	Uzbekistan	South Central Asia	S	Durum or macaroni
173	PI 654290	TJK2006:296	*T. turgidum subsp. durum*	Tazikistan	South Central Asia	S	Durum or macaroni
174	Cltr 13165	Langdon	*T. turgidum subsp. durum*	USA	Northern America	W	Durum or macaroni
175	PI 330553	189	*T. turgidum subsp. paleocolchicum*	UK	Northern Europe	S	Cultivated emmer
176	PI 211708	Egypt	*T. turgidum subsp. turanicum*	Egypt	Center of origin, Northern Africa	S	Khorasan or oriental
177	PI 166591	Ak	*T. turgidum subsp. turgidum*	Turkey	Center of origin	S	Rivet or cone
178	PI 167867	4314	*T. turgidum subsp. turgidum*	Turkey	Center of origin	S	Rivet or cone
179	PI 481591	IQ 223	*T. turgidum subsp. turgidum*	Iraq	Center of origin	S	Rivet or cone
180	PI 502933	Fo Shou Mai	*T. turgidum subsp. turgidum*	China	Largest producer	S	Rivet or cone
181	PI 208912	Zerdakia	*T. turgidum subsp. turgidum*	Iraq	Center of origin	S	Rivet or cone
182	PI 438971	AKMOLINKA 2	*T. turgidum subsp. turgidum*	Kazakhstan	South Central Asia	S	Rivet or cone
183	PI 427328	G2264	*T. urartu*	Iraq	Center of origin	S	Wild einkorn
184	PI 428183	G1759	*T. urartu*	Armenia	Center of origin	S	Wild einkorn
185	PI 428279	G3162	*T. urartu*	Lebanon	Center of domestication	W	Wild einkorn
186	PI 355707	69Z5.72	*T. zhukovskyi*	Georgia	Donor for GG genome and cross between *T. timopheevi* and *T. monococcum*	W	Cultivated hexaploid
187	PI 429099	6A-696	*Triticosecale sp.*	Germany	Cross between *T. dicoccum* & *S. cereale*; hexaploid	F	*Tritical* (Rye and durum cross)
188	PI 574284	ASVM4*4654	*T. hybrid*	USA	High cancer incidence	S	*Aegilops squarrosa/T. dicoccum*

A, *Aegilops*; F, Facultative; HRW, Hard red winter; HRS, Hard red spring; HWS, Hard white spring; HWW, Hard white winter; NA, not available; S, Spring; W, Winter; SRS, Soft red spring; SRW, Soft red winter; SWS, Soft white spring; SWW, Soft white winter; T, *Triticum*.

core collection is approximately 0.3% of the world barley holding, and the ICRISAT (International Crops Research Institute for the Semi Arid Crops, Hyderabad, India) sorghum core collection is about 1.5% of the domain size [31,32]. As many of the lines shown (Table 2) are wild accessions, data are not available on genetic and metabolic markers, agronomic and morphological characteristics, thus limiting the descriptive information provided.

Future direction
Having established this core collection and obtained grain for each line from GRIN, the next step in the

identification of distinct wheat lines with enhanced biomedical activity is the interrogation of these lines via phytochemical profiles using LC-TOF-MS analysis of wheat grain extracts according to our recently published procedures [9]. The chromatographic data that result will be subjected to advanced multivariate regression techniques that plot multidimensional relationships to define the chemical diversity within the core collection. The same extracts used for metabolic profiling will then be subjected to *in-vitro* biological analysis to assign a relative value for anticancer activity to each wheat line. For wheat lines with the greatest *in-vitro* activity, *in-vivo* testing in appropriate animal cancer models will be conducted. For wheat lines with *in-vivo* anticancer activity, the genetic and metabolomic traits that account for protection will be identified and appropriate experiments conducted to determine the extent to which environmental factors impact the stable expression of the traits of interest [10].

Conclusion

While there has been an active discussion of adding value to wheat through the enhancement of its human health benefits, no systematic approaches have been establish to advance this effort. The work reported herein constitutes the first essential step needed to examine wheat germplasm resources in order to identify health benefits that may exist and to develop them fully for the benefit of the consuming public.

Availability of supporting data

The datasets supporting the results of this article are available in the Germplasm Resources Information Network (GRIN) repository from the United States Department of Agriculture (USDA), http://www.ars-grin.gov/npgs/index.html; in the Food and Agriculture Organization of the United States (FAOSTAT) repository from the World Health Organization, http://faostat3.fao.org/home/index.html#COMPARE; and in the GLOBOCAN repository from the International Agency for Research on Cancer (IARC), http://globocan.iarc.fr.

Abbreviations
ASR: Age-standardized rate per 100,000 individuals; CC: Core collection; GRIN: Germplasm Resources Information Network; mt: Metric tons; PI: Plant introduction.

Competing interests
The authors declare that they have no competing interests.

Authors' contributions
MS directed the selection of specific species from GRIN, SM participated in data analysis and germplasm selection, and HT provided overall direction and specific guidance relative to interpretation of the cancer prevalence data. All authors participated in writing the manuscript. All authors read and approved the final manuscript.

Authors' information
MS is a Research Associate in the Department of Soil and Crop Sciences, SM is a doctoral candidate in the Cell and Molecular Biology Program, and HT directs the Cancer Prevention Laboratory at Colorado State University.

Acknowledgments
The authors would like to thank Dr. Harold Bockelman for providing wheat germplasm from GRIN, the International Agency for Research on Cancer for allowing us to use the GLOBOCAN cancer map, and Stephanie MacLeaod and John McGinley for their assistance in the preparation of this manuscript.

References
1. Arthur AE, Peterson KE, Rozek LS, Taylor JM, Light E, Chepeha DB, Hebert JR, Terrell JE, Wolf GT, Duffy SA: **Pretreatment dietary patterns, weight status, and head and neck squamous cell carcinoma prognosis.** *Am J Clin Nutr* 2013, **97:**360–368.
2. Cohen LA, Zhao Z, Zang EA, Wynn TT, Simi B, Rivenson A: **Wheat bran and psyllium diets: effects on N-methylnitrosourea-induced mammary tumorigenesis in F344 rats.** *J Natl Cancer Inst* 1996, **88:**899–907.
3. Fung TT, Hu FB, Pereira MA, Liu S, Stampfer MJ, Colditz GA, Willett WC: **Whole-grain intake and the risk of type 2 diabetes: a prospective study in men.** *Am J Clin Nutr* 2002, **76:**535–540.
4. Gil A, Ortega RM, Maldonado J: **Wholegrain cereals and bread: a duet of the Mediterranean diet for the prevention of chronic diseases.** *Public Health Nutr* 2011, **14:**2316–2322.
5. Jacobs DR, Marquart LF, Slavin J: **Whole grain intake and cancer: an expanded review and meta analysis.** *FASEB J* 1998, **12:**A875.
6. Pereira MA, Jacobs DR Jr, Pins JJ, Raatz SK, Gross MD, Slavin JL, Seaquist ER: **Effect of whole grains on insulin sensitivity in overweight hyperinsulinemic adults.** *Am J Clin Nutr* 2002, **75:**848–855.
7. Schatzkin A, Mouw T, Park Y, Subar AF, Kipnis V, Hollenbeck A, Leitzmann MF, Thompson FE: **Dietary fiber and whole-grain consumption in relation to colorectal cancer in the NIH-AARP Diet and Health Study.** *Am J Clin Nutr* 2007, **85:**1353–1360.
8. Tabung F, Steck SE, Su LJ, Mohler JL, Fontham ET, Bensen JT, Hebert JR, Zhang H, Arab L: **Intake of grains and dietary fiber and prostate cancer aggressiveness by race.** *Prostate Cancer* 2012, **201:**323296.
9. Matthews SB, Santra M, Mensack MM, Wolfe P, Byrne PF, Thompson HJ: **Metabolite profiling of a diverse collection of wheat lines using ultraperformance liquid chromatography coupled with time-of-flight mass spectrometry.** *PLoS One* 2012, **7:**e44179.
10. Thompson MD, Thompson HJ: **Biomedical agriculture: a systematic approach to food crop improvement for chronic disease prevention.** *Adv Agron* 2009, **102:**1–54.
11. Thompson HJ: **Vegetable and fruit intake and the development of cancer: a brief review and analysis.** In *Bioactive Foods in Promoting Health.* Edited by Watson RR, Preedy VR. Oxford: Academic; 2010:19–36.
12. Centers for Disease Control: *Health, United States, 2007.* Atlanta, GA: Centers for Disease Control; 2007.
13. World Health Organization: *Chronic Disease.* Geneva: WHO; 2008.
14. Marshall S: **Role of insulin, adipocyte hormones, and nutrient-sensing pathways in regulating fuel metabolism and energy homeostasis: a nutritional perspective of diabetes, obesity, and cancer.** *Sci STKE* 2006, **2006:**re7.
15. Hirsch HA, Iliopoulos D, Joshi A, Zhang Y, Jaeger SA, Bulyk M, Tsichlis PN, Shirley Liu X, Struhl K: **A transcriptional signature and common gene networks link cancer with lipid metabolism and diverse human diseases.** *Cancer Cell* 2010, **17:**348–361.
16. Tacutu R, Budovsky, Yanai H, Fraifeld VE: **Molecular links between cellular senescence, longevity and age-related diseases - a systems biology perspective.** *Aging* 2011, **3:**1178–1191.
17. Ezzati M, Riboli E: **Can noncommunicable diseases be prevented? Lessons from studies of populations and individuals.** *Science* 2012, **337:**1482–1487.
18. Eyre H, Kahn R, Robertson RM, Clark NG, Doyle C, Hong Y, Gansler T, Glynn T, Smith RA, Taubert K, Thun MJ: **Preventing cancer, cardiovascular disease, and diabetes: a common agenda for the American Cancer Society, the American Diabetes Association, and the American Heart Association.** *Circulation* 2004, **109:**3244–3255.
19. WCRF/AICR: *Food, Nutrition, Physical Activity, and the Prevention of Cancer: a Global Perspective.* Washington, DC: AICR; 2007.
20. Spring B, Moller AC, Coons MJ: **Multiple health behaviours: overview and implications.** *J Public Health (Oxf)* 2012, **Suppl 1:**i3–i10.
21. International Agency for Research on Cancer: *Cancer Incidence, Mortality and Prevalence Worldwide in 2008.* Lyon: IARC; 2008. http://globocan.iarc.fr/factsheets/cancers/all.asp.
22. FAOSTAT: *Production of the world's most important grain crops.*; 2008. http://faostat.org.
23. Carver BF: *Wheat Science and Trade.* Ames, IA: Wiley-Blackwell; 2009.
24. Gooding MJ, Davies WP: *Wheat Production and Utilization.* London: CAB International; 1997.

25. Stallknecht GF, Gilbertson KM, Ranney JE: **Alternative wheat cereals as food grains: Einkorn, emmer, spelt, kamut, and triticale.** In *Progress in new crops*. Edited by Janick J. Alexandria, VA: ASHS Press; 1996:156–170.
26. National Plant Germplasm System; 2012. http://www.ars-grin.gov/cgi-bin/npgs.
27. Brown AHD: **Core collections: a practical approach to genetic resources management.** *Genome* 1989, **31**:818–824.
28. Hancock JF: *Plant evolution and the origin of crop species*. Wallingford: CABI Pub; 2004.
29. Sun QM, Zhou RH, Gao LF, Zhao GY, Jia JZ: **The characterization and geographical distribution of the genes responsible for vernalization requirement in Chinese bread wheat.** *J Integr Plant Biol* 2009, **51**:423–432.
30. Preston JC, Kellogg EA: **Discrete developmental roles for temperate cereal grass VERNALIZATION1/FRUITFULL-like genes in flowering competency and the transition to flowering.** *Plant Physiol* 2008, **146**:265–276.
31. Van Hintum TJL: **Core collections of plant genetic resources.** In *PGRI Technical Bulletin No.3*. Edited by Brown AHD, Spillane C, Hodgkin T. Rome: International Plant Genetic Resources Institute; 2000:6–51.
32. van Hintum TJL: **The Core Selector, a system to generate representative selections of germplasm accessions.** *Plant Genet Resour Newsl* 1999, **118**:64–67.

5

Crop adaptation to climate change in the semi-arid zone in Tanzania: the role of genetic resources and seed systems

Ola T Westengen[1,2*] and Anne K Brysting[2]

Abstract

Background: Rural livelihoods relying on agriculture are particularly vulnerable to climate change. Climate models project increasingly negative effects on maize and sorghum production in sub-Saharan Africa. We present a case study of the role of genetic resources and seed systems in adapting to climatic stress from the semi-arid agroecological zone in Tanzania.

Results: Crop adaptation, switching to more drought-tolerant crop species or varieties, is an important adaptation strategy within a diverse portfolio of livelihood responses to climatic stress. Crop adaptation involves the adoption of improved maize varieties combined with continued use of local varieties of both maize and sorghum. Regression modelling shows that households receiving the extension service and owning livestock are more likely to switch to drought-tolerant varieties as a response to climatic stress than those without access to these assets. The seed system in the study area consists of both formal and informal elements. The informal channels supply the highest quantities of both sorghum and maize seeds. Recycling of improved varieties of maize is common and the majority of households practice seed selection. Detailed assessment of the three different categories of genetic resources – local, improved and farmer-recycled varieties – reveals that drought tolerance is more frequently reported as a reason for growing local varieties than for growing improved varieties of maize and sorghum. The significantly later maturity reported for local varieties compared to the improved varieties bred to have a short growing cycle indicates that households distinguish between drought-tolerance and drought-avoidance traits.

Conclusions: Seed system perspectives on crop adaptation offer insights into the complex ways crop adaptation is realized at the livelihood level. The integration of informal and formal seed system elements is important for the adaptive capacity of agriculture-based livelihoods. Our findings highlight the value and importance of location-specific information about crop variety use for arriving at realistic recommendations in impact and adaptation studies.

Keywords: Adaptation, Agriculture, Seed systems, Genetic resources, Maize, Sorghum

Background

Adaptation to climate change is a major issue in the current food security discourse [1,2]. Livelihoods in developing countries depending on agriculture are particularly vulnerable to changes in the mean and variability of climate and the need for adaptation is highlighted in crop impact studies from sub-Saharan Africa (SSA) [3,4]. Adaptation options in agriculture involve changes at the farm management level as well as changes in the policy and institutional decision environment [5]. In the portfolio of common on-farm and non-farm livelihood adaptation strategies, crop adaptation (changing to crop species or varieties that are resistant to climatic stress) is among the most cited adaptation measures [3-10]. The important role that crop adaptation has achieved in the discourse is illustrated by a United Nations General Assembly resolution from 2009, which 'underlines the importance of … making crops more tolerant to environmental stress, including drought and climate change' [11]. Despite the general agreement about the importance of crop adaptation there are diverging views in the adaptation literature

* Correspondence: ola.westengen@sum.uio.no
[1]Centre for Development and the Environment, University of Oslo, Blindern, P.O. Box 1116, NO-0317 Oslo, Norway
[2]Centre for Ecological and Evolutionary Synthesis (CEES), Department of Biosciences, University of Oslo, Blindern, P.O. Box 1066, NO-0316 Oslo, Norway

about how this adaptation option can be realized at the livelihood level.

The need to adapt crops to changing environmental conditions is not new, rather it is the most fundamental co-evolutionary relationship between crops and humans since the dawn of agriculture [12,13]. Crop adaptation in the Darwinian sense is the evolution of crops to become better suited to their environments. In traditional agriculture, adaptation is an interplay between natural selection and selection by farmers. In modernized agriculture, which has been rapidly spreading in the developing world with the Green Revolution since the 1960s, the science of plant breeding has largely replaced the farmer's role in crop development [14]. Conventional plant breeding can briefly be defined as crossing plants with desired characteristics and selecting offspring combining those desirable characteristics to produce so-called improved varieties. The technology and the political economic context are different, but genetic diversity is the raw material for adaptation both in on-farm crop development and in professional plant breeding. This role of crop diversity is reflected in the term *genetic resources*, which encompasses seeds, plants and plant parts useful in crop breeding, research or conservation for their genetic attributes [15]. In traditional agriculture, genetic resources for adaptation are sourced from the farmer's own field or through gifts and trade with other farmers, and sometimes through gene flow from other varieties or wild relatives of the crop. In modernized agriculture, the plant breeders act as intermediaries between the genetic resources and the farmer, and the genetic resources used to breed new varieties are normally sourced from genebanks and genetic stocks.

The terms *informal* and *formal* seed systems are used to distinguish between the two different sources of genetic resources [16,17]. In SSA's smallholder agriculture, most seeds are sourced through the informal seed system and only a small proportion of the seeds planted every year are sourced through formal market channels with direct links to plant breeding [18,19]. However, farm saving and recycling of improved varieties is common [20-22]. Socio-economic work from Mexico has demonstrated that recycling and hybridization between landraces and improved varieties of maize is a deliberate strategy used by small-holder farmers to combine desirable traits from improved and local varieties, a phenomenon known as creolization [23,24]. Thus, the distinction between the formal and informal seed systems is not clear-cut. Development initiatives aiming at replacing the informal seed system with formal seed systems modelled after those found in industrialized countries have been questioned and challenged in recent seed system literature [25-28].

Vermeulen *et al.* [8], in a review of options to support agriculture and food security under climate change, point out that while we know much about what regions and crops are likely to be sensitive to climate change, there is limited scientific knowledge about how current farming systems can adapt. This gap arguably reflects methodological and epistemological differences between two scholarly traditions dealing with agriculture in a climate change context. The first is impact oriented and the other studies the vulnerability and capacity of affected livelihoods [29,30]. While there is no clear-cut boundary between the two literatures, their difference is apparent in the Intergovernmental Panel on Climate Change (IPCC) report from Working Group II on *Impacts, Adaptation and Vulnerability* [31] where the two literatures are reviewed by different expert groups in different chapters [6,32]. The impact-oriented literature projects yield loss in the largest food crops and focuses on the consequences for regional and national food security, often with a view to recommending targeting of adaptation measures [3,6,33,34]. The literature on capacity and vulnerability commonly applies a livelihood and poverty perspective on adaptation [32,35,36] and typically treats coping and adaptation as objects for empirical research. The difference in starting point (crop impact vs socio-economic impact) often leads to differences in recommendations with regard to how crop adaptation can be realized in developing countries. The impact-oriented literature projects shifts in crop climates that are likely to make the current crop varieties unsuitable and commonly arrives at recommending breeding and dissemination of stress-tolerant and climate-ready varieties. On the other hand, the focus on local conditions and safety nets in the vulnerability and capacity studies often leads to an emphasis on the adaptive importance of local crops and traditional knowledge.

We here present a study of crop adaptation in an area where households experience climate stress and cultivate crops for which climate change models project adverse effects on future yield. This paper is a contribution towards bridging the gap between different scholarly traditions on climate change adaptation. We apply a combination of the livelihood approach and seed system perspectives when addressing the following research questions: (a) Is crop adaptation an important adaptation strategy? (b) What livelihood factors are associated with practicing on-farm adaptation activities in general and crop adaptation in particular? (c) What is the role of genetic resources and seed systems in crop adaptation?

Methods

Study site and impact projections

The current study was carried out during the harvest season in 2010 in two Tanzanian villages: Mangae in the Morogoro district and Laikala in the Dodoma district (Figure 1). The agricultural sector is the main source of employment and livelihood for 77.5% of the population in Tanzania [37]. Despite food self-sufficiency at the national

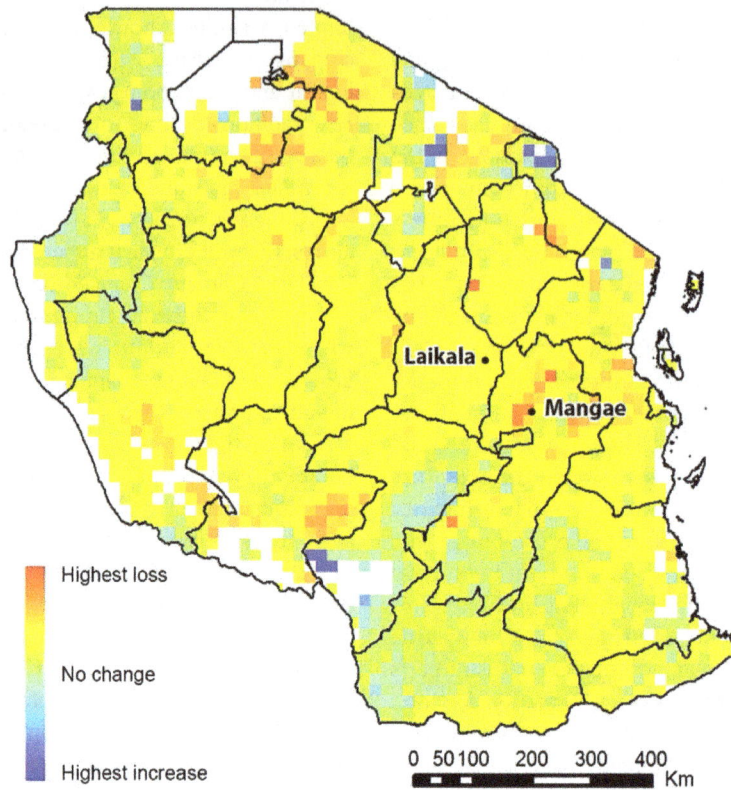

Figure 1 Map of study sites and maize production impacts. Tanzania and the villages Mangae in the Morogoro district and Laikala in the Dodoma district. Simulated change in percentage of maize yield in 2050 compared to current conditions in 10 arc-minute grids based on the mean value from the CERES-Maize model run on downscaled data from two GCMs and two SRES scenarios (HadCM3 and ECHam 4 models, A1 and B1 scenarios). Map based on data from Thornton et al. [4].

level, one third of the population is unable to meet their dietary energy requirements [38]. Maize is the largest crop, produced on 58% of the total cereal area in 2010, and sorghum is the third most produced crop nationally after maize and rice [39]. We used three criteria to select study sites: (1) the sites are located in areas with projected negative effects of climate change on maize and sorghum production, (2) agriculture is today climatically stressed and (3) maize and sorghum are major crops in the agricultural part of people's livelihood.

Impact projections for maize and sorghum yields vary according to the scale and model used. At the regional SSA scale, Schlenker and Lobell [40] coupled historical crop production data with predictions for temperature and precipitation changes from 16 climate change models under the A1b emission scenario of the IPCC, and projected aggregate production losses of 17% for sorghum and 22% for maize by mid-century. For Tanzania, the nationally projected impact of a 2°C seasonal temperature increase is an 8.8% reduction of sorghum yield and 13% for maize in the same period [41]. Thornton et al. [4] ran the biophysical crop model CERES-Maize [42] with a fine spatial resolution (10 arc-minute grids or ~18 km

resolution) across East Africa, and predicted that the semi-arid region in Tanzania is one of the areas where maize yields are likely to be reduced by 20% or more. The villages in the current study are located in areas where maize production is predicted to be adversely affected (Figure 1).

Theoretical framework and statistical methods

We used a livelihood approach [43,44] to study adaptation and assess the relative importance of crop adaptation in light of the institutional context and other adaptation options accessible to the households. Drawing on the seed system literature [17,25,27], we study seed systems as the institutions that mediate access to genetic resources. Furthermore, we analyze the association between different categories of genetic resources and a range of production and consumption variables, drawing on the socio-economic literature on the benefits of different types of crop varieties [24,45].

We used both qualitative and quantitative methods when collecting and analyzing the data. We conducted key informant interviews of actors in the formal seed system in the area of research as well as with villagers and authorities in the villages included in our case study. The quantitative analysis is based on a random sample of 320 households in two villages, who were interviewed using a structured questionnaire with both closed and open-ended questions. The questionnaire consisted of four sections: a section on asset status, a section on stress factor ranking, a section on coping activities in times of stress and a section on the seed system and varieties used. In the section on coping activities we asked if households were practicing any of the activities on a list of coping activities and if the reason was climatic stress or other stressors. The list was based on coping activities documented in the literature [22,46,47] and elicited in the initial qualitative phase of data collection. We used logistic regression models with the common on-farm coping activities (practiced by more than 15% of the households due to climatic stress) as response variables and a set of livelihood factors as explanatory variables. In these models the log odds of the activities were modeled as a linear combination of the explanatory variables. The explanatory variables included in our model were chosen based on hypotheses of either positive or negative associations with the response variables based on findings in other quantitative adaptation studies [10,48,49]. To account for asset status we included human capital variables (household size and the sex, age and education of the household head), financial capital variables (annual income and a dummy variable for livestock ownership), an institutional dummy variable on the extension service on cropping and a village effect dummy variable. The modeling was coded in R [50] and the models were evaluated with a Wald test from the Analysis of Overdispersed Data package [51].

The seed system part of the survey was formulated first by asking about sources of seeds and planting material for all crop species cultivated and second by asking in detail about area allocation and quality aspects of the different varieties of maize and sorghum. First, we classified the different seed sources as either formal or informal based on the seed system typology outlined in Sperling *et al.* [17]. Second, we classified the different varieties as belonging to one of the three genetic resource categories: local, improved or farmer-recycled. The classification was based on information provided by the household on the variety cultivated and cross-checked with information on variety names and history from key informant interviews with farmer groups, extension workers and plant breeders. The term 'local varieties' is used to distinguish the varieties said to have a long history in the area from the improved varieties produced in the formal seed system. For sorghum we could have used other terms such as 'landrace' or 'traditional variety', but in the outcrossing crop maize, which is subject to gene flow from recently introduced varieties, these terms are problematic and we chose to use the same terminology for both crops. Farm-saved improved varieties re-used on a farm in two seasons or more were classified as farmer-recycled. We tested the correlation between crop

switching and the area cultivated to the different categories using the non-parametric correlation test Spearman's rho (ρ). Finally, we analyzed the deviance between local and improved varieties with regard to a range of production and consumption variables using a chi-square test.

Results and discussion
Livelihoods under stress
The households ranked drought, conflict or competition over water and the unreliable onset of the rainy season as the three worst stress factors (Figure 2). Thus, problems of availability and access to water and seasonal variability are considered worse than biotic stress caused by crop diseases, destruction of fields by wild animals, problems with market access or floods caused by excessive rains. Other studies from the area have similarly found that climatic stress is the major factor behind the reduction in agricultural productivity and that the perception that there is a need to adjust livelihoods to an increased risk for drought and changes in rainfall and temperature is common in the semi-arid zone in Tanzania [52,53].

The livelihood approach is a lens for studying complex rural development questions and is widely used to study risk responses in rural livelihoods [43,44]. Ellis [43] defines a livelihood as: 'the assets, the activities and the access to these that together determine the living gained by the individual or household'. Households in Laikala and Mangae practice a wide range of coping activities and climatic stress is a major cause behind the diverse livelihood strategy portfolio (Figure 3). We found that receiving food aid is the most common way to cope with climatic stress among all activities recorded. The other three non-farm climatic stress responses undertaken by more than 15% of the households are: cutting back on the number of meals per day, incurring debt and using informal community support networks. Assessments of climatic stress responses of smallholders in developing countries often find the distinction between coping and adaptation to be blurred [54-56]. The common non-farm responses recorded in this study are short-term measures used to cope with stress while most of the on-farm strategies fit well with the IPCC definition of adaptation: 'the process of adjustment to actual or expected climate and its effects, in order to moderate harm or exploit beneficial opportunities' [31]. Seven on-farm activities are practiced by more than 15% of the villagers: shifting the cropping area, switching to drought-tolerant varieties, shifting planting dates, switching to drought-tolerant species, diversifying crops, extending farmland and diversifying livestock. These activities involve forward planning to anticipate climatic stress and are not merely ways to cope and survive during unexpected stress events. The diversity of livelihood activities is typical for rural households facing climatic stress in the region [47,52,57-60] and the importance of on-farm strategies recorded here

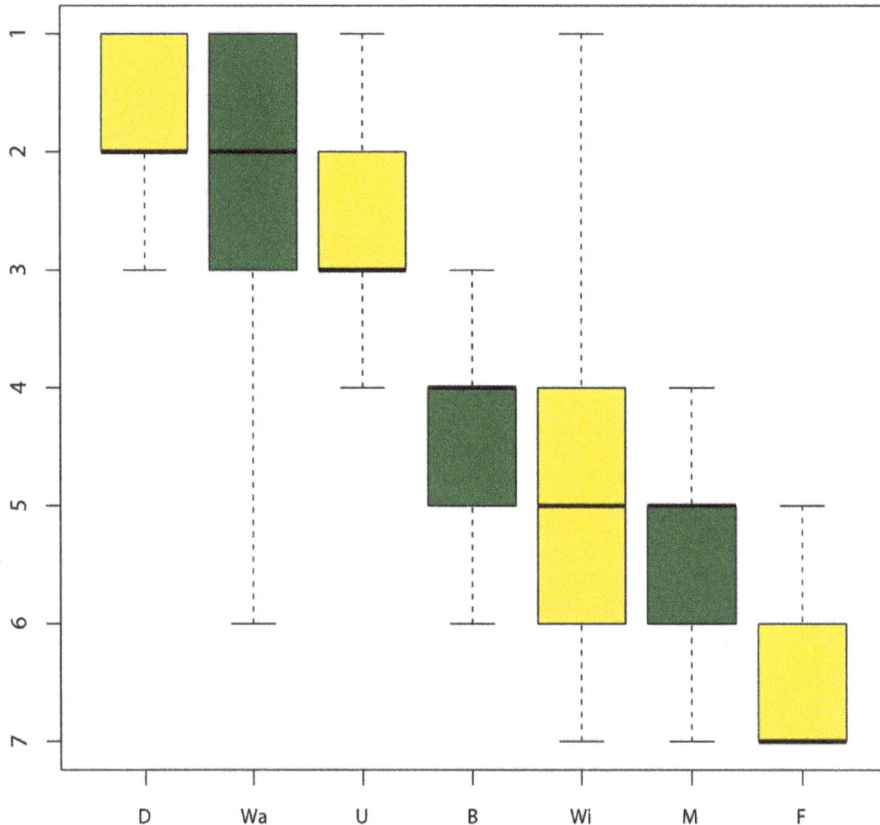

Figure 2 Stress factors ranked by households. Box plot with mean (bold line), quartiles (boxes) and variability outside the upper and lower quartiles (whiskers). Stress factors ranked from 1 (worst) to 7 by households in Mangae and Laikala ($n = 320$): drought (D); conflict or competition over water (Wa); unreliable onset of the rainy season (U); biotic stress (B); wild animals (Wi); market access (M); floods (F).

resonates with findings in other studies of rural livelihood strategies in Morogoro and Dodoma [46,58,59]. In this livelihood framing of crop adaptation we find that the two crop adaptation activities assessed are among the most important responses to climatic stress, with 46% and 40% of all respondents having switched to drought-tolerant varieties and drought-tolerant crop species, respectively. Thus, crop adaptation is a central part of the livelihood response to climatic stress in the area.

Determinants of adaptation

To determine what livelihood factors are associated with undertaking on-farm adaptation activities in general and crop adaptation in particular we modelled the seven most important on-farm activities as a linear combination of a set of explanatory variables representing household asset status. The explanatory variables included in our models have a joint significance in six of the adaptation activity models tested (Table 1). The only model for which the selected variables were not jointly significant was the activity 'shift planting dates' ($P = 0.12$). Only two explanatory variables representing asset status are significantly correlated with crop adaptation, namely livestock ownership and

receiving the extension service. Both switching to drought-tolerant varieties and switching to drought-tolerant species are positively correlated with livestock ownership, indicating that livestock owners are more likely to practice crop adaptation than those without livestock. Receiving the extension service has a significant positive effect on switching to drought-tolerant varieties and extending farmland. There is a significant village effect with households in Laikala, the driest village, being less likely to switch to drought-tolerant varieties and more likely to switch to drought-tolerant species than those in Mangae.

Before further discussion of the livelihood factors associated with crop adaptation, it is useful to consider what kind of genetic resources are used by the respondents who answered that they had switched to a drought-tolerant variety as a response to climatic stress. While the intention was to capture all kinds of variety switches, it appears that for maize the respondents understood the question as whether they have switched to an improved drought-tolerant variety. Two findings support this conclusion: (1) the significant positive association with households receiving the extension service in the regression model and (2) the significant positive correlation between this activity and the area allocated

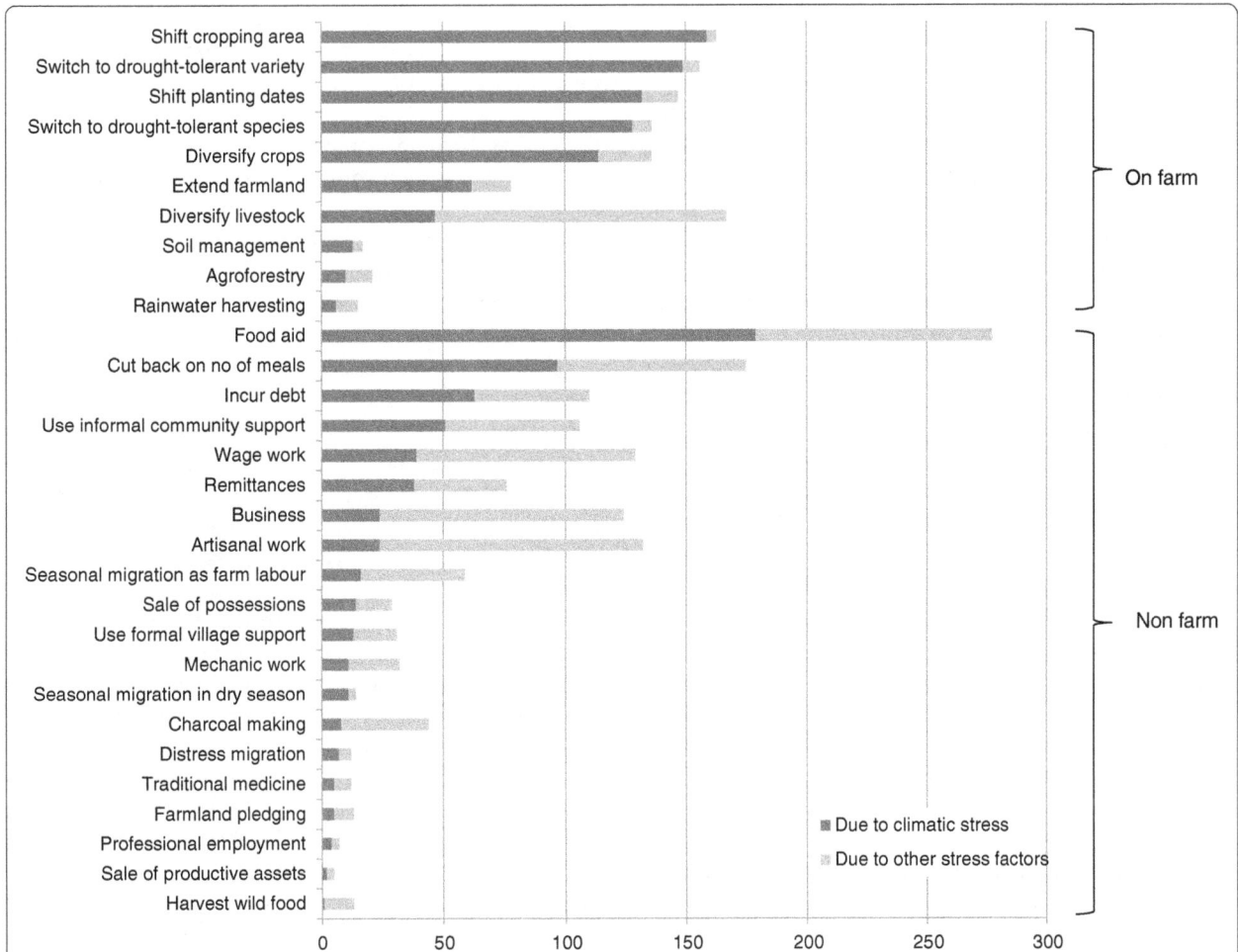

Figure 3 Coping activities in times of stress. The number of households in Mangae and Laikala reporting that they practice different coping activities (grey bars), and the share doing so because of climatic stress (dark grey part of bars).

Table 1 Regression models for on-farm coping activities practiced by households with livelihood factors as explanatory variables[a]

Activity	Shift cropping area	Switch to drought-tolerant variety	Switch to drought-tolerant species	Diversify crops	Extend farmland	Diversify livestock
Sex of household head	8.145×10^{-1}**	-3.489×10^{-2}	-9.470×10^{-2}	-4.035×10^{-1}	6.545×10^{-1}*	9.329×10^{-1}**
Age of household head	-6.419×10^{-3}	-2.961×10^{-3}	-1.244×10^{-3}	-8.348×10^{-3}	-6.811×10^{-3}	-2.073×10^{-3}
Education	4.804×10^{-1}	-2.739×10^{-1}	1.001×10^{-1}	-1.753×10^{-1}	9.645×10^{-2}	-9.479×10^{-2}
Household size	1.203×10^{-2}	-9.705×10^{-3}	4.775×10^{-2}	2.682×10^{-2}	-1.095×10^{-2}	1.589×10^{-1}**
Livestock ownership	9.324×10^{-1}*	7.656×10^{-1}*	1.037**	1.069×10^{-1}	2.780×10^{-1}	2.520***
Annual income	-2.906×10^{-8}	1.925×10^{-8}	8.799×10^{-11}	5.494×10^{-8}	7.887×10^{-8}	7.698×10^{-8}
Extension service	2.388×10^{-3}	8.177×10^{-1}**	2.184×10^{-1}	8.109×10^{-2}	6.252×10^{-1}*	-1.855×10^{-1}
Village effect	7.638×10^{-1}**	-6.596×10^{-1}*	8.010×10^{-1}**	5.785×10^{-1}*	2.861×10^{-2}	9.071×10^{-1}**
Wald test	$P = 0.00017$	$P = 0.011$	$P = 8.8 \times 10^{-5}$	$P = 0.017$	$P = 1.7 \times 10^{-13}$	$P = 1.8 \times 10^{-7}$

[a]Significance: ***$P < 0.0001$; **$P < 0.001$; *$P < 0.01$.

to improved varieties ($\rho = 0.19$, $P < 0.001$). In comparison, the area allocated to local varieties is negatively correlated with the same activity ($\rho = -0.17$, $P < 0.01$). Thus, rather than capturing all kinds of adaptive switches, the question apparently mainly captures the adoption of improved varieties.

In this study we did not find the typical human capital factors usually associated with adoption of improved varieties, such as household size and the sex, education and age of the household head [61], to be significantly positively associated with crop adaptation. Out of the two financial capital variables included, annual income and possession of livestock, only the latter positively affected the probability of undertaking crop adaptation. Interestingly, the same pattern was found for wealth indicators in a scoping study for drought-tolerant maize done by the International Maize and Wheat Improvement Centre (CIMMYT) in East Africa [62]. The same study found that neither the length of education nor the sex of the household head affected the purchasing of improved varieties in Tanzania, while these factors significantly enhanced the likelihood in other countries in SSA. A study from Ethiopia [10] found that larger households headed by older males with more education and higher income were more likely to undertake crop adaptation. A study from Malawi [63] similarly found that larger households headed by males and experiencing climate-related shock were more likely to purchase seeds. The lack of positive association between the household's human capital and crop adaptation in this study indicates that institutional factors play a larger role. This is confirmed by the positive association between receiving the agricultural extension service and switching to drought-tolerant varieties. The government's extension service is promoting improved varieties of both sorghum and maize in the villages studied and some of the varieties are promoted as drought-tolerant. Prior to the growing season, during which this study was undertaken, improved maize varieties were sold through the extension service at subsidized rates in Mangae and improved sorghum was distributed by the extension service in Laikala under a seed aid program. The subsidized and free distribution of improved varieties is supposed to target the poorest and most needy households. This probably explains why receiving the extension service seems to 'trump' other factors that normally influence the likelihood of cultivating improved seeds. The important role

of participation in government programs is observed in several studies of determinants of adoption of improved varieties from other areas [45,64].

The role of the seed system

The local seed system consists of both formal and informal elements and the largest proportion of seeds is sourced through informal seed channels both for maize and sorghum (Table 2). Among the four informal seed sources most seeds are sourced from farmers' own harvest for both crops and the local seed market is the second most important channel. For maize, 24% of seeds are sourced through seed sources classified as formal seed system channels, while in sorghum only 8% are from these channels. In the terminology of the livelihood approach, seeds are important assets for agriculture-based livelihoods and seed systems embody the institutions that mediate access to this asset. This assessment thus demonstrates that the informal seed system is more important than the formal system, measured in the quantity of seeds accessed, but also the formal system provides a considerable quantity of maize seeds in the study area.

Distinguishing between the three genetic resource categories (local, improved and farmer-recycled varieties), we see a marked difference between maize and sorghum (Figure 4). For sorghum, local varieties dominate both in terms of the number of households cultivating them and in terms of the area allocated. For maize, improved varieties are somewhat more commonly cultivated than local varieties and farm saving and recycling of seeds is quite common. A relatively limited number of households cultivates two or three categories of maize, but most cultivators of improved sorghum also cultivate local varieties. The dominance of the informal seed system for maize outlined in the previous section is thus not coupled with a similar dominance of local varieties. The informal maize seed system is open to an influx of improved varieties originating in the formal seed system. Farmers' seed use is more complex than a mere choice between different varieties off-the-shelf from the informal or the formal seed system. A large proportion of the respondents reported that they select seeds for next year's planting. Out of the 275 sorghum-growing households in our survey, 77% practice selection and among the 310 maize-growing households, 78% practice selection. The combination of recycling of improved open-pollinated maize varieties and seed selection suggests that improved varieties

Table 2 Proportion of maize and sorghum seeds sourced from different supply channels in Mangae and Laikala

Crop	Seed source (percentage)[a]							Formal (percentage)	Informal (percentage)
	Own	Barter	Gift	Local market	Shop	Government	Other		
Maize	42	6	4	21	9	15	3	24	73
Sorghum	61	9	7	11	5	3	5	8	87

[a]Seed source share is given as percentage of total weight of seeds reported for the crop. Seeds from own harvest, through barter, gift and local market are classified as informal seed channels. Seeds from shop and government provision are classified as formal channels.

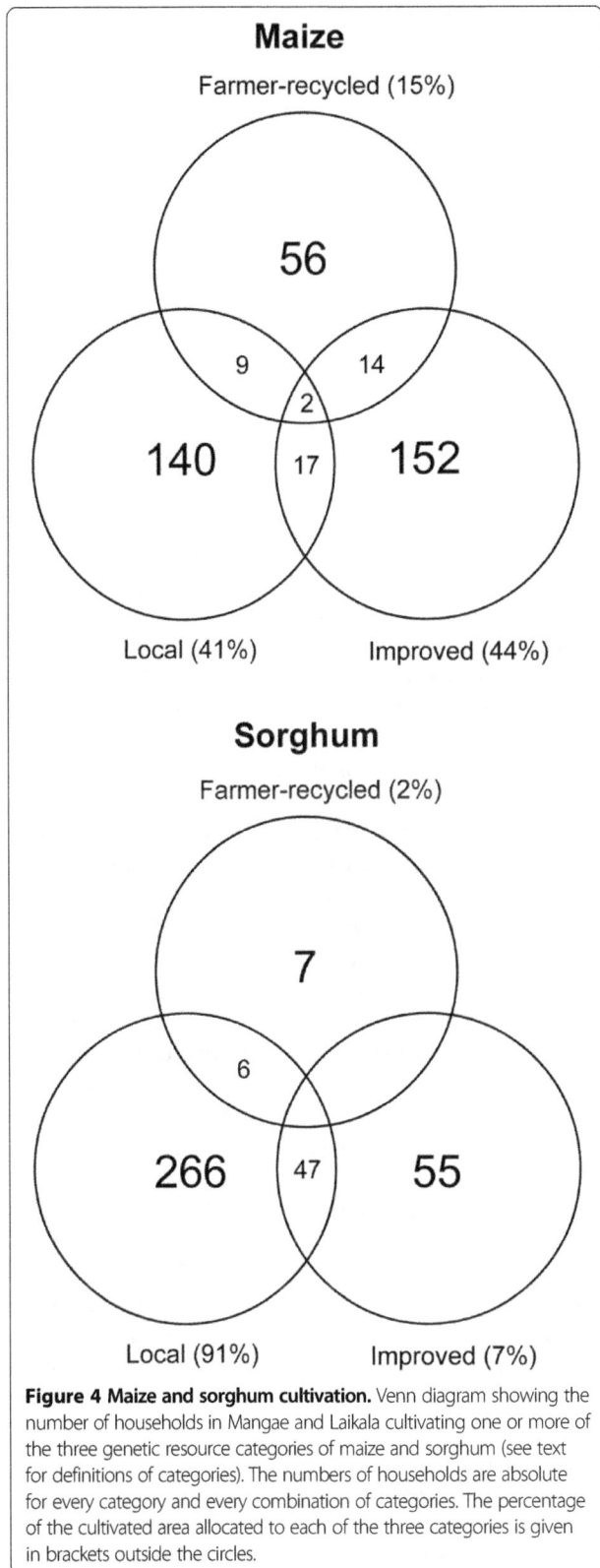

Figure 4 Maize and sorghum cultivation. Venn diagram showing the number of households in Mangae and Laikala cultivating one or more of the three genetic resource categories of maize and sorghum (see text for definitions of categories). The numbers of households are absolute for every category and every combination of categories. The percentage of the cultivated area allocated to each of the three categories is given in brackets outside the circles.

are used as genetic resources with beneficial creolization as a potential outcome. In another paper we explore the consequences of this seed management in more depth using single nucleotide polymorphisms to study genetic diversity and differentiation in the local maize seed system in one of the villages in this study [65].

The assessment of households' reasons for cultivating the different genetic resource categories and their perceptions of consumption and production qualities casts more light on the complexity of crop adaptation (Table 3). The average field size allocated to local varieties compared to the size of fields planted with improved varieties is significantly larger both for maize and sorghum, but the difference is more marked for sorghum. Drought tolerance is the most frequently reported reason for growing both maize and sorghum. We found that 57% of the local maize cultivators and 81% of the local sorghum cultivators reported drought tolerance as a major reason for cultivating the local varieties. Yield receives the second highest score among the cultivators of local varieties of both crops and is the most reported reason for cultivating improved maize. Biotic stress resistance receives a low to zero score as a major reason for cultivation across all genetic resource categories. This is in line with the results of the stress factor ranking and confirms that abiotic stress is perceived as a more serious problem than biotic stress by households in this area. The higher proportion of households reporting drought tolerance as a reason for cultivating local sorghum than for cultivating local maize reflects that sorghum is a considerably more drought-tolerant crop species than maize [18]. The shift to improved varieties due to climatic stress indicated by the regression model is not reflected in the assessment of production variables. On the contrary, a significantly higher proportion of households report cultivating local varieties of both crops because of their drought tolerance compared to those reporting cultivation of improved varieties for this reason. Among the two consumption variables assessed, tastiness and storability, we did not find significant differences between improved and local varieties.

The scores given by households to the two production variables, maturity and drought resistance, reveal a complexity in the perception of traits relevant for withstanding climatic stress. Local varieties of both sorghum and maize are reported to have a significantly longer maturity period than improved ones. Since a short time for maturation is generally considered a valuable drought-resistance trait [62,66], this is apparently contradictory to the finding of no significant difference between the average household scores given to local and improved varieties for this trait. A possible explanation can be found in the scientific definition of drought resistance, which is commonly defined as encompassing both drought avoidance and drought tolerance. Varieties that do not avoid drought and need a long growing period to mature might still be relatively drought resistant because they are able to tolerate drought.

Table 3 Production and consumption variables reported by households in Mangae and Laikala for each field[a]

Variable		Maize			Sorghum		
		Local (n = 156)	Improved (n = 189)	Farmer-recycled (n = 60)	Local (n = 290)	Improved (n = 56)	Farmer-recycled (n = 7)
Field size	Acres	3.73*	3.27	3.36	3.28***	1.42	3.36
Reason for growing[b]	Drought tolerance	0.57*	0.41	0.46	0.81*	0.57	0.57
	Yield	0.52	0.53	0.71	0.35	0.33	0.57
	Biotic stress resistance	0	0.01	0	0.01*	0	0
Consumption	Storability[c]	1.26	1.24	1.22	1.56	1.22	1
	Tastiness[c]	1.02	1.08	1.07	1.11	1.13	1
Production	Maturity[d]	1.97***	1.31	1.34	2.90***	1.39	1.86
	Drought resistance[e]	1.94	1.96	1.77	1.37	1.57	1.57

[a]The mean values for the variables are reported with chi-square tests for the difference between the local and improved categories. Significance: ***$P < 0.0001$; **$P < 0.001$; *$P < 0.01$; [b]Multiple responses for households listing up to three factors (1 if a major reason, 0 if not); [c]1 = Good, 2 = Ok, 3 = Not so good; [d]1 < 100 days, 2 = 100 to 120, 3 > 120; [e]1 = Very good, 2 = Good, 3 = Not good.

The local varieties of sorghum in the study area take 5 to 7 months to mature and the local varieties of maize take more than 3 months. The improved varieties of sorghum and maize have considerably shorter maturation periods: from 120 days down to only 90 days for the earliest maturing varieties of both crops. The maize variety kito, which has a short maturation period, was recommended by the extension service due to the late onset of rains in the growing season when this study was conducted. However, short maturation period varieties are not necessarily the most robust varieties if drought strikes before flowering and grain filling [67,68]. Because of the unpredictable rains early in the growing season, drought escaping varieties such as kito might actually be less drought resistant than other improved and local varieties. This can explain why varieties with other physiological traits that are important for withstanding drought are favored by the households in Mangae and Laikala. The most commonly improved maize varieties are staha and TMV1, which were released by the public Tanzania Agricultural Research Organization. Neither are short maturation period varieties according to CIMMYT, which classifies staha as a very late maturing variety and TMV1 as an intermediate to late maturing variety [68]. The observation that short maturation period not necessarily is a reason for choosing a variety to cope with abiotic stress has also been observed in other studies of farmers' perceptions. In a study from Chiapas, Mexico, Bellon and Taylor [69] observed that while a short growing period was perceived as a positive trait associated with improved maize varieties, resistance to drought was a positive trait only attributed to local varieties. In a later study from the same area, drought resistance was ranked as one of the more important traits and one that was given a significantly higher rank for local compared to improved varieties [24].

The complexity in the perception of drought resistance is also apparent for sorghum in the study area. The sorghum variety wahi is, like the maize kito, an example of an improved short maturation period variety promoted by the formal seed system with very limited adoption among farmers. A short maturation period is a breeding target for sorghum in SSA because the day-length sensitivity common in local varieties is considered a production constraint [18]. The day-length-sensitive local varieties shift from vegetative to reproductive growth when the days shorten to a critical period, regardless of the date of planting. This trait has mixed implications for coping with and adapting to climatic stress. Its primary function is to ensure that grain matures under dry conditions, which is important for avoiding grain rot, mold and other diseases [18]. Thus, day-length sensitivity is an adaptation to the temporal habitat of the local varieties, and together with local preferences in taste and end-product use, this adaptation probably goes a long way to explaining the low adoption of improved sorghum among smallholders in the region. However, day-length sensitivity makes local varieties vulnerable to shifts in the normal seasonality. The photoperiod stays the same even if the climate conditions change and this raises questions about the adaptive potential of the local genetic resources as well as about the ability of the seed system to deliver this adaptation in a timely manner. The spatial scope of informal seed systems in providing farmers with appropriate genetic resources to adapt is beginning to attract research informed by climate projections [70]. Our results suggest that the temporal scope of seed systems warrants more research in this context.

Conclusions
Seed system perspectives and adaptation
There is mounting evidence that agriculture in SSA will have to adapt to the adverse effects of climate change and much attention from development practitioners and scholars is directed towards the question of the means of adaptation. It is generally agreed that the genetic resources of important food crops are key assets for adaptation in

rural households, but different research outlooks lead to different conclusions about what kind of genetic resources best allows farmers to adapt. In an agricultural modernization perspective, crop adaptation is commonly framed as a question of the public and commercial development of improved varieties, and farmers' crop adaptation options are framed as the adoption of new technology. This framing of crop adaptation does not represent the current reality in subsistence agriculture in SSA, where most of the seeds planted are uncertified and sourced through informal seed system channels.

Framing crop adaptation with a livelihood approach captures the complex resource access situation faced by rural households in the semi-arid zone in Tanzania. The results presented here confirm that climatic stress is already a major stress factor in the livelihoods of people living in the area and that crop adaptation is among the more important responses used by households. Crop adaptation involves adoption of improved maize varieties combined with continued use of local varieties of both maize and sorghum, which despite their late maturity are valued for their drought tolerance. The seed system in the study area consists of both formal and informal elements, but the informal elements are the most important supply channels. All genetic resource categories are subject to on-farm selection and hybridization, and farmer selection may lead to incremental on-farm crop adaptation. Our findings support the view that the integration of informal and formal seed system elements will be important for agriculture-based livelihoods in meeting the challenges ahead [25-27].

This study highlights two important methodological challenges in impact and adaptation studies. First, the use of generic variety data in modeling crop impact on a large geographical scale may be problematic because of the diversity between crop varieties with regard to drought avoidance and drought tolerance traits. Biophysical crop models often fail to capture that varieties sourced through informal seed channels dominate smallholder agriculture in SSA. No single variety can represent the diverse situations in a single village, let alone in studies making projections at the national or regional level. The second methodological challenge is relevant for the small, but growing body of regression model-based adaptation studies in which a 'switch to a drought-tolerant variety' is used as a response variable. People's understanding of what a drought-tolerant variety constitutes is likely to be influenced by national and international development actors' promotion of improved varieties under this banner. In this study households' perceptions of the characteristics of and reasons for growing the different genetic resource categories reveal that local and recycled varieties play a role in adaptation to drought stress that was left undetected in the regression modeling. The perceptions of the varieties observed here are not objective measurements, but they are nevertheless important

in understanding actual decision-making by farmers [24]. Different attributes of the varieties influence the use and allocation of land to local, improved and farmer-recycled varieties. Our findings highlight the value and importance of location-specific information about crop variety use for arriving at realistic recommendations in impact and adaptation studies. Seed system perspectives of crop adaptation offer insights into the complex ways crop adaptation is realized at the livelihood level and can contribute to increase knowledge of how farming in SSA can and will adapt to a changing climate.

Abbreviations
CIMMYT: International Maize and Wheat Improvement Centre; IPCC: Intergovernmental Panel on Climate Change; SSA: sub-Saharan Africa.

Competing interests
The authors declare that they have no competing interests.

Authors' contributions
OTW designed the study, analyzed the data and drafted the manuscript. Both authors contributed to and approved the final manuscript.

Acknowledgements
This work was supported through a grant from the University of Oslo. We are grateful to Jans Bobert and Khamaldin Mutabazi from the project Resilient Agro-landscapes to Climate Change in Tanzania for invaluable support and advice during the fieldwork. We thank Philip K Thornton for providing the yield projection data used in Figure 1. We acknowledge Sokoine University of Agriculture for granting the research permit. We are grateful for the key information and advice we obtained from academic staff at Sokoine University, as well as the many actors in the formal seed sector. Special thanks go to the late maize breeder Alfred Moshi, who provided key insights into breeding and the adoption of improved maize in the study area. Many thanks are also due to Philip Daninga and the excellent group of MSc students who assisted with the fieldwork. Finally we would like to express our gratitude to the people and leaders of Mangae and Laikala villages for their willingness to participate in this study. This manuscript was improved by critical and insightful comments from Trygve Berg, Desmond McNeill and two anonymous reviewers.

References
1. HLPE: *Food security and climate change. A report by the high level panel of experts on food security and nutrition of the Committee on World food Security.* Rome: FAO; 2012.
2. Godfray HCJ, Beddington JR, Crute IR, Haddad L, Lawrence D, Muir JF, Pretty J, Robinson S, Thomas SM, Toulmin C: **Food security: the challenge of feeding 9 billion people.** *Science* 2010, **327**(5967):812–818.
3. Lobell DB, Burke MB, Tebaldi C, Mastrandrea MD, Falcon WP, Naylor RL: **Prioritizing climate change adaptation needs for food security in 2030.** *Science* 2008, **319**(5863):607–610.
4. Thornton PK, Jones PG, Alagarswamy G, Andresen J: **Spatial variation of crop yield response to climate change in East Africa.** *Glob Environ Change: Human Policy Dim* 2009, **19**(1):54–65.
5. Howden SM, Soussana JF, Tubiello FN, Chhetri N, Dunlop M, Meinke H: **Adapting agriculture to climate change.** *Proc Natl Acad Sci USA* 2007, **104**(50):19691–19696.
6. Easterling W, Aggarwal P, Batima P, Brander K, Erda L, Howden M, Kirilenko A, Morton J, Soussana J, Schmidhuber J: **Food, fibre, and forest products.** In *Climate Change 2007: climate change impacts, adaptation and vulnerability, contribution of working group II to the fourth assessment report of the Intergovernmental Panel on Climate Change.* Edited by Parry ML, Canziani OF, Palutikof JP, van der Linden PJ, Hanson CE. Cambridge, UK: Cambridge University Press; 2007:273–313.
7. McIntyre BD, Herren HR, Wakhungu J, Watson RT: *Agriculture at a crossroads. International Assessment of Agricultural Knowledge, Science and Technology for Development (IAASTD): Global report.* Washington, DC, USA: Island Press; 2009.

8.   Vermeulen SJ, Aggarwal P, Ainslie A, Angelone C, Campbell BM, Challinor A, Hansen JW, Ingram J, Jarvis A, Kristjanson P: Options for support to agriculture and food security under climate change. *Environ Sci Pol* 2012, **15**(1):136–144.

9.   Bryan E, Deressa TT, Gbetibouo GA, Ringler C: Adaptation to climate change in Ethiopia and South Africa: options and constraints. *Environ Sci Pol* 2009, **12**(4):413–426.

10. Deressa TT, Hassan RM, Ringler C, Alemu T, Yesuf M: Determinants of farmers' choice of adaptation methods to climate change in the Nile Basin of Ethiopia. *Glob Environ Change: Human Policy Dim* 2009, **19**(2):248–255.

11. United Nations general assembly resolution a/RES/64/197. [http://www.un.org/en/ga/64/resolutions.shtml]

12. Darwin C: *The variation of animals and plants under domestication.* Middlesex, UK: The Echo Library; 2007.

13. Harlan JR: *Crops and man.* Madison, WI, USA: American Society of Agronomy; 1975.

14. Murphy DJ: *People plants and genes: the story of crops and humanity.* Oxford, UK: Oxford University Press; 2007.

15. Fowler C, Hodgkin T: Plant genetic resources for food and agriculture: assessing global availability. *Annu Rev Environ Resour* 2004, **29**:143–179.

16. Almekinders CJM, Louwaars NP, Debruijn GH: Local seed systems and their importance for an improved seed supply in developing countries. *Euphytica* 1994, **78**(3):207–216.

17. Sperling L, Cooper HD, Remington T: Moving towards more effective seed aid. *J Dev Stud* 2008, **44**(4):586–612.

18. DeVries J, Toenniessen G: *Securing the harvest: biotechnology, breeding and seed systems for African crops.* New York, USA: CABI; 2002.

19. Langyintuo AS, Mwangi W, Diallo AO, MacRobert J, Dixon J, Banziger M: Challenges of the maize seed industry in eastern and southern Africa: a compelling case for private-public intervention to promote growth. *Food Policy* 2010, **35**(4):323–331.

20. Lunduka R, Fisher M, Snapp S: Could farmer interest in a diversity of seed attributes explain adoption plateaus for modern maize varieties in Malawi? *Food Policy* 2012, **37**(5):504–510.

21. Gibson R, Lyimo N, Temu A, Stathers T, Page W, Nsemwa L, Acola G, Lamboll R: Maize seed selection by East African smallholder farmers and resistance to maize streak virus*. *Ann Appl Biol* 2005, **147**(2):153–159.

22. Mortimore M, Adams W: Farmer adaptation, change and 'crisis' in the Sahel. *Glob Environ Change: Human Policy Dim* 2001, **11**(1):49–57.

23. Bellon MR, Adato M, Becerril J, Mindek D: Poor farmers' perceived benefits from different types of maize germplasm: the case of creolization in lowland tropical Mexico. *World Dev* 2006, **34**(1):113–129.

24. Bellon M, Risopoulos J: Small-scale farmers expand the benefits of improved maize germplasm: a case study from Chiapas, Mexico. *World Dev* 2001, **29**(5):799–811.

25. Louwaars NP, de Boef WS: Integrated seed sector development in Africa: a conceptual framework for creating coherence between practices, programs, and policies. *J Crop Improve* 2012, **26**(1):39–59.

26. Scoones I, Thompson J: The politics of seed in Africa's green revolution: alternative narratives and competing pathways. *IDS Bull* 2011, **42**(4):1–23.

27. McGuire S, Sperling L: Making seed systems more resilient to stress. *Glob Environ Change: Human Policy Dim* 2013, **23**(3):644–653.

28. Sperling L, McGuire S: Fatal gaps in seed security strategy. *Food Sec* 2012, **4**(4):1–11.

29. Smit B, Wandel J: Adaptation, adaptive capacity and vulnerability. *Glob Environ Change: Human Policy Dim* 2006, **16**(3):282–292.

30. Vermeulen SJ, Challinor AJ, Thornton PK, Campbell BM, Eriyagama N, Vervoort JM, Kinyangi J, Jarvis A, Läderach P, Ramirez-Villegas J: Addressing uncertainty in adaptation planning for agriculture. *Proc Natl Acad Sci USA* 2013, **110**(21):8357–8362.

31. Parry ML, Canziani OF, Palutikof JP, van der Linden PJ, Hanson CE: *Climate change 2007: impacts, adaptation and vulnerability. Contribution of working group II to the fourth assessment report of the Intergovernmental Panel on Climate Change (IPCC).* Cambridge, UK: Cambridge University Press; 2007:976.

32. Adger WN, Agrawala S, Mirza MMQ, Conde C, O'Brien K, Pulhin J, Pulwarty R, Smit B, Takahashi K: Assessment of adaptation practices, options, constraints and capacity. In *Climate change 2007: impacts, adaptation and vulnerability. Contribution of working group II to the fourth assessment report of the Intergovernmental Panel on Climate Change.* Edited by Parry ML, Canziani OF, Palutikof JP, van der Linden PJ, Hanson CE. Cambridge, UK: Cambridge University Press; 2007:717–743.

33. Knox J, Hess T, Daccache A, Wheeler T: Climate change impacts on crop productivity in Africa and South Asia. *Environ Res Lett* 2012, **7**(3):034032.

34. Parry ML, Rosenzweig C, Iglesias A, Livermore M, Fischer G: Effects of climate change on global food production under SRES emissions and socio-economic scenarios. *Glob Environ Change: Human Policy Dim* 2004, **14**(1):53–67.

35. Adger WN, Huq S, Brown K, Conway D, Hulme M: Adaptation to climate change in the developing world. *Prog Dev Stud* 2003, **3**(3):179–195.

36. O'Brien K, Leichenko R, Kelkar U, Venema H, Aandahl G, Tompkins H, Javed A, Bhadwal S, Barg S, Nygaard L, West J: Mapping vulnerability to multiple stressors: climate change and globalization in India. *Glob Environ Change: Human Policy Dim* 2004, **14**(4):303–313.

37. Ministry of Agriculture, Food Security and Cooperatives: Ministry of agriculture, food security and cooperatives. [http://www.agriculture.go.tz/]

38. Haug R, Hella J: The art of balancing food security: securing availability and affordability of food in Tanzania. *Food Security* 2013, **5**:415–426.

39. FAOSTAT: [http://faostat3.fao.org/home/index.html]

40. Schlenker W, Lobell DB: Robust negative impacts of climate change on African agriculture. *Environ Res Lett* 2010, **5**(1):014010.

41. Rowhani P, Lobell DB, Linderman M, Ramankutty N: Climate variability and crop production in Tanzania. *Agr Forest Meteorol* 2011, **151**(4):449–460.

42. Ritchie J, Singh U, Godwin D, Bowen W: Cereal growth, development and yield. *Syst Appro Sustain Agri Dev* 1998, **7**:79–98.

43. Ellis F: *Rural livelihoods and diversity in developing countries.* Oxford, UK: Oxford University Press; 2000.

44. Scoones I: Livelihoods perspectives and rural development. *J Peasant Stud* 2009, **36**(1):171–196.

45. Bellon MR, Hellin J: Planting hybrids, keeping landraces: agricultural modernization and tradition among small-scale maize farmers in Chiapas, Mexico. *World Dev* 2011, **39**(8):1434–1443.

46. Ellis F, Mdoe N: Livelihoods and rural poverty reduction in Tanzania. *World Dev* 2003, **31**(8):1367–1384.

47. Paavola J: Livelihoods, vulnerability and adaptation to climate change in Morogoro, Tanzania. *Environ Sci Pol* 2008, **11**(7):642–654.

48. Fisher M, Chaudhury M, McCusker B: Do forests help rural households adapt to climate variability? Evidence from Southern Malawi. *World Dev* 2010, **38**(9):1241–1250.

49. Kristjanson P, Neufeldt H, Gassner A, Mango J, Kyazze FB, Desta S, Sayula G, Thiede B, Förch W, Thornton PK: Are food insecure smallholder households making changes in their farming practices? Evidence from East Africa. *Food Security* 2012, **4**(3):381–397.

50. R Project: [http://www.r-project.org/]

51. Analysis of overdispersed data. [http://cran.r-project.org/web/packages/aod/index.html].

52. Mary A, Majule A: Impacts of climate change, variability and adaptation strategies on agriculture in semi arid areas of Tanzania: the case of Manyoni district in Singida region, Tanzania. *Afr J Environ Sci Technol* 2009, **3**(8):206–218.

53. Slegers M: 'If only it would rain': Farmers' perceptions of rainfall and drought in semi-arid central Tanzania. *J Arid Environ* 2008, **72**(11):2106–2123.

54. Morton JF: The impact of climate change on smallholder and subsistence agriculture. *Proc Natl Acad Sci USA* 2007, **104**(50):19680–19685.

55. Mortimore M: Adapting to drought in the Sahel: lessons for climate change. *Wiley Interdiscip Rev Clim Chang* 2010, **1**(1):134–143.

56. Vermeulen SJ, Campbell BM, Ingram JS: Climate change and food systems. *Annu Rev Environ Resour* 2012, **37**(1):195–222.

57. Eriksen SH, Brown K, Kelly PM: The dynamics of vulnerability: locating coping strategies in Kenya and Tanzania. *Geogr J* 2005, **171**:287–305.

58. Liwenga ET: Adaptive livelihood strategies for coping with water scarcity in the drylands of central Tanzania. *Phys Chem Earth* 2008, **33**(8):775–779.

59. Naess LO: The role of local knowledge in adaptation to climate change. *Wiley Interdiscip Rev Clim Chang* 2013, **4**:99–106.

60. Goulden M, Naess LO, Vincent K, Adger WN: Accessing diversification, networks and traditional resource management as adaptations to climate extremes. In *Adapting to climate change: thresholds, values, governance.* Edited by Adger WN, Lorenzoni I, O'Brien K. Cambridge, UK: Cambridge University Press; 2009:514.

61. Feder G, Just RE, Zilberman D: Adoption of agricultural innovations in developing countries: a survey. *Econ Dev Cult Chang* 1985, **33**(2):255–298.

62. Erenstein O, Kassie GT, Langyintuo A, Mwangi W: *Characterization of maize producing households in drought prone regions of Eastern Africa.* Mexico: CIMMYT; 2011.

63. Nordhagen S, Pascual U: The impact of climate shocks on seed purchase decisions in Malawi: implications for climate change adaptation. *World Dev* 2013, **43**:238–251.

64. Chibwana C, Fisher M, Shively G: Cropland allocation effects of agricultural input subsidies in Malawi. *World Dev* 2012, **40**(1):124–133.

65. Westengen OT, Ring KH, Berg PR, Brysting AK: **Modern maize varieties going local in the semi-arid zone in Tanzania.** *BMC Evol Biol* 2014, **14**(1):1.

66. Bänziger M: *Breeding for drought and nitrogen stress tolerance in maize: from theory to practice.* Mexico: CIMMYT; 2000.

67. Lobell DB, Bänziger M, Magorokosho C, Vivek B: **Nonlinear heat effects on African maize as evidenced by historical yield trials.** *Nat Clim Chang* 2011, **1**(1):42–45.

68. CIMMYT: *Choosing the right open-pollinated maize in Southern Africa.* Mexico; 2010:3.

69. Bellon MR, Taylor JE: **'Folk' soil taxonomy and the partial adoption of new seed varieties.** *Econ Dev Cult Chang* 1993, **41**(4):763–786.

70. Bellon MR, Hodson D, Hellin J: **Assessing the vulnerability of traditional maize seed systems in Mexico to climate change.** *Proc Natl Acad Sci USA* 2011, **108**(33):13432–13437.

6

Exploration of 'hot-spots' of methane and nitrous oxide emission from the agriculture fields of Assam, India

Satyendra Nath Mishra[1†], Sudip Mitra[2*†], Latha Rangan[3], Subashisha Dutta[3] and Pooja Singh[2]

Abstract

Background: Agricultural soils contribute towards the emission of CH_4 (mainly from paddy fields) and N_2O (from N-fertilizer application), the two important greenhouse gases causing global warming. Most studies had developed the inventories of CH_4 and N_2O emission at the country level (larger scale) for India, but not many studies are available at the local scale (e.g. district level) on these greenhouse gases (GHGs). Assam is an important state in the North Eastern region of India. In addition to being the regional economic hub for the entire region, agriculture is the major contributor to the state's gross domestic product. In Assam about three-fourths of the area is under paddy cultivation and rice is the staple food. With this background, a district wise inventory of CH_4 and N_2O emission in the North Eastern state of Assam, India was carried out using different emission factors, viz., IPCC, Indian factors and others, to highlight the discrepancies that arose in the emission estimation of these important GHGs while used at the smaller scale i.e. district level. This study emphasizes the need for better methodologies at the local level for GHGs inventories. This study also reiterates the fact that no emission factor is universally applicable across all regions. The GHGs like CH_4 and N_2O are highly site and crop specific and the factors required for their inventory are driven by cultural practices, agronomic management, soil resources and socio-economic drivers.

Material and methods: In this study, Intergovernmental Panel on Climate Change (IPCC) methodology was used for the estimation of CH_4 and N_2O emission. In case of N_2O emission, both direct and indirect emission from agricultural soil was estimated for the various districts of Assam.

Results: The CH_4 (base year 2000–2001) and N_2O (base year 2001–2002) emission was estimated to be 121 Gg and 1.36 Gg from rice paddy and agricultural fields of Assam state respectively.

Conclusions: This study is the first report on the estimation of the GHG emission at the district level from the entire state of Assam, agriculturally one very important state of North Eastern India. This state is also considered as remote due to its geographical location. The study clearly elucidates that there is large variation in the emission inventory of CH_4 and N_2O at the district level (local scale) when different emission factors are used. This calls for detailed and comprehensive data collection and mapping at the micro level for accurate inventory of greenhouse gases in future from agriculture fields.

Keywords: Agriculture, Paddy fields, Methane, Nitrous oxide, Assam, India

* Correspondence: sudipmitra@yahoo.com
†Equal contributors
[2]School of Environmental Sciences, Jawaharlal Nehru University, New Delhi 110067, India
Full list of author information is available at the end of the article

Background

Increasing industrialisation and developmental activities across the globe have led to stress on the Earth's resources. One of the major damaging impacts of this increasing developmental activity (especially industrial and agricultural) is increasing concentration of the greenhouse gases (GHGs), namely, carbon dioxide (CO_2), methane (CH_4), nitrous oxide (N_2O), chlorofluorocarbon (CFCs), etcetera, which are potential causes of global warming due to the enhanced greenhouse effect. CO_2, CH_4 and N_2O are key GHGs that contribute toward global warming at 60%, 15% and 5% respectively [1,2]. Concentrations of these gases in the atmosphere are increasing at 0.4%, 0.3% and 0.22% per year respectively [1,3]. On average, the agricultural sector emits about 47% and 58% of total global anthropogenic emissions of CH_4 and N_2O respectively. Although CH_4 and N_2O emission constitutes only about 20% of the total GHG emissions, they play a significant role in global warming due to their higher values of global warming potential (GWP) of 21 and 310 respectively. It has been estimated that about 74% of the agricultural GHG emissions are from non-annex[1] countries [4]. The amount of CH_4 emission from paddy fields (about 50 to100 Tg yr^{-1}) accounts for about 10% to 20% of total CH_4 emission around the world. Huang et al. [5] projected that the CH_4 emission from rice fields may increase to 145 Tg yr^{-1} by 2025. Industrial nitrogen fixation for use in agriculture had increased from less than 10 Tg yr^{-1} in the 1950s to over 80 Tg yr^{-1} by the year 2000. Nitrogen applied in agricultural systems is emitted in different forms like dinitrogen, ammonium, dissolved organic nitrogen or NO_x. Of all these N_2O, which is increasing in the atmosphere at the rate 0.2% to 0.3% per year, is of particular concern [6].

The anthropogenic emission of CO_2, CH_4 and N_2O in India was about 1,398,700 Gg, 20,560 Gg and 240 Gg respectively in the year 2007 [7]. In 2007, the agriculture sector was the largest source of CH_4 emission, accounting for about 65% of the total. Of this, livestock, paddy cultivation and onsite burning of crop residues represented shares of 48%, 16% and 1% respectively. In view of this, Indian scientists have placed special emphasis in recent times on the exploration of CH_4 emission from paddy fields [8-12]. In 2007, the agricultural sector accounted for about 65% of total N_2O emission in India. The main source of direct and indirect N_2O emission in agriculture was the application of nitrogen fertilizer[2] [7]. In 2008, the Government of India (GoI) came up with the National Action Plan on Climate Change. Of the eight national missions mentioned in this plan, one deals with a national mission for sustainable agriculture [13]. After the energy sector, being a dominant and dynamic source of GHG emission, the agricultural sector got special attention for studies and management to abate GHG

emissions [14]. It has been easy to change the technology to reduce GHG emission from the energy sector either by using regulatory norms or good backstop technology. However, this is not the case with the agricultural sector, as agriculture directly deals with the cultural, socioeconomic matrix of society and farmers, and local setup. So, it is not easy to assess the emissions, due to the random distribution of variables on which emissions depend, or to develop a proper mitigation plan at field level. Few studies have been taken at experimental level to estimate GHG emissions at the local level [15,16].

National level data provides lots of information and inputs for national and international level planning and negotiation. However, data and information at district level would be imperative and very important in the near future for local level decision-making, and for upcoming district and regional planning activities initiated at local and regional level in the age of the decentralised planning approach. It is unanimously accepted across the scientific and policy-making bodies that while climate change and global warming is a global phenomenon, its solution lies at the level of local planning and adaptation. Considering this background, a local level study was carried out by estimating district-wise emission of CH_4 and N_2O from rice paddy and agricultural fields respectively for the state of Assam, India, as per the Intergovernmental Panel on Climate Change (IPCC) guidelines and using other available emission factors[3]. Assam is the gateway to the north-eastern part of India, situated between 90° to 96° longitude east and 24° to 28° latitude north. Assam is bordered in the north and east by the Kingdom of Bhutan and Arunachal Pradesh state. The states, namely, Nagaland, Manipur and Mizoram are situated in the South of Assam, and West Bengal state and Bangladesh to the west. Meghalaya state lies to the south-west of Assam.

Material and methods

Extensive data collection and investigation was carried out to make the inventory of CH_4 and N_2O emissions at the district[4] level (smaller administrative unit) in one of the major paddy-growing states of India, namely, Assam in the north-eastern part of India. Assam was chosen for this study as paddy is the major crop in this state, is grown three times a year, and is major source of agricultural gross domestic product (GDP)[5]. Emission of CH_4 from paddy fields and N_2O from agricultural fields (as per the Indian emission factor and IPCC standard) were represented geographically on the map of Assam using Geomatica GIS software.

Data collection

The CH_4 emission inventory from the paddy fields of all the districts of Assam was calculated by taking into consideration that the paddy area was under the high yielding variety(HYV). Information about paddy fields under different water management systems was obtained from the *Statistical Handbook of Assam 2003* [17]. District-wise irrigation potential utilized in Assam during the Kharif, Rabi, and pre-Kharif seasons in 2000 to 2001 was taken as the paddy field area under a continuous irrigation system. Paddy is the major crop in Assam, is grown in all three seasons of the year, and demands a huge amount of water. It was assumed that all the irrigation potentials were used only for the paddy cultivation. Information about the district-wise area under a rain-fed ecosystem was obtained by subtracting the area under HYV of paddy for the year 2000 to 2001, with that of the area under continuous irrigation. Emission factors for these water management practices were followed as per the IPCC report.

Estimation of N_2O emission from agricultural fields was done using the district-wise data of N-fertilizer use in the year 2001 to 2002 [17]. N_2O emission due to animal manure was calculated using livestock data from the 1997 livestock census [17]. During calculation, two factors that affect direct N_2O emission are not taken into consideration due to non-availability of data, namely, N_2O emission due to the N- fixed by the crops biologically and the amount of emission contributed by the burning of N-fixing and non-N fixing crop residue in the State (Tables 1 and 2). The choice of different base years for CH_4 and N_2O was made on the basis of data availability at that point of time when the study was conducted. Multiple years of data may provide better accuracy in the inventory. Considering the fact that this study is the first attempt towards a district-level inventory of the Assam state, even one year of data provides a reasonable idea about the GHG emission potentials at the district level and the importance of developing site-specific emission factors.

Methodology for the emission inventory
Methane
The district-wise inventory of CH_4 emission from Assam was calculated on the basis of the IPCC formula that was issued in the revised guideline in 1996 [19]. Also three different emission factors were used to estimate the CH_4 emission from the districts of Assam. This was to highlight the differences in the CH_4 emission and the need for a proper local-level emission inventory database for better planning and mitigation strategies in the future. The details of the four different factors that were used are as follows.

Emission factor as reported by Gupta et al
The seasonal integrated flux of 46 g m^{-2} was used for the calculation of CH_4 emission in this case. This

emission flux was from the Jorhat experimental farms in which the HYV *Mahsuri* cultivar of paddy was grown in the year 1991. The seasonal integrated flux was much higher than the other experimental stations in India due to the irrigated water regime, addition of organic amendment in the experimental field [32], and higher content of organic carbon in the soil of North Eastern states (about 5%) of India (about 1%) [8].

IPCC emission factor based on Bhatia et al
The IPCC emission factor for the base year 1994 to 1995 was calculated by dividing the methane emission of Assam as calculated in the study reported by Bhatia *et al.* using the IPCC emission factor. The emission factor was calculated as follows:

Emission factor in g m^{-2} = total emission from paddy field of Assam [8]/total area under paddy cultivation [17] = the factor comes out to be 12.81 g m^{-2}.

Indian emission factor based on Bhatia et al
The Indian emission factor for base year 1994 to 1995 was calculated by dividing the methane emission of Assam as calculated by Bhatia *et al.* using the Indian emission factor.

Emission factor in g m^{-2} = total emission from paddy field of Assam [8]/total area under paddy cultivation [17] = the factor comes out to be 6.92 g m^{-2}.

Emission factor used for this study as per the IPCC standard equation
The emission factor and details used in this study for calculation of district-wise CH_4 emission for Assam was as per the IPCC formula. The assumptions made in the calculation of CH_4 emission were as follows: 1) paddy field irrigated under continuous flooding was taken based on the district-wise irrigation potential for the Kharif, Rabi, and pre-Kharif seasons in the year 2000 to 2001 [17]. Paddy is the major crop and is cultivated three times a year, so it was assumed that most of the irrigation was used for it; 2) paddy field under a rain-fed, flood-prone condition was obtained by subtracting the area under continuous flooding from the total area under paddy cultivation (HYV variety) for each district [17]. The formula used for the calculation of CH_4 emission as per the IPCC guideline [19] is as follows:

$$Emission\left(Tgyr^{-1}\right) = \Sigma_i\Sigma_jEF_{ij} * 10^{-12} \qquad (1)$$

where i = irrigation under continuous flooding system, j = rain-fed flood-prone, EF_j (seasonally integrated emission factor for rain-fed flood-prone) = 8 g m^{-2}, and EF_i (seasonally integrated emission factor for irrigation under continuous flooding) = 10 g m^{-2} [8].

Table 1 Details of factors used for the assessment of direct nitrous oxide (N_2O) emission

Primary factor	Break-up of primary factor	Details	Coefficient or value IPCC emission factor	Indian emission factor	Remark
F_{SN} (annual amount of synthetic N-fertilizer applied to soil adjusted for the amount that volatilizes as NH_3 and NO_x)	N_{FERT}	The total amount of synthetic fertilizer consumed annually [18]	–	–	–
	$Frac_{GASF}$	Fraction of fertilizer volatilize as NH_3 and NO_x	10.0% [19]	15.0% [8,20]	The difference in the emission factor was due to soil management practices, soil type, pH, climatic condition and also the methodology used for emission assessment [19]. Details of differences in the emission factor due to different methodology used for assessment have been discussed in the Sarkar study [20].
F_{AM} (annual amount of animal manure nitrogen applied to soils adjusted to account for volatilization of NH_3 and NO_x)	T	Each defined livestock	–	–	For this study four categories were taken, namely, cattle, buffalo, sheep and goat, based on the details available in the Assam statistical handbook [17].
	$N_{(T)}$	Number of animals in each category [17]	–	–	
	$N_{ex\,(T)}$	Annual average nitrogen excretion rate per head for each livestock	Recommended to use country specific factors [19]	Indian emission factor for each livestock category [21]	$N_{ex\,(T)}$ in g yr^{-1} = (wet dung excreted by livestock in g day^{-1})*(dry matter of livestock)* (nitrogen constant of livestock)*365
	$Frac_{GASM}$	Fraction of N that volatilizes in NH_3 and NOx	20.0% [19]	15.0% [8]	
	$Frac_{FUEL}$	Animal manure burnt for fuel	52.5% [22]	52.5% [22]	IPCC manual suggested to national study or official statistics of country or region [19]
	$Frac_{PRP}$	Fraction of animal manure deposited on soil by grazing livestock	Not used in this study	Not used in this study	No data were available
	$Frac_{COLLEC}$	Loss during the collection of dung	30.0% [23]	30.0% [23]	–
	$Frac_{FEED}$	Fraction of animal manure used as feed	0.0%	0.0%	Taken as zero, as animal manure is hardly used as feed in India [8]
	$Frac_{CONST}$	Fraction of animal manure used in construction	2.0% [22]	2.0% [22]	–
F_{BN} (amount of nitrogen fixed annually by nitrogen fixing crops)	$Crop_{BF}$	Seed yield of nitrogen fixing crops	Not used in this study	Not used in this study	If seen in terms of area under nitrogen fixing crop in Assam (about 1.23 lakh hectare was under pulses in 2000 to 2001, against gross cropped area of 38.43 lakh hectares) then F_{BN} contribution to total N_2O emission may be negligible [17]. However, it is imperative that to have comprehensive source and sink of GHG emission from agriculture sector, which would help in developing better mitigation strategy and policy in the future. This study could not estimate the emission of F_{BN} due to non-availability of data at district level.
	$Frac_{NCRB}$	Nitrogen content of grain and straw of legumes	Not used in this study	Not used in this study	

Table 1 Details of factors used for the assessment of direct nitrous oxide (N_2O) emission (Continued)

Factor	Symbol	Description			Remarks
F_{CR} (amount of nitrogen in crop residues returned to soil annually)	$Crop_{ST}$	Amount of straw of non-nitrogen fixing crops incorporated to the soil as residue	Not used in this study	Not used in this study	The gross cropped area in Assam - other than paddy and pulses -under spices, horticulture, vegetable, wheat etcetera, was about 12.74 lakh hectares in 2000 to 2001 [24]. Since crop residue in India is mostly used as fodder or as burning fuel, it is likely that the contribution to N_2O emission would not be substantial. However, it is always warranted that if data are made available, the emission inventory would have to be developed in the future.
	$Frac_{NCRST}$	Nitrogen content of residue of non-nitrogen fixing crops	Not used in this study	Not used in this study	
	$Crop_{BF}$	Amount of straw of nitrogen fixing crops incorporated to the soil as residue	Not used in this study	Not used in this study	
	$Frac_{NCRBF}$	Nitrogen content of residue of nitrogen fixing crops	Not used in this study	Not used in this study	
EF_1 (kg N_2O-N kg^{-1} N input)	–	The emission factor for N_2O-N emitted from various nitrogen additions in soil	0.0125 [19]	0.007 [8, 25, 26]	The N_2O emission through nitrification and denitrification in the field, applied with nitrogen fertilizer are strongly influenced by soil temperature, moisture, pH, and soluble organic matter availability [27]. It is to be noted that the IPCC emission factor is taken from the studies of Klemedtssoner al. [28] and Clayton et al. [27], as referenced in the IPCC manual [19]. These studies were done in Europe's peatland and clay loam grassland soil respectively. The Indian factor is based on the studies of Kumaret al. [25], Majumdar et al. [29] and Pathak et al. [26] which were done in India. It is to be noted that in European and Indian conditions the above-mentioned factors that influence the N_2O emission from soil differ markedly, which led to the differing values of emission factors.
F_{OS}	–	Area of organic soil harvested	Not used in this study	Not used in this study	Not an application for Indian conditions, as the organic content in Indian soil varies only from 1% to 5%, while organic soils are those having 12% to 18% organic carbon [8].
EF_2	–	Percent of N_2O emissions from organic soil	Not used in this study	Not used in this study	–

IPCC, Intergovernmental Panel on Climate Change; GHG, greenhouse gas.

Table 2 Details of coefficients used for the assessment of indirect nitrous oxide (N$_2$O) emission

Primary factor	Break-up of primary factor	Details	Coefficient or value		Remark
			IPCC emission factor	Indian emission factor	
N$_2$O$_{(G)}$ (N$_2$O emission from volatilization of applied nitrogen fertilizer and animal manure and its subsequent atmospheric deposition as NO$_x$ and NH$_4$)	N$_{FERT}$	The total amount of synthetic fertilizer consumed annually [17]	–	–	–
	The value of Frac$_{GASF}$, T, N$_{(T)}$, Nex$_{(T)}$, and Frac$_{GASM}$ were same as in Table 1.	–	–	–	–
	EF$_4$ [kg N$_2$O-N kg^{-1} NH$_4$-N and NO$_x$-N deposited]	Emission factor for N$_2$O emission from atmospheric NH$_3$ and NO$_x$	0.01 [19]	0.005 [8]	–
N$_2$O$_{(L)}$ (N$_2$O produced from leaching and runoff of applied nitrogen fertilizer and animal manure)	N$_{FERT}$	The total amount of synthetic fertilizer consumed annually [17]	–	–	–
	The value of T, N$_{(T)}$, and Nex$_{(T)}$, are same as in table 1.	–	–	–	–
	Frac$_{FUEL-AM}$	Animal manure burnt for fuel	52.5% [22]	52.5% [22]	IPCC manual suggests using official statistics of the nation or expert survey [19].
	Frac$_{PRP-AM}$	Fraction of animal manure that is deposited on to the soil by grazing animal	Not used in this study	Not used in this study	No data available.
	Frac$_{COLLEC}$	Loss during the collection of dung	30.0% [23]	30.0% [23]	–
	Frac$_{FEED-AM}$	Fraction of animal manure used as feed	0.0%	0.0%	Taken as zero, as animal manure is hardly used as feed in India [8].
	Frac$_{CONST-AM}$	Fraction of animal manure used in construction	2.0% [22]	2.0% [22]	IPCC manual suggests use of official statistics of the nation or expert survey [19].
	Frac$_{LEACH}$	Fraction of nitrogen lost through leaching	30.0% [19]	10.0% [8]	IPCC manual uses default value of 30% for Frac$_{LEACH}$. This default value was largely based on mass balance studies comparing agricultural N inputs to N recovered in rivers. The IPCC manual suggests the N that is deposited away from agricultural land, a lower value of Frac$_{LEACH}$ may be more appropriate based on regional or national studies [19]. Bhatia et al. [8] used an Indian emission factor of 10% based on the studies by Singh et al. [30] and Patel et al. [31].
	EF$_5$ [kg N$_2$O-N kg^{-1} leached and run off]	The emission factor for depositing N from leaching and run-off.	0.025 [19]	0.005 [8]	–

IPCC, Intergovernmental Panel on Climate Change.

Nitrous oxide

The district-wise emission inventory of N_2O from Assam was based on the formula given by the 1996 revised IPCC guideline [19]. It includes N_2O emitted as a result of the anthropogenic N-fertilizer input through the direct pathway of nitrification and denitrification from soil and also through indirect pathways, that include volatilisation losses, leaching and runoff from applied N-fertilizer, animal manure etcetera. Thus, the emission of N_2O was calculated in two steps, namely, direct N_2O emission from agricultural soil (net sown area) and indirect N_2O emission from agricultural soil (net sown area) as follows:

$$N_2O_{Total} = N_2O_{direct} - N + N_2O_{Indirect} - N \qquad (2)$$

Direct N_2O emission

The following equation (2.1) was used to estimate the direct emission of N_2O from the agricultural field (for details about IPCC and Indian factors used in this study see Table 1).

$$N_2O_{direct} - N = \{(F_{SN} + F_{AM} + F_{BN} + F_{CR}) * EF_1\} + (F_{OS} * EF_2) \qquad (2.1)$$

where EF_1 is percentage N_2O emission from the applied fertilizer, F_{OS} is the area of organic soil harvested, and EF_2 is percentage N_2O emitted from the organic soil.

F_{SN} denotes the annual amount of synthetic N- fertilizer applied to soil, adjusted to account for the amount that volatilizes as NH_3 and NO_x:

$$F_{SN} = N_{FERT} * (1 - Frac_{GASF}) \qquad (2.1.1)$$

where N_{FERT} denotes the total amount of synthetic fertilizer consumed annually and $Frac_{GASF}$ is the fraction of fertilizer that volatilizes as NH_3 and NO_x.

F_{AM} denotes the annual amount of animal manure N applied to soils, adjusted to account for the volatilization as NH_3 and NO_x:

$$F_{AM} = \Sigma_T (N_T * N_{ex(T)}) * (1 - Frac_{GASM})$$
$$* [1 - (Frac_{FUEL} + Frac_{PRP} + Frac_{COLLEC}$$
$$+ Frac_{FEED} + Frac_{CONST})] \qquad (2.1.2)$$

where T stands for each defined livestock category/species (in this study, four categories of livestock, namely, cattle, buffalo, sheep and goat have been taken), N_T is the number of animals in each category, $N_{ex(T)}$ is the annual average nitrogen excretion rate per head for each livestock category, $Frac_{GASM}$ is the fraction of N that volatilizes as NH_3 and NO_x, $Frac_{FUEL}$ denotes animal manure that is burnt for fuel, $Frac_{PRP}$ is the fraction of animal manure deposited on soil by grazing livestock, $Frac_{CONST}$ is the fraction of animal manure used as

construction, $Frac_{FEED}$ is the fraction of animal manure used as feed, $Frac_{-COLLEC}$ is the loss during collection of dung.

F_{BN} is the amount of N- fixed annually by N- fixing crop as:

$$F_{BN} = Crop_{BF} * Frac_{NCRBF} \qquad (2.1.3)$$

where $Crop_{BF}$ is the seed yield of N-fixing crops. Four crops, that is, gram, arhar, groundnut and soybean were taken into account for the calculation, and $Frac_{NCRBF}$ is the N content of grain and straw of legumes.

F_{CR} is the amount of N in crop residue returned to soil annually:

$$F_{CR} = (Crop_{ST} * Frac_{NCRST} + Crop_{SBF} * Frac_{NCRSBF}) \qquad (2.1.4)$$

Where, $Crop_{ST}$ is the amount of straw of non-N fixing crops incorporated into soil as residue, $Frac_{NCRST}$ is the N content of residue of non-N fixing crops, $Crop_{SBF}$ is the amount of straw of N-fixing crops incorporated to the soil as residue and $Frac_{NCRSBF}$ is the N content of residue of N-fixing crop.

Due to non-availability of data, F_{BN} and F_{CR} are not included in the calculation of N_2O-N $_{Direct}$ emission (see Table 1 for details and a note on its potential contribution to the emission inventory for N_2O).

Indirect N_2O emission

The following equation (2.2) was used for the calculation of indirect emission of N_2O ($N_2O_{indirect}$) from the agricultural fields as per the IPCC guideline (for details about IPCC and Indian factors used, see Table 2):

$$N_2O_{Indirect} = N_2O_{(G)} + N_2O_{(L)} \qquad (2.2)$$

where $N_2O_{(G)}$ is the N_2O produced from volatilization of applied N-fertilizer and animal manure and its subsequent atmospheric deposition as NO_x and NH_4. This is calculated by the formula (2.2.1) as below:

$$N_2O_{(G)} = [(N_{FERT} * Frac_{GASF}) + (\Sigma_T (N_{(T)} * N_{ex(T)} * Frac_{GASM})] * EF_4 \qquad (2.2.1)$$

where, N_{FERT} is the amount of fertilizer consumed annually, $Frac_{GASF}$ is the fraction of fertilizer that volatilizes as NH_3 and NO_x, $\sum_T (N_{(T)} * N_{ex(T)}$ is the amount of N in animal manure excreted annually, T is each defined livestock category, N_T is the number of animals in each category, $N_{ex(T)}$ is the annual N excretion rate per head for each livestock category and EF_4 is the emission factor for N_2O emission from atmospheric NH_3 and NO_x.

Table 3 Details of livestock categories and characteristic [21] used to calculate the nitrogen excretion rate per head ($N_{ex\ (T)}$)

Livestock characteristics	Livestock category			
	Cattle	Buffalo	Sheep	Goat
Wet dung excreted by cattle (kg per day)	8.335	10.380	1.430	0.625
Body weight (kg)	350	350	50 to 60	40
Urine (litres per day)	12.960	6.810	0.950	0.498
Dry matter content(%)	18	18	32	32
Nitrogen content (oven-dry,%)	1	1	1.87	1.87

N_2O (L) is N_2O produced from leaching and the runoff of the applied fertilizer and animal manure and calculated through the following formula (2.2.2):

$$N_2O_{(L)} = [N_{FERT} + \{\Sigma_T(N_{(T)} * N_{ex(T)}$$
$$* [1 - Frac_{FUEL-AM} + Frac_{PRP-AM}$$
$$+ Frac_{COLLEC} + Frac_{FEED-AM} + Frac_{CONST-M})]\}]$$
$$* Frac_{LEACH} * EF_5 \qquad (2.2.2)$$

where $Frac_{FUEL-AM}$ denotes animal manure that is burnt for fuel, $Frac_{PRP-AM}$ is the fraction of animal manure that is deposited onto the soil by grazing livestock, $Frac_{COLLEC}$ is the loss of dung during collection,

$Frac_{FEED-AM}$ is the fraction of animal manure that is being fed, $Frac_{CONST-AM}$ is the fraction of animal manure that is used as construction, $Frac_{LEACH}$ is the fraction of N-lost through leaching and EF_5 is the emission factor for deposited N from leaching and runoff.

The IPCC and Indian standard emission factor values were used to calculate the emission of N_2O-N_{Total} for Assam (see Table 1 and 2 for details). For the calculation of $N_{ext(T)}$ (in the F_{AM}), the annual average nitrogen excretion rate per head for each livestock category was calculated. The livestock categories used were cattle, buffalo, sheep and goat (see Table 3 for details). The values of annual average nitrogen excretion rate per head ($N_{ext(T)}$) for each livestock category was the same

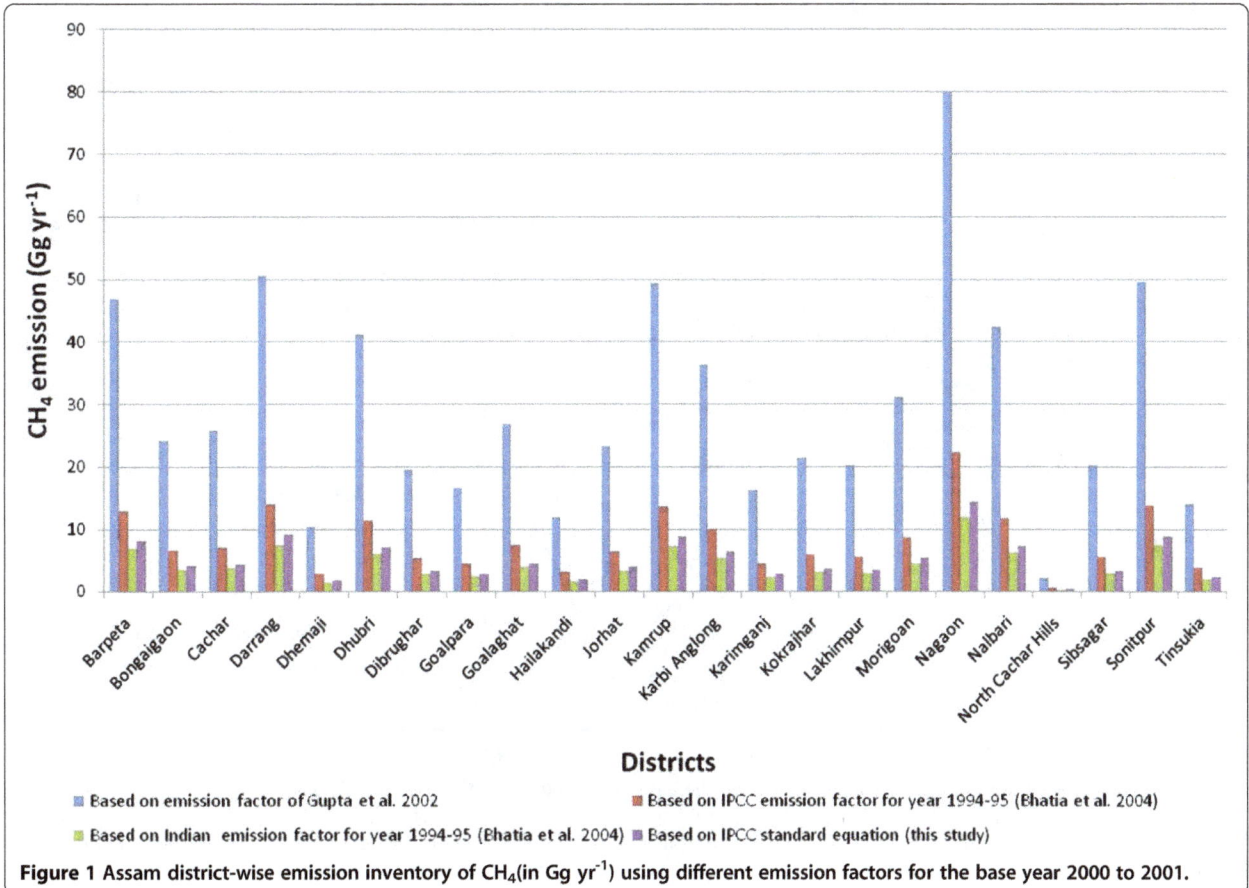

Figure 1 Assam district-wise emission inventory of CH_4(in Gg yr^{-1}) using different emission factors for the base year 2000 to 2001.

when calculating the N_2O emission as per IPCC and the Indian standard, since the IPCC guideline suggests using national or expert studies as the reference [19].

Results and discussion
Methane

The district wise emission of CH_4 from the rice paddy fields of Assam (as per the IPCC standard formula) was estimated to be 121 Gg for the base year 2000 to 2001. Nagaon district emitted the highest amount of CH_4 (14 Gg), followed by Darrang (9 Gg) and Sonitpur (8 Gg). The lowest amount of CH_4 emission was from the North-Cachar Hills district (0.47), which had lowest area under paddy cultivation. The CH_4 emission inventory of Assam State estimated from this study was compared with emission values obtained using three other emission factors (see Figure 1). CH_4 emission was highest (682 Gg) based on the 46 g m^{-2} emission factor of Gupta et $al.$ [33], followed by an emission of 190 Gg based on the 12.81 g m^{-2} emission factor of Bhatia et $al.$ [8], based on IPCC coefficients. The CH_4 emission of 102 Gg was lowest when the 6.92 g m^{-2} emission factor of Bhatia et $al.$ [8] was used, which was based on the

Indian emission factors. The CH_4 emission estimate based on our study comes out about 121 Gg for the base year 2000 to 2001.

It is interesting to note that while most of central plain districts of Assam along the banks of Brahmaputra had CH_4 emission in the range of 5 to 15 Gg, Nagaon district produced a much higher amount of CH_4 (see Figure 2). Nagaon district had a greaterarea under rice and also produced more rice in comparison to other central plain districts and thus the higher CH_4 emission estimation is very much in line with the anticipated results. It is a matter of concern for scientists, policy makers and planners when there is great difference between the highest (682 Gg) and lowest (102 Gg) values of CH_4 emission for a state. Discrepancies of such nature call for an immediate inventory of GHGs at local level for better mitigation planning in future. This variability calls for further strengthening of methodologies in the emission inventory of CH_4 from paddy fields at the micro level, to obtain better estimates.

Correlation analysis was carried out between N-fertilizer application and CH_4 emission values obtained using the Indian emission factor for the base year

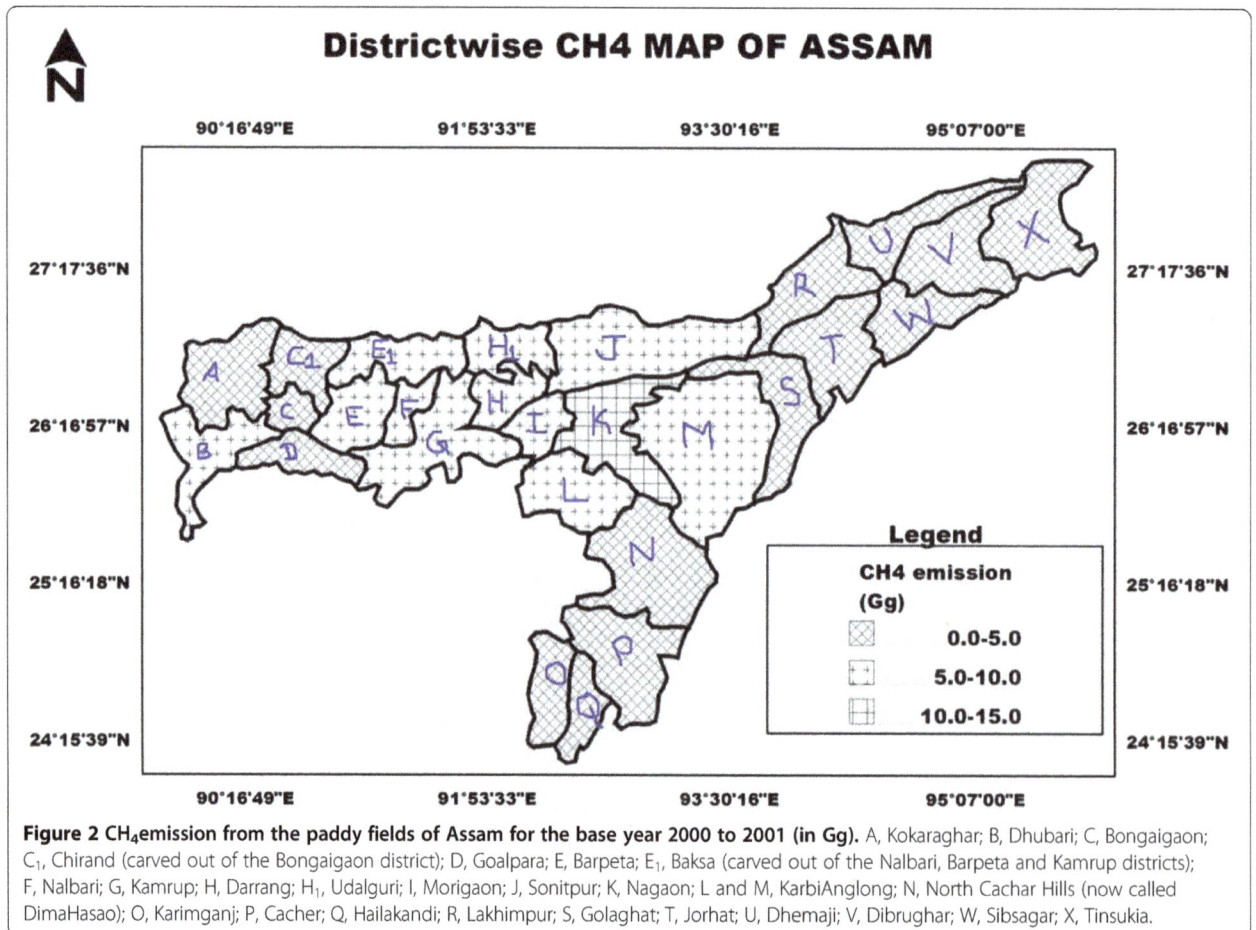

Figure 2 CH_4 emission from the paddy fields of Assam for the base year 2000 to 2001 (in Gg). A, Kokaraghar; B, Dhubari; C, Bongaigaon; C_1, Chirand (carved out of the Bongaigaon district); D, Goalpara; E, Barpeta; E_1, Baksa (carved out of the Nalbari, Barpeta and Kamrup districts); F, Nalbari; G, Kamrup; H, Darrang; H_1, Udalguri; I, Morigaon; J, Sonitpur; K, Nagaon; L and M, KarbiAnglong; N, North Cachar Hills (now called DimaHasao); O, Karimganj; P, Cacher; Q, Hailakandi; R, Lakhimpur; S, Golaghat; T, Jorhat; U, Dhemaji; V, Dibrughar; W, Sibsagar; X, Tinsukia.

2000 to 2001. A significant relationship ($r = 0.84$) was observed with 95% confidence. However, this correlation needs to be field-tested in Indian conditions, as there are contradicting claims about nitrogen fertilizer application and its impact on increasing [34] or decreasing [35] the CH_4 emission from paddy fields.

Nitrous oxide

Emission of N_2O-N using the IPCC standard was estimated at 1.36 Gg for the base year 2001 to 2002, while the emission estimated using the Indian standard was 0.43 Gg (Figure 3). This difference of about 216% between the lowest and highest values of the N_2O emission inventory was due to the different emission factors of the IPCC and the Indian standards used for the calculation of direct and indirect N_2O emission (see Table 1 and 2 for details). This variation in N_2O emission emphasises the need for micro-level planning for future data gathering and inventory management to have an accurate and dependable database of GHG emissions from agricultural fields. This may lead to better agriculture management and mitigation planning in future. For N_2O also, the central plain districts of Assam emissions were much higher than for rest of the state (0.04 Gg to 0.18 Gg; see Figure 4). Unlike CH_4, in the case of N_2O, the different ranges of emission are widely spread over the state.

Correlation analysis was performed to find out the relationships between the N_2O emissions and application of N-fertilizer. Correlation analysis was carried out for the emission values obtained from the IPCC and Indian standard for the year 2001 to 2002. It was found that there is a significant correlation ($r = 0.98$ at 95% confidence level) between N_2O emission (Indian standard) with that of N-fertilizer used in that year. Similar strong correlation ($r = 0.98$ at 95% confidence level) was found between N_2O emission (using an IPCC emission factor) and N-fertilizer used in that base year. These significant correlations reinstate the earlier findings that N-fertilizers are the major sources of N_2O emitted from agricultural soils.

Conclusions

District-wise emission of CH_4 (for the base year 2000 to 2001) from the paddy fields of Assam is estimated to be 121 Gg based on this study. However, there is a large difference in the highest (682 Gg) and lowest (102 Gg) value of CH_4 emission when different emission factors were used. The district-wise N_2O emission (for the base year 2001 to 2002) for Assam State was estimated to be 1.36 Gg and 0.43 Gg using the IPCC and Indian factors respectively.

The study clearly shows that there is large variation in the emission inventory of CH_4 and N_2O at the district level when different emission factors are used. This

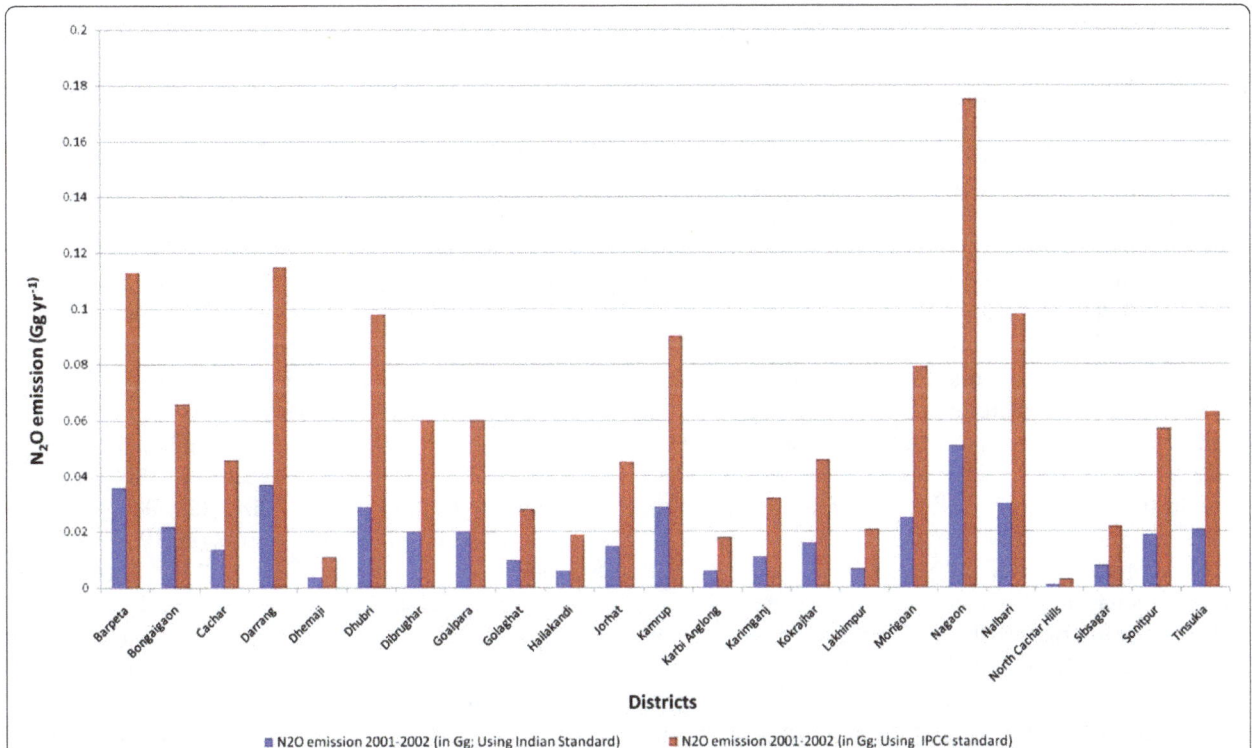

Figure 3 Assam district-wise emission of nitrous oxide (N_2O) in Gg yr^{-1} using different emission factors for the base year 2001 to 2002.

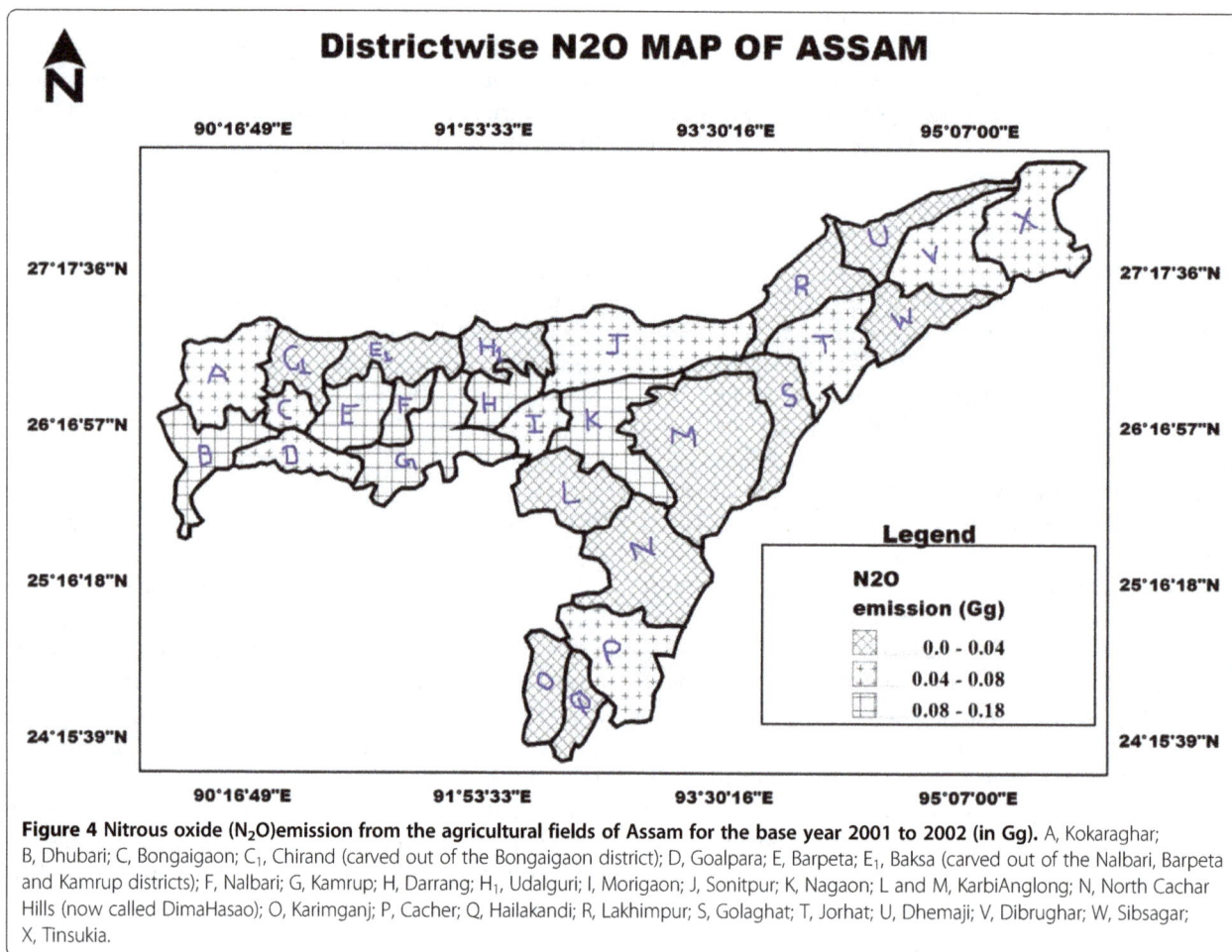

Figure 4 Nitrous oxide (N₂O)emission from the agricultural fields of Assam for the base year 2001 to 2002 (in Gg). A, Kokaraghar; B, Dhubari; C, Bongaigaon; C₁, Chirand (carved out of the Bongaigaon district); D, Goalpara; E, Barpeta; E₁, Baksa (carved out of the Nalbari, Barpeta and Kamrup districts); F, Nalbari; G, Kamrup; H, Darrang; H₁, Udalguri; I, Morigaon; J, Sonitpur; K, Nagaon; L and M, KarbiAnglong; N, North Cachar Hills (now called DimaHasao); O, Karimganj; P, Cacher; Q, Hailakandi; R, Lakhimpur; S, Golaghat; T, Jorhat; U, Dhemaji; V, Dibrughar; W, Sibsagar; X, Tinsukia.

indicates that for an accurate inventory of CH_4 and N_2O, universal or regional emission factors cannot be applied at the smaller scale (for example,at district level). The emission of CH_4 and N_2O from the agricultural sector is a function of specific crop and site, and is influenced by factors such as cultural practice, agronomic management, soil type and socio-economic drivers. This calls for detailed and comprehensive mapping and data collection at the micro-level for accurate inventory of CH_4 and N_2O in the future. This is pertinent in the present context when more emphasis is being given to the localisation of issues to avert the negative effect of climate change, and the bottom-up approach to planning. In the Indian context, this is even more important as power and planning are being devolved more at the bottom-most level of governance, namely, the Panchayats and District Planning Committee as per the Indian Constitution (article 243ZD[3-a]) on the issue of environmental planning and conservation [36]. So, it is important to have strong and able local-level planning to mitigate the CH_4 and N_2O emission in the future, on the basis of sound estimates of these gases. This requires little extra effort from various stakeholders to raise the level of awareness among researchers, policy makers and farmers by providing them adequate training and sensitisation to local issues.

Endnotes

[a]Non-annex countries are a group of 152 countries classified by the IPCC, belonging to the low-income group, with very few classified as a middle-income group. For detail see <http://unfccc.int/parties_and_ observers/parties/non_annex_i/items/2833.php>.

[b]Since 1950, the area under cultivation in India had increased by only 27.43 times (96.6 million hectares in 1951 to 1952 to 123.1 million hectares in 1999 to 2000), while the application of N-fertilizer (55, 000 tonnes in 1950 to 1951 to 11,592,500 tonnes in 1999 to 2000) increased by about 20,977 times [18,37,38]. This increasing use of N-fertilizer on the limited land is not only posing a threat in term of N_2O emission causing global warming but also reducing N-use efficiency of the plant.

[c]The emission of CH_4 and N_2O was calculated for different base years, namely, 2000 to 2001 and 2001 to 2002 respectively. This was due to the availability of relevant data for the calculations.

[d]When the study was carried out in 2004, the data were available only for 23 districts of Assam state. In 2012, Assam has total of 27 districts. For this study we have done the calculation for 23 districts. While representing the emission on the GIS map we mentioned the name of the new districts and also the name of the parent district from which the new district was carved out.

[e]In the year 2000 to 2001, of the total gross cropped area of 38.43 lakh hectares in Assam, about 26.46 lakh hectares was under paddy (autumn, winter and summer paddy cultivation) cultivation [24,39].

Abbreviations

CFCs: Chlorofluorocarbon; CH_4: Methane; CO_2: Carbon dioxide; GDP: Gross domestic product; Gg: Gigagram; GHG: Greenhouse gas; GIS: Geographic information systems; GoI: Government of India; GWP: Global warming potential; HYV: High yielding variety; IPCC: Intergovernmental Panel on Climate Change; N_2O: Nitrous oxide; Tg: Terragram.

Competing interests

The authors declare that they have no competing interests.

Authors' contributions

SNM and SM conceived the idea of study and participated in its initial design, data collection, data analysis, and initial drafting of the research work. LR, SD and PS participated in further data analysis and drafting the final manuscript. All authors read and approved the final manuscript.

Author details

[1]Institute of Rural Management, Post Box No. 60, Anand 388001Gujarat, India. [2]School of Environmental Sciences, Jawaharlal Nehru University, New Delhi 110067, India. [3]Indian Institute of Technology-Guwahati, Assam 781039, India.

References

1. Metz B, Davidson OR, Bosch PR, Dave R, Meyer LA: (Eds): *Climate change - contribution of working group III to the fourth assessment report of the intergovernmental panel on climate change*. Cambridge: Cambridge University Press; 2007.
2. Watson RT, Zinyowera MC, Moss RH: **Climate change 1995- impacts, adaptations and mitigation of climate change. Scientific-technical analyses**. In *Contribution of working group II to the second assessment report of the Intergovernmental Panel on Climate Change*. Cambridge: Cambridge University Press; 1996:289–324.
3. Battle M, Bender M, Sowers T, Tans PP, Butler JH, Elkins JW, Ellis JT, Conway T, Zhang N, Lang P, Clarke AD: **Atmospheric gas concentrations over the past 20 century measured in air from firn at the south pole**. *Nature* 1996, **383**:231–235.
4. United States Environmental Protection Agency (USEPA): *Global Anthropogenic Non-CO_2Greenhouse Gas Emissions: 1990-2020*. Washington DC: 2006. http://www.epa.gov/nonco2/econ-inv/downloads/GlobalAnthroEmissionsReport.pdf.
5. Huang Y, Wang H, Huang H, Feng ZW, Yang ZH, Luo YC: **Characteristic of methane emission from wetland rice-duck complex ecosystem**. *Agriculture, Ecosystems and Environment* 2005, **105**:181–193.
6. Ahrens TD, Beman JM, Harrison JA, Jewett PK, Matson PA: **A synthesis of nitrogen transformations and transfers from land to the *sea in the Yaqui Valley agricultural region* of northwest Mexico**. *Water Resources Research* 2008, **44**:W00A05. doi:10.1029/2007WR006661.
7. Government of India: *Interim report of the expert group on low carbon strategies for inclusive growth*. New Delhi: 2011.
8. Bhatia A, Pathak H, Aggrawal PK: **Inventory of methane and nitrous oxide emission from agricultural soils of India and their global warming potential**. *Current Science* 2004, **87**:317–324.
9. Sinha SK: **Global methane emission from rice paddies: excellent methodology but poor extrapolation**. *Current Science* 1995, **68**:643–646.
10. Parashar, *et al*: **Methane budget from paddy fields in India**. *Chemosphere* 1996, **33**:737–757.
11. Jain MC, Kumar S, Wassmann R, Mitra S, Jain MC, Singh SD, Singh JP, Singh R, Yadav AK, Gupta S: **Methane emissions from irrigated rice fields in northern India (New Delhi)**. *Nutrient Cycling in Agroecosystems* 2000, **58**:75–83.
12. Khosa MK, Sidhu BS, Benbi DK: **Methane emission from rice fields in relation to management of irrigation water**. *Journal of Environmental Biology* 2011, **32**:169–172.
13. Government of India (GoI): *National Action Plan on Climate Change*. New Delhi: 2008.
14. Sharma SK, Choudhury A, Sarkar P, Biswas S, Singh A, Dadhich PK, Singh AK, Majumdar S, Bhatia A, Mohini M, Kumar R, Jha CS, Murthy MSR, Ravindranath NH, Bhattacharya JK, Karthik M, Bhattacharya S, Chauhan R: **Greenhouse gas inventory estimates for India**. *Current Science* 2011, **101**:405–415.
15. Ghosh S, Majumdar D, Jain MC: **Methane and nitrous oxide emissions from irrigated rice of North India**. *Chemosphere* 2003, **51**:181–195.
16. Gogoi B, Baruah KK: **Nitrous oxide emission from tea (Camellia sinensis (L.) O. kuntze)- planted soils of North East India and soil parameters associated with the emission**. *Current Science* 2011, **101**:531–535.
17. Government of Assam (GoA): *Statistical Handbook-Assam, Directorate of Economics and statistics*. Guwahati: 2003.
18. Centre for Monitoring Indian Economy (CMIE): *Agriculture Production in India - State wise and Crop wise data*. Mumbai: 1988:19–26.
19. Inter-Governmental Panel on Climate Change (IPCC): *Good Practice Guidance and Uncertainty Management in National Greenhouse Gas Inventories*. New York: Cambridge University Press; 1996.
20. Sarkar MC, Banerjee NK, Rana DS, Uppal KS: **Field measurements of ammonia volatilization losses of N from Urea applied to wheat**. *Fertilizer News* 1991, **36**:25–29.
21. Gaur AC, Neelakanthan S, Dargan KS: *Organic Manures*. New Delhi: Indian Agriculture Research Institute; 1984.
22. Tandon HLS: *Fertilizer, Organic manures, Recyclable wastes and Biofertilizers*. New Delhi: FDCO; 1992.
23. TERI: *The Energy Research Institute: Energy Data Directory and year Book 2000–01*. New Delhi: TERI Press.
24. Assam Small Farmers' Agri-Business Consortium: *Assam agriculture at glance*. Guwahati: 2005. http://assamagribusiness.nic.in/ASAGRI.pdf.
25. Kumar U, Jain MC, Kumar S, Pathak H, Majumdar D: **Effects of moisture levels and nitrification inhibitors on N_2O emission from a fertilized clay loam soil**. *Current Science* 2000, **79**:224–228.
26. Pathak H, Bhatia A, Prasad S, Jain MC, Kumar S, Singh S, Kumar U: **Emission of nitrous oxide from soil in rice-wheat systems of Indo-Gangetic plains of India**. *Environ. Monit. Assess* 2002, **77**:163–178.
27. Clayton H, McTaggart IP, Parker J, Swan L, Smith KA: **Nitrous oxide emissions from fertilised grassland: a 2-year study of the effects of fertiliser form and environmental conditions**. *Biology and Fertility of Soils* 1997, **25**:252–260.
28. Klemedtsson L, Klemedtsson AK, Escala M, Kulmala A: **Inventory of N_2O emission from farmed European peatlands**. In *Approaches to Greenhouse Gas Inventories of Biogenic Sources in Agriculture*. Edited by Freibauer A, Kaltschmitt M. Lökeberg: Sweden: Proceedings of the Workshop at Lökeberg; 1998:79–91.
29. Majumdar D, Kumar S, Pathak H, Jain MC, Kumar U: **Reducing nitrous oxide emission from rice field with nitrification inhibitors**. *Agric. Ecosyst. Environ* 2000, **81**:163–169.
30. Singh B, Singh Y, Khind CS, Meelu OP: **Leaching losses of urea-N applied to permeable soils under different hydrological situations**. *Fertilizer Research* 1991, **28**:179–184.
31. Patel SK, Pamda B, Mohanty SK: **Relative ammonia loss from urea based fertilizers applied to rice under different hydrological situations**. *Fertilizer Research* 1989, **19**:113–119.
32. Bhattacharya S, Gupta PK, Parashar DC: *Greenhouse gas emission in India: methane budget estimates from rice fields based on data available upto 1995* Centre for Global Change, National Physical Laboratory, Scientific Report No. 21; 1996.
33. Gupta PK, Sharma C, Mitra AP: *Methane measurements from rice fields in India*. New Delhi: Centre on Global Change, National Physical Laboratory; 2002.
34. Lindau C, Bollich P, Delaune R, Patrick W, Law V: **Effect of urea fertilizer and environmental factors on methane emission from a Louisiana USA rice field**. *Plant Soil* 1991, **136**: 195–203.

35. Zou J, Huang Y, Jiang J, Zheng X, Sass RL: **A 3-year field measurement of methane and nitrous oxide emissions from rice paddies in China: effects of water regime, crop residue, and fertilizer application.** *Global Biogeochem. Cycles* 2005, **19**GB2021. doi:10.1029/2004GB002401.
36. Kashyap SC: *Our constitution – an introduction to India' constitution and constitutional law*. New Delhi: National Book Trust; 2001.
37. Centre for Monitoring Indian Economy (CMIE): *Agriculture*. Mumbai; 2004.
38. The Fertilizer Association of India (FAI): *Fertilizer Statistics*. New Delhi; 2000.
39. Barah BC, Betne R, Bhowmick BC: **Status of rice production system in Assam: a research perspective**. In *Prioritization of strategies for agriculture development in north-eastern India: October 2001; New Delhi*. Edited by Barah BC. New Delhi: National centre for Agriculture Economics and Policy Research; October 2001:50–68.

7

Underutilized wild edible plants in the Chilga District, northwestern Ethiopia: focus on wild woody plants

Mekuanent Tebkew[1*], Zebene Asfaw[2] and Solomon Zewudie[3]

Abstract

Background: Ethiopia encompasses an extraordinary number of ecological zones and plant diversity. However, the diversity of plants is highly threatened due to lack of institutional capacity, population pressure, land degradation and deforestation. An adequate documentation of these plants also has not been conducted. The farmers in Ethiopia face serious and growing food insecurity caused by drought, land degradation and climate change. Thus, rural communities are dependent on underutilized wild edible plants to meet their food and nutritional needs. Hence, this study was conducted to examine the distribution, diversity, role, management condition and associated traditional knowledge of underutilized wild edible plants with a focus on woody plants in the Chilga District, northwestern Ethiopia.

Methods: A questionnaire survey, semi-structured interviews, preference and direct matrix rankings, a market survey and focused group discussion methods were employed for data collection. Data were collected from 96 respondents. A plant inventory was also conducted on 144 quadrates in two agroecologies and in three uses. Both quantitative and qualitative data analysis methods were used. Statistical Analysis System (SAS) version 9.0 was used for statistical analysis. Analysis of Variance ($P < 0.05$) was used to compare diversity indices and species richness between agroecologies and among kebeles.

Results: Thirty-three underutilized wild edible plants were recorded in the study area. Of the recorded plants, 45% were trees. Fruits (76%) were the most frequently used plant parts. More than half of the respondents (56.3% in the midland and 66.7% in the lowland area) consumed underutilized wild edible plants for supplementing staple food. Underutilized wild edible plant citation of the poor was significantly higher ($P < 0.05$) than medium and rich classes. Underutilized wild edible plants in the study area were threatened by agricultural expansion, overharvesting for fuel wood and construction, and by overgrazing. However, these plants have been given minimum conservation attention.

Conclusions: Thirty-three underutilized wild edible plants were recorded in the study area. The community consumes underutilized wild edible plants for supplementing staple food, filling food gaps and for recreation. The local community applies only some management practices to some wild edible plants. Therefore, special management is needed to sustain the benefits of these plants.

Keywords: farmlands, herbaceous species, natural forest, riverine forests, woody species

* Correspondence: mekuannttebkew@yahoo.com
[1]Department of Natural Resource Management, University of Gondar,
P.O. Box 196, Gondar, Ethiopia
Full list of author information is available at the end of the article

Introduction

Nearly 40,000 to 100,000 plant species have been regularly used for food, shelter and medicines in the world [1]. However, only a small number of plants are widely used. The remaining plant diversity is underutilized [1,2]. Underutilized plants contribute immensely to family food security and serve as means of survival during times of drought, famine, shocks and risks [3]. They can also supplement nutritional requirements due to their better nutritional value [4,5].

Ethiopia encompasses an extraordinary number of ecological zones [6] and plant flora [3,7-9]. Currently, however, the biodiversity of Ethiopia faces various threats. The main threats are government institutional capacity, population growth, land degradation, deforestation and weak management [6,10]. Similarly, in the northwestern region of Ethiopia, which is endowed with sub-humid, dry and lowland areas, plenty of wild and semi-wild edible plants are present. However, the diversity of underutilized wild edible woody plants is not well known.

The lives of farmers in Ethiopia face several challenges including deforestation, drought, land degradation, and climate change [11], all of which contribute to serious food insecurity among households [12]. Thus, in most cases, rural communities depend on underutilized edible plants [13] due to their easily accessibility [5]. Understanding underutilized wild edible woody plants (UWEP) along with their threats forms a base for local decision-making and helps with the application of appropriate management. In addition, it assists us with the selection of a species that can adapt in different land use systems. Nevertheless, little documentation has been conducted in Ethiopia (that is, only 5% of Ethiopian districts have been studied) [13]. Moreover, information is lacking on the ecological distribution and potential of UWEPs for agroforestry practices and for maintaining food security in other districts of Ethiopia including the Chilga district.

The term 'underutilized wild edible plants' in this study refers to all wild edible plants (a) that are locally abundant, but globally rare; (b) that are undervalued, that is, their current public and private value is below its potential; and (c) for which there is a lack of research and information. Therefore, this study will (1) identify the distribution and quantify the diversity of UWEPs in different land uses, kebeles and agroecologies; (2) determine the role of UWEPs in livelihood diversification; (3) document the knowledge associated with utilization and management of UWEPs; and (4) identify the factors that threaten future uses and conservations of UWEPs in the Chilga District.

Materials and methods
Site description
The study was conducted in the Chilga district, North Gondar Zone of the Amhara Regional State. The Chilga District is located 12˚55" N and 37˚06" E (Figure 1). It comprises 43 administrative kebeles (KA), which are the lowest administrative units next to districts. The altitude of the district ranges from 900 to 2,267 meters above sea level (m.a.s.l.) There are two agroecologies: midland (1,500 to 2,267 m.a.s.l) and lowland (900 to 1,500 m.a.s.l). About 33% of the area in the district is midland, while 67% is lowland agroecology. There are rivers and streams traversing the district, and these often serve as sources of water for the population (CDOA, 2012). The major soil covers of the Chilga district are 45% Cambisols, 40% Vertisols, and 15% Nitosols [14]. The natural vegetation of Chilga is mainly composed of various lowland and midland species [14]. The temperature of the district ranges from 11 to 32°C with a mean annual rainfall between 995 to 1,175 mm. It had a total population of 241,712 and a total area of 3,181 km^2 The livelihood of the local people mainly based on subsistence mixed agriculture (crop-livestock production).

Selection of study sites
The study was conducted in four kebeles of the Chilga District from 8 October to 20 December 2012. District and kebele experts were contacted for general information. In addition, secondary archived materials were reviewed from CDOA to get further information.

The district contains midland and lowland agroecologies that are classified based on altitude, annual rainfall and temperature differences. Moreover, the sociodemographic and biophysical characteristics of the two agroecologies are not the same. Thus, based on accessibility for data collection and availability of UWEPs, two KAs from each agroecology and two villages from each KA were selected (see Table 1).

Selection of key informants and households
For this study, key informants (KI) are defined as knowledgeable persons about underutilized wild edible woody plants and local conditions. After selecting two villages at each KA, 24 KI (three from each village) were selected by using a snowball method to collect preliminary data and for questionnaire development following the method of Bernard [15]. Three farmers were randomly asked to call five knowledgeable persons in the village. Then, the three most knowledgeable KI were selected out of fifteen KI in each village with a total of 24 KI being selected from the study villages. Then, semi-structured interviews and questionnaires were prepared to interview the KI and household (HH), respectively. A simple stratification of HHs was conducted by age (≤40 and >40) and wealth (poor, medium and rich), which were commonly used in assessing the local knowledge and plant utilization [16]. Hence, 12 HH classified by age category (≤40 and >40 in 1:1 ratio) and wealth (4 HH for each wealth class) were taken in

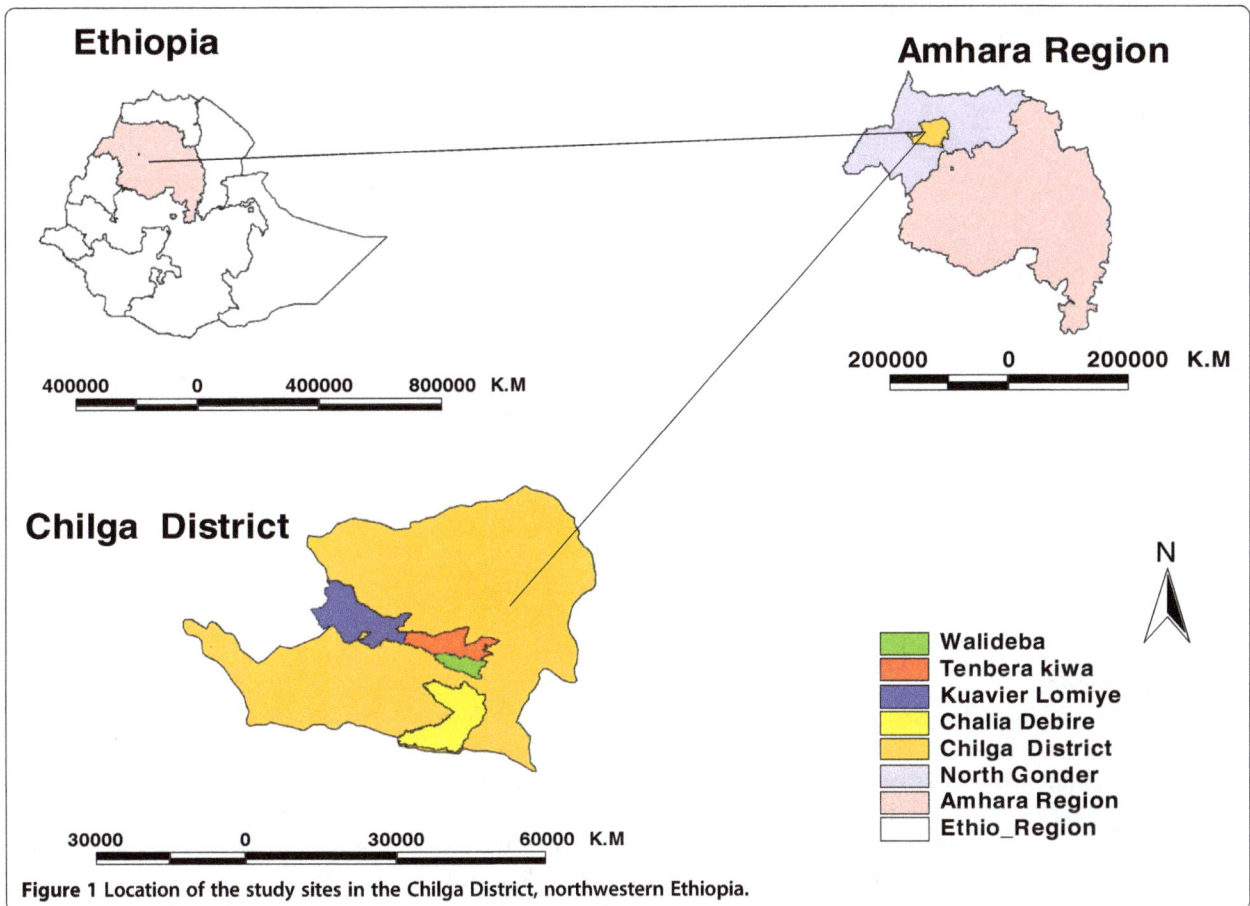

Figure 1 Location of the study sites in the Chilga District, northwestern Ethiopia.

each study village (Table 2). Thus, 96 HHs (78 males and 18 females) from four KA (24 HHs from each) were interviewed for the whole study, thereby constituting 5% of the population.

Data collection
Questionnaire survey and key informant interview
Questionnaires and semistructured interviews were prepared, pretested and administered to HHs and KIs,

Table 1 Sampled kebeles, villages, households and altitude in the Chilga District, northwestern Ethiopia

Study kebeles	Sampled villages	Number of respondents	Altitude (m.a.s.l.)
Quavier Lomiye (N = 24)	Achera	12	Below 1,500
	Bele Wuha	12	
Tenbera Kiwa (N = 24)	Gint	12	Below 1,500
	Kilel	12	
Walideba (N = 24)	Bete Skangie	12	Above 1,500
	Mehalgie	12	
Chalia Debire (N = 24)	Ateraho	12	Above 1,500
	Awugiber	12	

respectively (16, 17). All interviewees were met on a 'one-to-one' basis and asked the same standard (open- and closed-ended) questions using the local language (Amharic) based on their consent, including expansions or clarifications as needed. Consent was given by the informants to the district agricultural office. Information including vernacular names, parts used and consumption role of the plants was gathered. In addition, traditional management practices, other uses, and threats of UWEPs were also recorded.

Field observation and focused group discussion
Repeated field observations were conducted in the study sites by walking transects where most of the UWEPs are grown/cultivated. The purpose of the field observation was to obtain actual information of presence, growth habit, habitat characteristics and identification of edible plant species mentioned during the interviews. A focused group discussion of KI was conducted at each study site to verify the data and identification of underutilized edible plants. All underutilized wild edible plants listed in the socioeconomic survey were verified and idiosyncratic ideas were removed from the data.

Table 2 Importance value index (IVI) of the top ten woody species in three land uses and two agroecologies in Chilga District, northwestern Ethiopia

Scientific name	LLA			MLA			Average
	NF	R	FL	NF	R	FL	
Anogeissus leiocarrpa	39.88^b	39.51^b	17.25	32.80^a	16.66^c	13.68	26.63^a
Syzygium guineense	-	0.95	-	28.72^b	86.63^a	32.99^c	24.88^b
Focus sycomorus	22.73	8.19	15.01	7.92	7.72	42.22^b	17.30
Diospyros abyssinica	3.22	80.93^a	8.81	2.11	8.41	1.07	17.31
Flueggea virosa	35.80^c	12.85	32.31^b	3.74	-	1.03	14.29
Terminalia laxiflora	42.26^a	1.77	25.94^c	8.71	1.77	5.10	13.96
Croton macrostachus	-	-	4.91	18.00^c	11.13	47.83^a	13.64
Maytenus arbutifolia	2.28	13.27	1.41	14.48	23.94^b	25.86	13.54
Acacia polyachanta	21.56	15.31	39.45^a	-	-	-	12.72
Acalypha sp.	1.51	21.84^c	-	13.06	3.83	3.19	7.24
Others	130.76	105.38	154.91	170.46	139.91	127.03	138.49

LLA, lowland agroecology; MLA, midland agroecology; NF, natural forest; R, riverine forest; FL, farmland; Bold numbers with superscripts of a, b and c in each column are top three IVI values from highest to lowest order.

Direct matrix ranking

Direct matrix ranking was used to compare selected multipurpose species based on service categories. A direct matrix ranking method was exercised for commonly reported multipurpose UWEPs to assess their relative importance to local people and the extent of the existing threats related to their use values following the method of [16,17]. Six KI from each agroecology were selected and asked to assign a use value (5 to 0) for seven species in the lowland agroecology and for five species in the midland agroecology. The frequency of citation as multipurpose species was used for ranking UWEPs. Use categories for the comparison included construction, medicine, fruit/food, fuel wood, shade, farm and household implements, and fences, as used in the work of Cotton (16) and Martin (17).

Preference ranking

Preference ranking of selected UWEPs was conducted using taste and use criteria for each study kebele to assess the perception of the community. The most preferred underutilized wild edible plants in each study kebele were selected by KIs, and ranking (4-most preferred, 3-commonly preferred, 2-preferred but not so common, and 1-occasionally used) was conducted by all respondents, who followed the method of Jain et al. [17]. Similarly ranking of conservation demand for selected underutilized wild edible plants was conducted to assess conservation status of most preferred species following the method of Jain et al. [18] (4-for the species whose conservation is highly demanded, 3-conservation urgently demanded, 2-conservation required but not so urgent, and 1-conservation not required at present). Finally, such ranking of the species was summed up, and average ranking was employed at the site level.

Threats to underutilized edible

The major human and natural factors that possibly threaten the survival of underutilized wild edible plants were identified through preliminary assessment. Thus, based on the relative importance of the threatening factors, priority ranking was conducted by eight KIs using the method of [16]. One to five scores were assigned where one was for the least while five was for the most destructive threat. Then, all ranks were summed up, and total ranking was conducted to determine the main threats.

Market survey

Assessments of Gint, Negadie Bahire, and Chandeba local markets (nearest market places to the study sites) were conducted to judge market prices of the recorded underutilized edible plants.

Plant identification

All encountered plants were identified and recorded by their vernacular names. Later, these were converted to their botanical names using *Flora of Ethiopia and Eritrea* [19-25] and experience. Plant specimens were collected and taken to the National Herbarium of Addis Ababa University for plant identification of plants that were not identified in the field.

Vegetation inventory Field inventory of all UWEPS (woody) was employed with key informants to obtain information on the type of plants and their species diversity.

To conduct this, the selected study villages were stratified into three land uses (forms), that is, (1) natural forest (NF), (2) farmland (FL), and (3) riverine forests (R). Then, representative populations in each land use/form were selected. Six quadrates were laid in each land use per village following two transects running parallel along the gradient. The two transect lines were 200 m apart and quadrates along the transect lines were also 200 m apart both in the natural forest and farmland. For riverine areas, two transect lines were established for each village following the water flow and three quadrate were taken at 200-m intervals from each transect line. The first quadrate

Table 3 List of all underutilized edible plants encountered in the Chilga District, northwestern Ethiopia

Scientific name (Family)	Vernacular name (Amh)	Habit	Habitat	Added values	PU	MD	Voucher number
Acanthus sennii Chiov. (Acanthaceae)	Kushashile	SH	BN, N,R	FE, FU,	FL	F	MT-001
Balanites aegyptiaca (L.) . (Balanitaceae)	Kudekuda	T	N,R,F	FE, FU,SC,	F	F	MT-002
Boletus edulis Bull. Ex Fries. (Bolentaceae)	Enguday	H	F, N	-	ST	P	MT-003
Carissa Spinarum L. (Apocynaceae)	Agam	SH	N,R,	FE, M, FU, FT	F	F	MT-004
Corchorus olitorius L. (Tiliaceae)	Kudra	H	F	FD, Rope	L	P	MT-005
Cordia africana L. (Boraginaceae)	Wanza	T	H	T,FU, FE, FU, SH, FD, M	F	F	MT-006
Dichrostachys cinerea Wight & Am (Fabaceae)	andera	S/ T	N	-	F	F	MT-007
Dioscorea prahensilis Benth (Dioscoreaceae)	Senssa	C	N,R	-	R	P	MT-008
Diospyros abyssinica (Hiem) F. Wite (Ebenaceae)	Serkin	T	F,R	FU,T,CO,SH, FE,	F	F	MT-009
Diospyros mesiliformis Hochst ex.A.DC. (Ebenaceae)	Gurmacha	T	F,R	T,FU,SH,CO,FE,SC	F	F	MT-010
Dovyalis abyssinica (A. Rich.) Warburg. (Flacourtiaceae)	Koshim	T	N,R	FU, FE,	F	F	MT-011
Ficus sur Forssk. (Moraceae)	Shola	T	R,F,N	CO,FE,SC,SH,FU	F	F	MT-012
Ficus sycomorus L. (Moraceae)	Bamba	T	F,N,R,H	CO,FE,SC,FD,SH, HB	F	F	MT-013
Ficus vallis-choudae Del. (Moraceae)	Bambula	T	N	FU, FE	F	F	MT-014
Ficus vasta Forssk. (Moraceae)	Warka	T	N,R,F,H	FU, CO, FE, SC,	F	F	MT-015
Flueggea virosa Guill. & Perr. (Euphorbiaceae)	Shasha	SH	F,N,R	FE, FU,CO,FU	F	F	MT-016
Gardenia ternifolia Schumach and Thonn. (Rubiaceae)	Gambilo	S/T	F,N,R	FD, FU,	F	F	MT-017
Gloriosa superba (Liliaceae)	Yemariam twa	H/sh	F	FD, WASHING	F	F	MT-018
Hibiscus cannabinus L. (Malvaceae)	Yeberha Wayika	H	F	FD	F	D	MT-019
Hibiscus esculentus L. (Malvaceae)	Wayika	H	F	M, FD,	L		MT-020
Maytenus senegalensis Forssk (Celastraceae)	Koshikosh	T	N,F,R	FE,FT,	F	F	MT-021
Mimusops kummel Bruce ex A.DC. (Sapotaceae)	Ishe	T	R,F	FE,FU,HB,CO,	F	F	MT-022
Morusmeso zygia (Moraceae)	Injori	C	N, R	-	F	F	MT-023
Pittosporum viridiflorum Sims. (Pittosporaceae)	Dengay Seber	S	N,F, R	FU	F	F	MT-024
Rhus glutinosa A. Rich. subsp. Abyssinica (Oliv.) M. Gilnert (Anacardiaceae)	Qamo	S	N, R	Fe, FU	F	F	MT-025
Rosa abyssinica Lindley (Rosaceae)	Qega	SH	N	FE,FU,	F	F	MT-026
Saba comorensis (Bo).) Pichon (Apocynaceae)	Ashama	C	R	CO, SH,	F	F	MT-027
Sporobolus africanus (Poir) Robyns and Tournay (Poaceae)	Muriye	H	N,FLD,R	FD	S	D	MT-028
Syzygium guineense (Willd.) DC. (Myrtaceae)	Dokima	T	R,F	M, FE, FU, SH, CO	F	F	MT-019
Tamarindus indica L. (Fabaceae)	Kumer	T	R,N,F	FE,FU,CO,SC	S	D	MT-030
Ximenia americana L. (Olacaceae)	Enkoye	Sh	N	FU,M	F	F	MT-031
Ziziphus abyssinica Hoschst. (Rhamnaceae)	Abetere	T	F,N,R,	Fu,Fe	F	F	MT-032
Ziziphus spina-christi Willd. (Rhamnaceae)	Arka	S/T	N	Fe,Fu, co,Fd,sc	F	F	MT-033

Key to abbreviations: Habit; Sh, shrub; T, tree; H, herb; C, climber; **Habitat:** N, natural forest; F, farmland; R; riverine and valley; H, home garden; Fld, field. BN, boundary; **Added value**: Fu, fuel wood; CH, charcoal; M, medicinal; CO, construction; Fe, fencing; SC, soil and water conservation; FD, fodder; Sh, shade; HB, production; T, timber; FT, farm and household tools. PU (**Parts used):** F, fruit; R, root; Fl, flower nectar; S, seed; W, whole part. **Mode of utilization** (MD): F, fresh; P, prepared/cooked; D, dried and prepared.

in each land use was laid out randomly from the east of the sampled population. The quadrate sizes for natural forest and riverine forest was 20 m × 20 m (400 m^2) for trees (main plot), 10 m × 10 m for shrubs and saplings at the center of main plot, and 1 m × 1 m for herbs and seedlings with an 'X' design that followed the method of [26,27]. For farmland, a quadrate size of 50 m × 40 m (main plot) for trees, 25 m × 20 m for saplings and shrubs, and five 2 m × 2 m in an 'X' design for seedlings and herbs was taken.

In the smallest subplots (1 m^2) or 4 m^2, all herbs and seedlings were identified and counted (including all plants less than 50 cm in height). In the subplots of 10 m × 10 m or 25 m × 20 m, diameter at breast height (DBH) of all saplings and shrubs (DBH <10 cm) was measured with a caliper. The DBH of trees (DBH ≥10 cm) were measured in the entire plot area. Trees forked below breast height were measured independently and average DBH was taken. One hundred forty-four main quadrates were taken for the whole study.

Data analysis

Both quantitative and qualitative data analyses were conducted after the necessary data collection. Statistical Analysis System (SAS) 9.0 version software was used for descriptive and statistical analysis. Descriptive analyses (informant consensus, direct matrix ranking and preference ranking) were presented in the form of percentages, figures and means. An analysis of variance (ANOVA) of diversity indices (Shannon Weiner diversity index, Simpson's diversity index and evenness), species richness and wealth classes was conducted to compare diversity between agroecologies and among kebeles. A T test mean separation (P <0.05) of diversity indices and wealth classes was conducted using least significance difference (LSd). The Spearman Rank Correlation test was also employed to evaluate whether there is significant (P <0.05) correlation between age and UWEPs list.

Species richness

It is defined as the number of species per quadrate, area or community. In this particular case, the numbers of observed species across the whole sample quadrates of each land use in each agroecology and Kebele were used as a representation of species richness. However, quadrate species richness is used as a comparison of species richness between agroecologies and kebeles.

Shannon diversity index (H')

It is very important when comparing diversity among samples and habitats to use an index that is more sensitive to richness. The value of H' is usually found to fall between 1.5 and 3.5 and only rarely surpasses 4.5 [28,29].

Therefore, the species diversity of the midland and lowland agroecologies was estimated as follows:

$$H' = -\sum_{1}^{s} Pi \ln pi \tag{1}$$

where H' = is Shannon diversity and Pi = proportion of individual species.

Simpson's diversity index

Simpson's diversity index is the most widely used method for estimating the richness diversity of community and is used to compare different communities or habitats [29]. The Simpson diversity is less sensitive to richness and more sensitive to evenness. The Simpson's diversity is calculated as follows:

$$D = 1 - \sum_{n=1}^{s} Pi^2 \tag{2}$$

Evenness

It is a widely used and understood method for estimating evenness of the communities [30]. The most commonly used index of estimating evenness diversity is the Pielou index, which is determined as follows:

$$J' = H'/H' \max \tag{3}$$

where J' = Pielou evenness index; H' = the observed value of Shannon index; H'max = lnS, and S = total number of species.

Importance value index and relative frequency

Estimation of the importance value index (IVI), abundance, frequency and relative frequency was conducted to assess the importance of each species to the survey sites using the method of [31] and [32] and these estimates are determined with the following:

$$\text{IVI} = \text{Relative Frequency} + \text{Relative Density} + \text{Relative Dominance} \tag{4}$$

Table 4 Underutilized woody edible plants species diversity for three land uses in lowland agroecology (LLA) and midland agroecology (MLA) of the study area, northwestern Ethiopia

Agroecology	Species richness	Simpson_1-D	Shannon	Evenness
Lowland				
NF	14	0.69	1.73	0.66
R	14	0.79	1.93	0.73
FL	12	0.76	1.78	0.71
Midland				
NF	12	0.78	1.77	0.71
R	14	0.79	2.05	0.78
FL	12	0.82	1.92	0.77

NF, natural forest; R, riverine forest; FL, farmland.

Table 5 Underutilized woody edible plants species diversity for three land uses in the study sites, northwestern Ethiopia

Study sites	Species richness			Simpson_1-D			Shannon _H			Evenness		
	NF	R	FL	NF	R	FL	N	R	FL	NF	R	FL
Quavere Lomiye	10	13	6	0.74	0.8	0.67	1.7	1.95	1.33	0.74	0.76	0.74
Tenbera	12	7	11	0.63	0.73	0.7	1.55	1.55	1.67	0.62	0.8	0.7
Walideba	12	13	10	0.76	0.88	0.76	1.78	2.34	1.72	0.71	0.91	0.75
Chalia Debire	7	10	4	0.67	0.66	0.6	1.37	1.57	1.05	0.7	0.68	0.76

LLA, lowland agroecology; MLA, midland agroecology; NF, natural forest; R, riverine forest; FL, farmland.

$$\text{Frequency} = \text{number of occurence of a given species in the plots studied} \quad (5)$$

$$\text{Relative Frequency (\%)} = \frac{\text{Frequency of any species}}{\text{Total frequency of all species}} * 100 \quad (6)$$

$$\text{Relative Density (\%)} = \frac{\text{Number of individual of each species per ha}}{\text{Total number of individuals of all species per ha}} * 100 \quad (7)$$

$$\text{Relative Dominance (\%)} = \frac{\text{Basal area of each species}}{\text{Total basal area of all species}} * 100 \quad (8)$$

$$\text{Basal area} = \frac{\prod D^2}{4}, \quad D = \text{Diameter of trees at breast height} \quad (9)$$

Results

Importance value index

The importance value index (IVI) of woody species is dominated by four species (86.12%), namely *Anogeissus leiocarrpa*, *Syzygium guineense*, *Diospyros abyssinica* and *Ficus sycomorus* (Table 2 and Additional file 1). The species distribution varies in land uses and in agroecologies. In the lowlands (LLA), *Terminalia laxiflora*, *Diospyros abyssinica* and *Acacia polyachanta* are the top ranking species for NF, riverine and FL, respectively. *Anogeissus leiocarrpa*, *Syzygium guineense and Ficus sycomorus* are also the first ranked species in the midland (MLA) for NF, riverine and FL land uses, respectively. Some of the woody plants in the study area were habitat specific. For example, *Syzygium guineense* was recorded only from riverine land use of the LLA.

Diversity of underutilized wild edible plants

Thirty-three UWEPs (28 woody and five herbaceous) species were recorded in the study area (Table 3). The family Moraceae had five species; the families Malvaceae, Fabaceae, and Euphorbiaceae had two species each; and the remaining families had one species each.

Table 6 Means (±std) of woody underutilized edible plant diversity for the three land uses in the study area, northwestern Ethiopia

Diversity indices	Agroecology	NF	R	FL
Richness	MLA	$2.87^a \pm 0.29$	$3.71^a \pm 0.38$	$2.00^a \pm 0.22$
	LLA	$3.42^a \pm 0.22$	$3.17^a \pm 0.35$	$1.87^a \pm 0.33$
	Over all mean	3.15 ± 0.18	3.44 ± 0.26	1.93 ± 0.19
Simpson	MLA	$0.44^a \pm 0.05$	$0.55^a \pm 0.04$	$0.31^a \pm 0.05$
	LLA	$0.55^a \pm 0.03$	$0.46^a \pm 0.06$	$0.27^a \pm 0.06$
	Overall mean	0.49 ± 0.03	0.51 ± 0.03	0.29 ± 0.04
Shannon	MLA	$0.77^a \pm 0.09$	$1.02^a \pm 0.09$	$0.51^a \pm 0.09$
	LLA	$0.97^a \pm 0.07$	$0.85^a \pm 0.12$	$0.46^a \pm 0.11$
	Overall mean	0.87 ± 0.06	0.93 ± 0.08	0.48 ± 0.07
Evenness	MLA	$0.68^b \pm 0.06$	$0.80^a \pm 0.06$	$0.50^a \pm 0.06$
	LLA	$0.82^a \pm 0.03$	$0.66^a \pm 0.08$	$0.43^a \pm 0.09$
	Overall mean	0.75 ± 0.03	0.73 ± 0.05	0.56 ± 0.09

[a,b]Means with the same letter ordered vertically within each diversity parameter are not significant ($P < 0.05$).
FL, farmland; LLA, lowland agroecology; MLA, midland agroecology; NF, natural forest; R, riverine forest.

Table 7 Mean (±std) of underutilized woody wild edible plant diversity for the three land uses in lowland agroecology (LLA) of the Chilga District, northwestern Ethiopia

Diversity indices	Kebeles	NF	R	FL
Richness	Quavere Lomiye	$3.08^a \pm 0.31$	$4.25^a \pm 0.43$	$1.92^a \pm 0.29$
	Tenbera	$3.75^a \pm 0.30$	$2.08^b \pm 0.34$	$1.83^a \pm 0.61$
	Overall mean	3.42 ± 0.22	3.17 ± 0.35	1.87 ± 0.33
Simpson	Quavere Lomiye	$0.52^a \pm 0.06$	$0.59^a \pm 0.06$	$0.34^a \pm 0.07$
	Tenbera	$0.57^a \pm 0.05$	$0.34^b \pm 0.09$	$0.19^a \pm 0.09$
	Overall mean	0.55 ± 0.03	0.46 ± 0.06	0.27 ± 0.06
Shannon	Quavere Lomiye	$0.90^a \pm 0.10$	$1.15^a \pm 0.13$	$0.53^a \pm 0.13$
	Tenbera	$1.05^a \pm 0.09$	$0.55^b \pm 0.16$	$0.39^a \pm 0.18$
	Overall mean	0.98 ± 0.07	0.85 ± 0.12	0.46 ± 0.11
Evenness	Quavere Lomiye	$0.83^a \pm 0.04$	$0.78^a \pm 0.07$	$0.59^a \pm 0.13$
	Tenbera	$0.82^a \pm 0.05$	$0.55^a \pm 0.14$	$0.27^a \pm 0.12$
	Overall mean	0.82 ± 0.03	0.66 ± 0.08	0.43 ± 0.09

[a,b]Means with the same letter ordered vertically within each diversity parameter are not significant (P <0.05).
FL, farmland; NF, natural forest; R, riverine forest.

Species richness of underutilized wild edible woody plants (UWEPs) recorded in the study area is presented in Tables 4 and 5. Overall, mean value difference in species richness and diversity indices of UWEPs in two agroecologies were in the order of: R > NF > FL (Table 6). Evenness in the LLA was significantly (P <0.05) higher than MLA for NF land use. Comparison of diversity within each agroecology shows species richness of riverine land use at Quavier Lomiye Kebele was significantly higher than at Tenbera in the LLA (Table 7). Within the MLA, for NF land use, Simpson and evenness diversity indices of Chalia Debire Kebele were significantly higher than Walideba Kebele Simpson and evenness diversity (Table 8).

Underutilized wild edible plants habits and parts used

Growth forms of UWEPs comprise trees, shrubs, herbs and climbers. Of all (n = 33) UWEPs, most (45%) were trees followed by shrubs (about 27.3%) (Figure 2a). Trees were also the dominant growth forms at each study site, representing about half at Walideba, Chalia Debire and Quabier Lomiye and 46.4% at Tenbera Kebele. The study also indicated that fruits were the most commonly

Table 8 Mean (±std) of woody wild edible species diversity for three land uses in midland agroecology (MLA) in northwestern Ethiopia

Diversity Indices	Kebele	NF	R	FL
Richness	Walideba	$2.67^a \pm 0.56$	$3.08^a \pm 0.57$	$2.33^a \pm 0.38$
	Chalia Debire	$3.08^a \pm 0.19$	$4.33^a \pm 0.47$	$1.67^a \pm 0.19$
	Overall mean	2.87 ± 0.29	3.71 ± 0.38	2.00 ± 0.22
Simpson	Walideba	$0.35^b \pm 0.078$	$0.52^a \pm 0.08$	$0.36^a \pm 0.08$
	Chalia Debire	$0.54^a \pm 0.03$	$0.58^a \pm 0.03$	$0.27^a \pm 0.07$
	Overall mean	0.44 ± 0.05	0.55 ± 0.04	0.32 ± 0.06
Shannon	Walideba	$0.64^a \pm 0.15$	$0.92^a \pm 0.17$	$0.61^a \pm 0.15$
	Chalia Debire	$0.91^a \pm 0.06$	$1.11^a \pm 0.09$	$0.40^a \pm 0.11$
	Overall mean	0.78 ± 0.09	1.02 ± 0.09	0.51 ± 0.09
Evenness	Walideba	$0.54^b \pm 0.10$	$0.79^a \pm 0.12$	$0.59^a \pm 0.13$
	Chalia Debire	$0.83^a \pm 0.04$	$0.82^a \pm 0.04$	$0.53^a \pm 0.14$
	Overall mean	0.68 ± 0.06	0.80 ± 0.06	0.56 ± 0.09

[a,b]Means with the same letter ordered vertically within each diversity parameter are not significant (P <0.05).
NF, natural forest; R, riverine forest; FL, farmland.

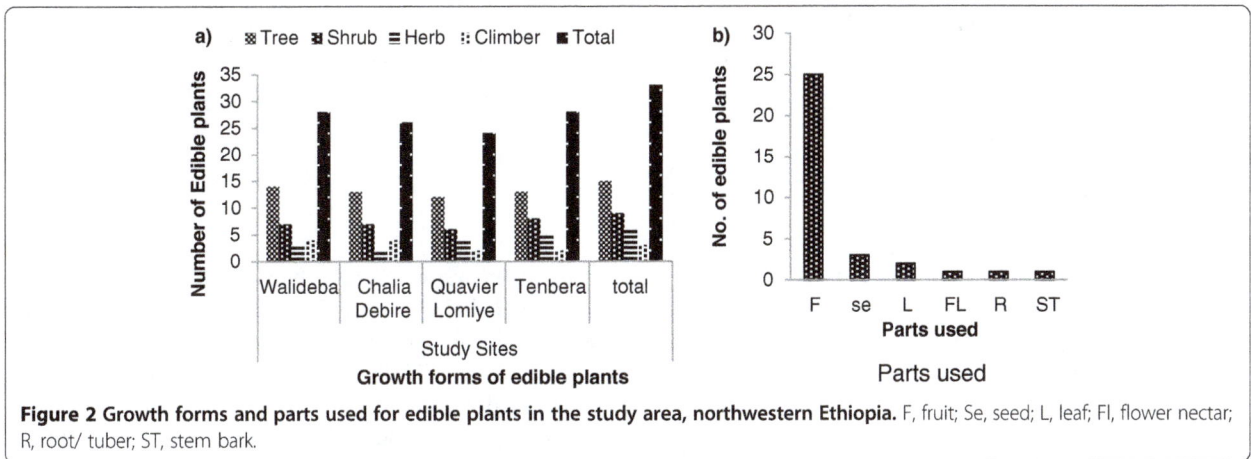

Figure 2 Growth forms and parts used for edible plants in the study area, northwestern Ethiopia. F, fruit; Se, seed; L, leaf; Fl, flower nectar; R, root/ tuber; ST, stem bark.

used parts of UWEPs (76%), while flowers, roots, and stems were the least used parts (Figure 2b).

Utilization and role of underutilized wild edible plants

The results of this study showed that UWEPs play an important role in the household livelihood diversification, and these results are summarized below:

1. Diversity of uses.

   UWEPs offer various uses such as fuel wood, fencing, construction, medicinal, fodder, timber, honey production and detergent (Figure 3). More than three-fourths of UWEPs in the study sites are used for fuel wood and fencing purposes. The community utilized these plants for various reasons (Table 9). About 56% and 33.3% of the respondents in MLA mentioned that UWEPs were used to supplement their staple foods and for refreshment, respectively. Also 21% of the respondents consume products to fill food gaps. In the LLA, 67% of the respondents mentioned that UWEPs were used in

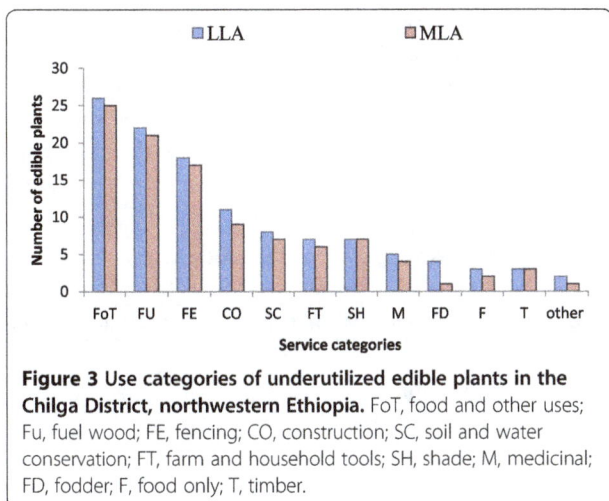

Figure 3 Use categories of underutilized edible plants in the Chilga District, northwestern Ethiopia. FoT, food and other uses; Fu, fuel wood; FE, fencing; CO, construction; SC, soil and water conservation; FT, farm and household tools; SH, shade; M, medicinal; FD, fodder; F, food only; T, timber.

times of normal diet. UWEP utilization by wealth class using the free-list exercise approach is shown in Table 10.

UWEP citation in the poor HHs was significantly higher ($P < 0.05$) than for medium and rich wealth classes in the Walideba and Chalia Debire kebeles. Similarly, UWEP citation in the poor HHs was significantly higher than for the richest groups in Quavier Lomiye Kebele. Plant citation in Tenbera Kebele was significantly different among wealth classes in the following order: poor > medium > rich.

2. Income generating role of underutilized wild edible plants.

   UWEPs could generate income for HHs through either sales to domestic market or exporting to neighboring countries, mainly the Sudan (Table 11). Twelve marketable UWEPs were recorded with the mean unit price of products ranging from 0.63 to 5.6 ETB (Ethiopian Birr). Youngsters in the local market sell most of these marketable UWEPs (eight plants). Of the marketable UWEPs, *Tamarindus indica* and *Hibiscus cannabinus* were the most highly priced species. About 83% and 17% of the marketable UWEPs are marketed in the local markets and for export, respectively (Figure 4a). Of all marketable UWEPs, fruits represent the highest proportion of edible parts (about 75%) (Figure 4 b).

3. Direct matrix ranking for underutilized wild edible plants.

   Seven UWEP species in the LLA and five in the MLA were selected, and direct matrix ranking was conducted based on the key informants' preferences of use category. Thus, food and fuel use categories were the first and second ranked categories, respectively, for the selected multipurpose species (Table 12) for lowland and (Table 13) for MLA. When overall categories were considered, species with top score ranking were ordered as follows:

Table 9 Consumption roles of underutilized wild edible plants in the study area, northwestern Ethiopia

Agroecology	Supplemental		Recreational		Fill food gaps		All	
	Informants	%	Informants	%	Informants	%	Informants	%
MLA	27	56.3	16	33.3	10	20.8	10	20.8
LLA	32	66.7	14	29.2	6	12.5	11	22.9

LLA, lowland agroecology; MLA, midland agroecology.

Diospyros mespiliformis > Syzygium guineense > Tamarindus indica for LLA; *Syzygium guineens > Diospyros abyssinica > Ximenia americana* for MLA. *Carisa spinarum* received the list score from both agroecologies.

Informant consensus and preference of underutilized wild edible plants

Species preference ranking of UWEPs was conducted based on the informants' consensus approach to find out the relative importance of plants to the local community (Table 14). The most preferred species in decreasing order were *S. guineense, D. abyssinica, Carissa spinarum, Mimusops kummel, D. mesiliformis, Ximenia americana* and *Cordia africana*. However, species preference varies from kebele to kebele. *S. guineense, D. mespiliformis and M. kummel* were the most preferred edible plants, cited by 95.8% of respondents in Walideba Kebele. Similarly, *S. guineense and M. kummel* were the best species cited by all respondents in Chalia Debire Kebele. Again, *D. mesiliformis* was the preferred species in Quavier Lomiye and Tenbera kebeles (species cited by ≥29 informants of the total respondents).

Selection and ranking of the most preferred candidate UWEPs, based on their taste and use, was conducted in each kebele. *S. guineense* is the most preferred species in Walideba and Chalia Debire kebeles (Figure 5 a and b), while *Saba comorensis was* the most preferred in Quavier Lomiye and Tenebera (Figure 5 c and d). In terms of conservation demand of species, *M. kummel* and *X. americana* were the most highly preferred ones in the MLA and LLA study sites, respectively.

Underutilized wild edible plant list exercise

With regard to list length for free plants, variation exists between age groups. Respondents younger than 40 years

of age cited more UWEPs than older people did across the study kebeles (Figure 6a). The Spearman correlation test has also shown a significant negative correlation between age and UWEPs list (r = -0.326, P <0.05).

The local community gains knowledge about UWEPs utilization, processing and management through experience (Figure 6b). However, the major acquisition/transfer method was from parents, friends and relatives, and neighbors in both agroecologies. Sources of knowledge or acquisitions for the majority of respondents (87.5%) in MLA and 73% in LLA were the parents.

Traditional management practices and threats of underutilized edible plants

The results of this study indicated that there was a lack of training to local community to improve management, conservation and utilization of UWEPs (Table 15). More than 77% of the respondents in MLA and 87.5% in LLA indicated that training on conservation and utilization of UWEPs was not given from any concerned body. This indicated that the local community utilizes, manages and conserves these plant resources only through experience and traditional knowledge.

Identification and preference ranking of the major threats of UWEPs based on their destructive effects were conducted (Table 16). It confirmed that illegal charcoal production, fuel wood collection, construction, agricultural land expansion, overgrazing and fire were the dominant threats.

Local communities practice different traditional management, which includes planting around the home garden, pruning, pollarding, fencing, and preventing growth of the most important plants by local culture in both the LLA and MLA (Table 16). Of the management practices; fencing, pollarding and local culture are the dominant practices in the LLA. In similar fashion, planting around

Table 10 Mean (±std) of underutilized wild edible plants free-list exercise by wealth class in the study sites, northwestern Ethiopia

Wealth category	Walideba (Mean ± std)	Chalia Debire (Mean ± std)	Quavier Lomiye (Mean ± std)	Tenbera (Mean ± std)
Poor (n = 32)	15.50[a] ± 2.14	14.25[a] ± 0.65	12.38[a] ± 1.19	14.13[a] ± 0.81
Medium (n = 32)	9.75[b] ± 0.70	9.75[b] ± 0.26	10.13[ab] ± 0.92	9.63[b] ± 0.73
Rich (n = 32)	7.75[b] ± 0.82	9.75[b] ± 0.80	8.25[b] ± 0.88	6.63[c] ± 0.98
Over All (96)	11.00 ± 1.03	11.25 ± 0.68	10.25 ± 0.66	10.13 ± 0.81

[a,b]Means with different letters ordered vertically in each Kebele were significant (P <0.05).

Table 11 List of marketable underutilized wild edible plants in the Chilga District, northwestern Ethiopia

Species name	Parts marketed	Unit	Mean price[a]	Number of respondents	Seller group[b]	Market category	Rank
Carisa spinarum	F	cup	0.83	6	younger	D	10
Corchorus olitorius	L	handful	2.2	10	all	D	8
Diospryos mesiliformis	F	cup	0.91	46	younger	D	2
Diospyros abyssinica	F	cup	0.8	27	younger's	D	6
Ficus sur	F	cup	0.63	2	younger	D	12
Hibiscus cannabinus	Se	cup	4.33	3	adult	Exp	11
Mimusops Kummel	F	cup	1	45	younger	D	3
Saba comorensis	F	number	0.7	35	all	D	4
Syzygium guineense	F	cup	1	64	younger's	D	1
Tamarindus indica	Se	kg	5.6	21	all	Exp	7
Ximenia Americana	F	cup	0.95	33	younger	D	5
Ziziphus spina- christi	F	cup	0.7	9	younger	D	9

F, fruit; Se, seed; L, leaf; L, local market; Exp, exported to other countries; a, Ethiopian birr (ETB); b, local classification of below 18-years old as younger and from 18- to 30-years of age as adult; Ranks were given by number of respondents.

home gardens and backyards and pollarding were the major practices in the MLA.

Fuel wood and illegal charcoal production and construction were the main threats in the MLA, whereas agricultural expansion, fuel wood collection, and charcoal making were dominant threats in the LLA.

Discussion

Importance value index

The importance value index (IVI) of all woody species indicates the ecological importance of each species to the area. In addition, it indicates the types of plant species on which the rural community is more reliant. The four top ranked species according to IVI value (*A. leiocarrpa, S. guineense, F. sycomorus* and *D. abyssinica*) in the whole area were the most important UWEPs. The high IVI of these species is probably due to their ability to produce a high number of seeds and maintain a persistent soil seed bank. There was variation in IVI in agroecology and land uses, probably due to variation of species adaptation to different agroecologies and human

disturbance [33]. Thus, high IVI confirmed the potential of these species to adapt to an area and to resist anthropogenic disturbance [33].

Floristic composition, distribution and diversity of underutilized edible plants

A good number of UWEPs (33 species) were recorded compared to other areas in the Amhara region [8]. These UWEPs include trees, shrubs and herbs. Most UWEPs were also documented elsewhere in Ethiopia; for instance, seven species are found in the semi-arid lowlands of southern Ethiopia [3,7], species in Derashe and Kucha, southern Ethiopia [7,20], species in Amhara region, northern Ethiopia (8); 20 species in northern Ethiopia; eight species in southeastern Ethiopia; five species in eastern Ethiopia; six species in southeastern Ethiopia [10]. The existence of these plants in different regions of the country indicates their ecological adaptation over a large geographical area and their edibility by different ethnic groups.

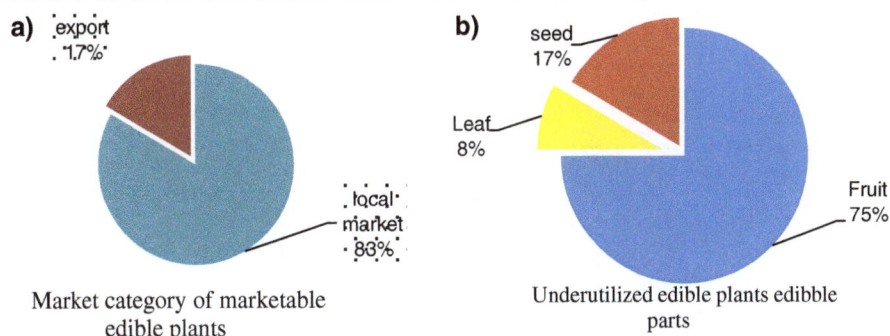

Market category of marketable edible plants

Underutilized edible plants edibble parts

Figure 4 Market categories of marketable edible plants (a) and parts sold (b) in the Chilga District, northwestern Ethiopia.

Table 12 Direct matrix average score of seven underutilized edible plants in lowland agroecology (LLA) in northwestern Ethiopia

Species	Use category								Total	Rank
	Food	Fuel wood	Fencing	Constru-ction	Shade	Medicinal	Home and farm tools	Timber		
Diospyros mespiliformis	5	3	2	4	3	0	2	2	20	1
Syzygium guineense	5	4	2	2	2	1	0	1	17	2
Tamarindus indica	5	4	2	2	2	0	2	1	17	2
Diospyros abyssinica	5	4	2	3	1	0	1	1	16	3
Ziziphus spina-christii	4	3	4	2	1	1	1	1	16	3
Ximenia Americana	5	4	3	0	0	2	2	0	15	4
Carisa spinarum	4	2	2	0	0	5	1	0	15	4
Total	33	24	18	13	9	8	7	5		
Rank	1	2	3	4	5	6	7	8		

Natural forests, riverine areas and farmlands were centers for UWEPs in the study sites in both MLA and LLA. Similar to this study UWEPs in other parts of Ethiopia were concentrated in farmlands, riverine areas and natural forests [7,8,10,33]. Forests of the world provide livelihoods and food for 300 million people in the form of non-timber forest products [34]. Again, their distributions in different agroecologies vary due to differences in species' physiological adaptations [7].

The study shows almost the same species richness among land uses. As Fentahun and Hager [8] explained, the species richness of wild edible plants is affected by altitude, personal preference and selective management of trees by farmers, especially for those found on farmlands. The Shannon diversity of land uses in both agroecologies is greater than 1.5. Thus, high diversity exists in both agroecologies. Significant differences among kebeles might be due to variation in the levels of human disturbance.

Utilization and socioeconomic implication
Trees, followed by shrubs, were the dominant growth forms of UWEPs in the study area. Although relatively small in number, herbaceous plants were also consumed. The report by Fentahun and Hager [8] in the Amhara region and in Tilahun and Mirutse [9] in the lower river valley of Debub Omo Zone was consistent with the present finding that trees were the leading growth forms. In contrast, Ermias et al. [13] indicated that shrubs were the dominant growth forms in Ethiopia, followed by trees, herbs and climbers. Fruits were the commonly utilized edible parts in the study area. In agreement with this finding, Ermias et al. [13] and Getachew et al. [35] found fruits as the widely used parts.

The results of this study indicate that UWEPs are used mainly to supplement staple food, to fill food gaps and for refreshment purposes. In agreement with the present study, other findings elsewhere [5,36,37] indicate their supplemental role. The greater number of plant citations by the poorest community in the study area indicates the consumption level and familiarity of the community with these plants. Seasonal food shortages, when household stocks were empty and the new crop was still in the field, were common times to focus on collecting, selling and consuming underutilized edible

Table 13 Direct matrix average score of five underutilized edible plants in midland agroecology (MLA) in northwestern Ethiopia

Use category	Syzygium guineense	Diospyros abyssinica	Ximenia americana	Mimusops kummel	Carissa spinarum	Total	Rank
Food	5	4	5	5	4	23	1
Fuel 2ood	4	4	4	4	3	18	2
Construction	3	4	3	3	0	12	3
Fencing	2	3	2	2	2	11	4
Shade	2	2	2	2	0	9	5
Medicinal	1	0	0	0	5	6	6
Home and Farm Tools	0	1	1	1	1	3	7
Timber	0	0	0	0	0	0	8
Total	17	17	16	16	14		
Rank	1	1	2	2	3		

Table 14 List of frequently cited edible plants in the study sites, northwestern Ethiopia

Edible species	Walideba	Chalia Debire	Quavier Lomiye	Tenbera	Total
Syzygium guineense	23	24	16	17	80
Carissa spinarum	20	20	19	17	76
Diospyros abyssinica	22	19	15	17	73
Diospyros mespiliformis	23	2	24	21	70
Ximenia Americana	11	21	18	18	68
Mimusops kummel	23	24	8	12	67
Cordia africana	19	22	7	15	63
Ficus sycomorus	12	13	17	11	53
Saba comorensis	2	14	22	14	52
Ficus sur	17	16	6	12	51
Balanites aegyptiaca	16	2	8	17	43
Dioscorea prahensilis	5	20	7	9	41
Ziziphus spina-christi	6	1	19	11	37
Tamarindus indica	0	0	19	10	29

plants [36,38]. Marginalized and poor communities are more vulnerable to drought, and thus, are more dependent on these plants [13,34,38,39].

The rural communities of this study, especially the lowland kebeles, preferred *Corchorus olitorius* more than other vegetables because they believe it has more nutritional value. Debela *et al.* [5] also reports on the nutritional richness of this plant. Thus, local people in the study area bypass nutritional insecurity by consuming UWEPs. Some species that were preferable in the study area (for example, T. indica, S. comorensis and C. spinarum) are also priority ranked species in Kenya [40].

Once more, most of UWEPs provide service in addition to food value. Different researchers elsewhere in Ethiopia have also noted multiple uses for UWEPs such as preparation of remedies, fuel wood, fencing, construction

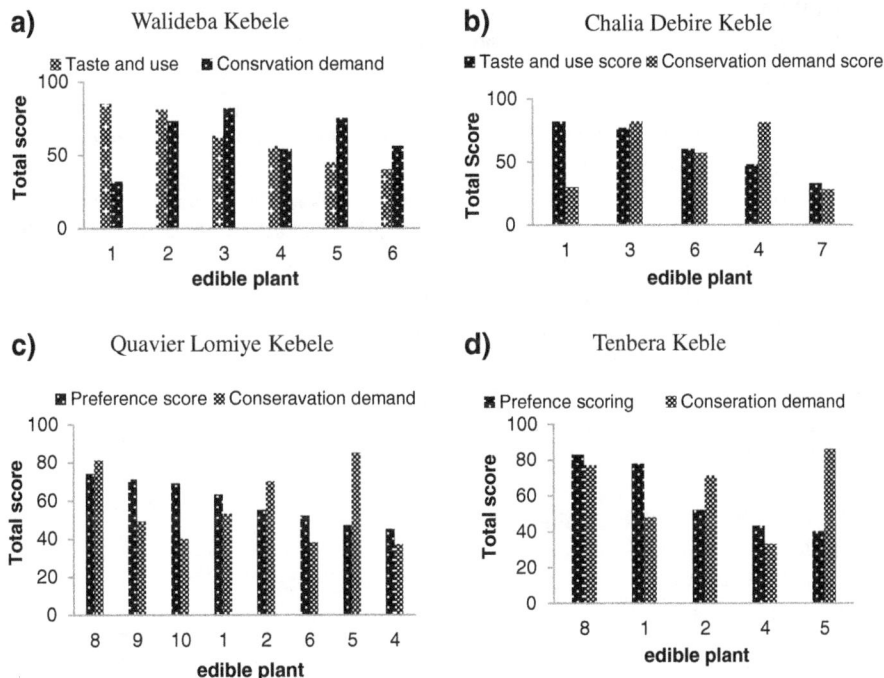

Figure 5 Preference ranking of edible plants in the study sites in the Chilga district, northwestern Ethiopia. Legend: 1 = *Syzygium guineense*; 2 = *Diospyros melisformis*; 3 = *Mimusops kummel*; 4 = *Diospyros abyssinica*; 5 = *Ximenia americana*; 6 = *Carissa spinarum*; 7 = *Cordia Africana*; 8 = *Saba comorensis*; 9 = *Corchorus olitorius*; 10 = *Tamarindus indica*.

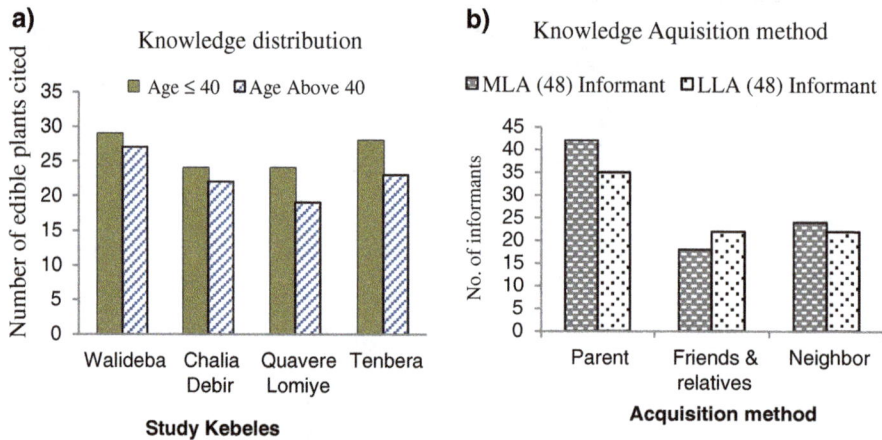

Figure 6 Knowledge distribution and acquisition methods of edible plant in the Chilga District, northwestern Ethiopia.

and timber, farm and household implements and livestock fodder [7-9]. Marketable wild edible plants such as *X. americana, S. guineense, T. indica, C. spinarum* and *M. kummel* also create household income opportunities. A few income-generating species like S. guineense, T. indica, C. spinarum and M. kummel were also reported in other areas [7,8]. Furthermore, income is gained from selling other parts such as stems and branches [3,7]. *C. Africana*, for instance, one of the most widely used edible plants in the study area, is preferred for timber production.

According to Bharucha and Pretty [34], there is no comprehensive global estimate of the economic value of UWEPs; therefore, quantitative analyses face methodological difficulties. For most studies, use of different valuation methods and diversity scales are rarely equivalent. Second, the sale of products is often illegal and therefore under-reported. This was observed in the study areas where various UWEPs plants sold at local market and even exported to Sudan were not valued correctly.

Knowledge distribution of underutilized edible plants

Ethnobotanical knowledge and practice within any culture varies depending on such factors as cultural attributes. In a free-list exercise, the differences in list length and content are measures of intraspecific cultural variation [41]. Younger respondents cited a higher number of UWEPs than elders in each study site. Fentahun and

Hager [8] also reported similar results, as younger respondents cited higher numbers than elders did. This might be because of the variation of informant's consumption preferences between adults and youngsters. Adults avoid eating most UWEPs because consumption of wild food plants is seen as a sign of poverty [42]. As a result, their list might be limited to only those species that they are accustomed to eating rather than what they know to be edible. The knowledge acquisition methods (parents are the major source) also proved the ignorance of some species by adults. In addition, younger people had an intimate association with plants in their day-to-day lives and had more experience with collection than the older people [8,43]. A lack of modern training in the study area is also an indicator of the acquisition of knowledge by experience and the effect of their interaction with plants in their daily life.

Table 16 Traditional management practices and threats of underutilized edible plants in the Chilga District, northwestern Ethiopia

CsI	LLA (48)		MLA (48)	
Management practices	Informants	%	Informants	%
Planting around home garden	9	18.8	18	37.5
Pruning	8	16.7	9	18.8
Pollarding	10	20.8	14	29.2
Fencing	10	20.8	11	22.9
Protected by culture	12	25	11	22.9
Threats				
Agricultural expansion	32	66.7	21	43.8
Construction	26	54.2	37	77.1
Fuel wood collection and charcoal making	32	66.7	40	83.3
Overgrazing	23	47.9	19	39.6
Fire	11	22.9	0	0

Table 15 Response of households on training gained about conservation and utilization of underutilized edible plants in the Chilga District, northwestern Ethiopia

Response	MLA (n = 48)		LLA (n = 48)	
	Informants	%	Informants	%
Yes	11	22.9	6	12.5
No	37	77.1	42	87.5
Total	48	100	48	100

Management practices and threats to underutilized edible plants

The farmers in the study area apply some management activities like pruning and pollarding for UWEPs in farmland use. In addition, the farmers plant some plants in their front yards and backyards. However, compared to other cultivated plants, the management practices were very small. The study by Fentahu and Hager [8] in other semi-arid areas of the Amhara region similarly depicted a lower level of management, and if any, limited to lopping and pollarding.

The threats to UWEPs in the study area are similar to threats that affect nonedible wild plants. Agricultural expansion, overgrazing, fire and other utilization-related factors such as fuel wood and charcoal, construction, and house and farm implements are the major threats in both agroecologies. Different researchers have documented these threats [3,7,13,38] to UWEPs. Furthermore, UWEPs are collected in natural environments, which are subjected to less management and exposed to anthropogenic threats [13].

Conclusions and recommendations

The Chilga District, located in the Amhara region, northwestern Ethiopia is endowed with diverse UWEPs and ethnobotanical knowledge. Thirty-three UWEPs (22 families) were recorded. These plants are distributed in natural forests, riverine forests and farmlands.

UWEPs are consumed for supplementing staple food, filling food gaps and recreational value. This utilization is significant in the poorest communities. Most of the encountered UWEPs provide other services including medicinal value, fuel wood and charcoal, construction, timber, and farm and household implements, thereby generating income from the sale of the products and their parts. Thus, UWEPs are the solution to diversification of rural household livelihoods. The utilization of some UWEPs (such as *S. guineense, D. abyssinica, C. spinarum, M. kummel, D. mesiliformis, X. americana and C. africana)* is more popular than for others.

UWEPs are suffering from the threats of agricultural expansion, overharvesting (for fuel wood, construction, and fencing) and overgrazing in both agroecologies. However, the management activities are also lower.

The local community utilizes UWEPs without gaining organized training, which is a threat to the sustainability of these plants. Thus, provision of training from the district, zonal offices, and NGOs for different management activities and the application of this training to projects that seek to maximize the value of UWEPs to local community are needed. This study focuses on diversity, role, threats and associated plant-use knowledge. Further study on the nutritional analysis and economic valuation of UWEPs is needed.

Abbreviations

FL: farmland; HH: household; IVA: importance value index; KA: kebele; KI: key informants; LLA: lowland agroecology; LSd: least significance difference; m.a. s.l: meters above sea level; MLA: midland agroecology; NF: natural Forest; R: riverine forest; UWEP: underutilized wild edible woody plant.

Competing interests

The authors declare that they have no competing interests.

Authors' contributions

MT: conception and design, data collection and analysis, manuscript writing and final approval of the manuscript. ZA: data analysis, critical revision and final approval of the manuscript. SZ: critical revision and final approval of the manuscript. All authors read and approved the final manuscript.

Acknowledgements

We thank to the Development Partnership in higher Education, Department for International development, (DeLPHE) for financial support of this research, which otherwise would have faced financial constraints. We are also thankful to the traditional healers and local people of Chilga District who generously shared their knowledge on underutilized edible plants. The Environmental Protection office and Agricultural office of Chilga District are also acknowledged for their provision of various basic data and support.

Author details

[1]Department of Natural Resource Management, University of Gondar, P.O. Box 196, Gondar, Ethiopia. [2]School of Forestry, Wondo Genet College of Forestry and Natural Resources, Hawassa University, P.O. Box 128, Shashemene, Hawassa, Ethiopia. [3]Department of Biology, Addis Ababa Science and Technology University, Addis Ababa, Ethiopia.

References

1. Magbagbeola JA, Adetoso JA, Owolabi OA: **Neglected and underutilized species (NUS): panacea for community focused development to poverty alleviation/poverty reduction in Nigeria.** *J Econ Intl Finance* 2010, 2:208–211.
2. Jaenicke H, Hoschele-Zeledon I: *Strategic Framework for Underutilized Plant Species Research and Development, with Special Reference to Asia and the Pacific, and to Sub-Saharan Africa.* Rome, Italy: International Centre for Underutilized Crops, Colombo, Sri Lanka and Global Facilitation Unit for Underutilized Species; 2006.
3. Assefa A, Abebe T: **Wild edible trees and shrubs in the semi-arid lowlands of southern Ethiopia.** *J Sci Dev* 2011, 1:5–19.
4. Van Andel T: *Non-timber forest products the value of wild plants.* Wageningen: Agrodok 39, Agromisa Foundation and CTA; 2006.
5. Hunde D, Njoka J, Zemede A, Nyangito M: **Wild edible fruits of importance for human nutrition in semiarid parts of east shewa zone, Ethiopia: associated indigenous knowledge and implications to food security.** *Pak J Nutr* 2011, 10:40–50.
6. IBC (Institute of Biodiversity Conservation): *Ethiopia: Second Country Report on the State of Plant Genetic Resources for Food and Agriculture (PGRFA) to FAO.* Addis Ababa, Ethiopia: 2007.
7. Kebu B, Fassil K: **Ethnobotanical study of wild edible plants in derashe and kucha districts, south Ethiopia.** *J Ethnobiol Ethnomed* 2006, 2:53.
8. Fentahun M, Hager H: **Wild edible fruit species cultural domain, informant species competence and preference in three districts of amhara region, Ethiopia.** *Ethnobotany Research & Applications* 2008, 6:487–502.
9. Tilahun T, Mirutse G: **Ethinobotanical study of wild edible plants of Kara and kwego semi-pastoralist people in lower Omo river valley, debub Omo zone, SNNPR, Ethiopia.** *J Ethnobiol Ethnomed* 2010, 6:23.
10. Demel T, Abeje E: **Status of indigenous fruits in Ethiopia. Review and Appraisal on the Status of Indigenous Fruits in Eastern Africa.** In *A*

synthesis report for IPGRI-SAFORGEN. Edited by Chikamai B, Eyog-Matig O. Kweka: 2004.

11. MoARD: *Horn of Africa Consultations of Food Security, Country Report.* Addis Ababa, Ethiopia: Ministry of Agriculture and Rural Development, Government of Ethiopia; 2007.

12. Sabates-Wheeler R, Tefera M, Bekele G: *Future-Agricultures, Working Paper 044.* Ethiopia: Future-Agricultures, Working Paper 044; 2012:43.

13. Lulekal E, Asfaw Z, Kelbessa E, Van Damme P: **Wild edible plants in Ethiopia: a review on their potential to combat food insecurity.** *Afrika Focus* 2011, **24**:71–121.

14. CDOA (Chilga District Office of Agricultural): *Chilga District Agricultural Office 2012 Annual Report.* Aykel, Ethiopia: 2012.

15. Bernard HR: *Research Methods in Anthropology: Qualitative and quantitative Approaches.* Walnut Creek CA: Altamira; 2002.

16. Cotton CM: *Ethnobotany: Principles and Applications.* Chichester, England: John Wiley and Sons Ltd; 1996.

17. Martin GJ: *Ethnobotany: A Methods Manual.* London, UK: Chapman and Hall; 1995.

18. Jain A, Sundriyal M, Roshnibala S, Kotoky R, Kanjilal PB, Singh HB, Sundriyal RC: **Dietary use and conservation concern of edible wetland plants at indo-Burma hotspot: a case study from northeast India.** *J Ethnobiol Ethnomed* 2011, **7**:29.

19. Hedberg I, Edwards S: **Pittosoraceae to Araliaceae.** In *Flora of Ethiopia and Eritrea Vol. 3.* Addis Ababa and Uppsala, Sweden: The National Herbarium Addis Ababa University; 1989:659.

20. Edwards S, Mesfin T, Hedberg I: **Canellaceae to Euphorbiaceae.** In *Flora of Ethiopia and Eritrea Vol. 2 Part 2.* Ethiopia and Upsala, Sweden: The National Herbarium Addis Ababa; 1995:456.

21. Edwards S, Sebsebe D, Hedberg I: **Hydrocharitaceae to Arecaceae.** In *Flora of Ethiopia and Eritrea, Volume 6.* Edited by Edwards S, Sebsebe D, Hedberg I. Addis Ababa and Uppsala: The National Herbarium; 1997:586.

22. Edwards S, Mesfin T, Sebsebe D, Hedberg I: **Magnoliaceae to Flacourtiaceae.** In *Flora of Ethiopia and Eritrea Vol. 2 Part 1.* Ethiopia and Upsala, Sweden: The National Herbarium Addis Ababa; 2000:532.

23. Mesfin T: **Asteraceae.** In *Flora of Ethiopia and Eritrea Vol. 4, Part 2.* Edited by Hedberg I, Friis I, Edwards S. Addis Ababa and Uppsala: The National Herbarium; 2004:408.

24. Hedberg I, Ensermu K, Edwards S, Sebessebe D, Persson E: **Plantaginaceae.** In *Flora of Ethiopia and Eritrea, Volume 5.* Edited by Hedberg I, Ensermu K, Edwards S, Sebessebe D, Persson E. Addis Ababa and Uppsala: The National Herbarium; 2006:690.

25. Hedberg I, Friis I, Person E: *General Part and Index to Vol 1–7. Flora of Ethiopia and Eritrea Volume 8.* Addis Ababa, Ethiopia and Uppsala, Sweden: The National Herbarium; 2009.

26. Stohlgren TJ, Falkner MB, Schell LD: **A modified Whittaker nested vegetation sampling method vegetation.** *J. Vegetatio* 1995, **117**:113–121.

27. Hoft M, Barik SK, Lykke AM: *Quantitative Ethnobotany. Applications of Multivariate and Statistical Analyses in Ethnobotany. People and Plants working paper 6.* Paris: UNESCO; 1999.

28. Clarke KR, Warwick RM: *Changes in Marine Communities: An Approach to Statistical Analysis and Interpretation, Primer-E.* 2nd edition. Plymouth: Springer verlag; 2001.

29. Maguran A: *Measuring Biological Diversity.* London: Blackwell science LtD, a Blackwell publishing company; 2004.

30. Pielou EC: **The measurement of diversity in different types of biological collections.** *J Theoret Biol* 1966, **13**:131–144.

31. Kent M, Coker P: *Vegetation Description and Analysis: A practical approach.* London: Belhaven Press; 1992.

32. Jha PK: **Environment plant resources and quantities estimation methods for vegetation analysis.** In *Ethnobotany for Conservation and Community Development.* Edited by Shrestha KK, Jha PK, Shengi P, Rastogi A, Rajbandary S, Joshi M. Nepal: Ethnobotanical Society of Nepal (ESO); 1997.

33. Adefires W: *Population Status and Socio-economic Importance of Gum and Resin Bearing Species in Borana Lowlands, Southern Ethiopia, M.Sc. Thesis.* Addis Ababa, Ethiopia: Addis Ababa University; 2006.

34. Bharucha Z, Pretty J: **The roles and values of wild foods in agricultural systems.** *Philoss Trans Royal Soc B* 2010, **365**:2913–2926.

35. Addis G, Urga K, Dikasso D: **Ethnobotanical study of edible wild plants in some selected districts of Ethiopia.** *J Human Ecol* 2005, **33**:83–118.

36. Fentahun M, Hager H: **Exploiting locally available resources for food and nutritional security enhancement: wild fruits diversity, potential and**

state of exploitation in the amhara region of Ethiopia. *J Food Sci* 2009, **1**:207–219.

37. Hunde D, Njoka J, Zemede A, Nyangito MM: **Seasonal availability and consumption of wild edible plants in semiarid Ethiopia: implications to food security and climate change adaptation.** *J Hort For* 2010, **3**:138–149.

38. Guinand Y, Dechassa Lemessa D: *Wild food plants in Southern Ethiopia: Reflections on the role of 'famine foods' at a time of drought.* Addis Ababa, Ethiopia: UNDP-EUE Field Mission Report; 2000.

39. Asfaw Z: **The future of wild food plants in southern Ethiopia: ecosystem conservation coupled with enhancement of the roles of key social groups.** *Acta Hortic* 2009, **806**:701–707.

40. Pauline M, Linus W: **Status of Indigenous Fruits in Kenya. Review and Appraisal on the Status of Indigenous Fruits in Eastern Africa.** In *A Synthesis Report for IPGRI-SAFORGEN.* FORNESSA-AFREA Logo; 2004.

41. Quailan R: **Consideration for collecting freelists in the field. Examples in ethnobotany.** *Field Methods,* **17**:1–16.

42. Garcia GSC: **The mother – child nexus: knowledge and valuation of wild food plants in Wayanad, Western Ghats.** *India J Ethnobiol Ethnomed* 2006, **2**:39.

43. Al-Qura'n SA: **Ethnobotanical and ecological studies of wild edible plants in Jordan.** *Libyan Agric Res Cent J Int* 2010, **1**:231–243.

8

Low potato yields in Kenya: do conventional input innovations account for the yields disparity?

Joseph Gichuru Wang'ombe[1,2*] and Meine Pieter van Dijk[1,3]

Abstract

Background: Potato yields in Kenya are less than half the amount obtained by some developed countries. Despite more acreage being dedicated to the crop, annual production has not improved. Kenya's low yields have been blamed on a failure to use clean seeds, fertilizers, fungicides and irrigation. The article examines the impact of adopting these innovations on enhancements of yields.

Results: The regression coefficients indicate that clean seeds have the greatest impact followed by irrigation, fungicides and fertilizers. However, clean seeds have the lowest adoption rate, with only 4.5% of the respondent sample using such seeds. Irrigation adoption was also low at 23% but there is widespread usage of fungicides and fertilizers at 92% and 96% respectively. Adoption of the four innovations more than doubled the yields but the absolute amount remained less than 50% of the 40 tons per hectare obtained by the leading world producers. The less than optimal gains can be attributed to the nonlinear relationships of the variables, which indicate the importance of more precise, proper application of inputs in order to obtain higher yields. Linear regression could only explain 10% of the variation but nonlinear regression improved R squared to 80%. The unexplained variables accounting for 20% appear to be essential for a further enhancement of yields, given the big difference between those currently achieved in Kenya and those in developed countries.

Conclusions: Whereas adoption of the inputs is important, there is a need to use precise, recommended application regimes in order to obtain better potato yields. Training, in the form of visits by innovation propagation agents, are shown to improve adoption rates although only about half (55%) of farmers reported receiving such visits in the preceding three years. This points to a need for the Ministry of Agriculture to lead in increasing the coverage of such visits. Taken together, the four innovations account for only a fraction of the yield variances highlighting the need for further research to identify other determinants of Kenya's low potato production.

Keywords: Potatoes, Innovation, Clean seeds, Fertilizers, Fungicides, Irrigation, Yields

Background

The most consumed food crop in the world is rice, followed by wheat, potatoes and maize in that order [1]. In Kenya, the potato is the second most important food crop after maize, which contributes 32% of overall dietary energy consumption and 68% of energy consumption from cereals [2]. The recurrent episodes of famine in periods of drought in recent years, coupled with Kenya's reliance on maize imports to meet its domestic needs suggest that the country has not thus far succeeded in realizing successful food security strategies. Indeed, contrary to other African countries, for example, Malawi, which have significantly reduced their dependence on cereal imports over the few years, in Kenya it has risen from 20.7% in 2000 to 2004, to 36.1% in 2007 to 2009 [3].

The potato has a demonstrated capacity to feed large populations. Nunn and Quin [4] showed how population and urbanization in Europe and America increased sharply during the eighteenth and nineteenth century following the introduction of the potato as a new food crop. The potato provides more food per hectare than other staples, given its short time to mature (80 to 120 days), which allows two crops per year. Now consumed in most regions of Kenya, the potato deserves consideration, therefore, as a potential focal crop in the

* Correspondence: gichuruwke@yahoo.com
[1]Maastricht School of Management, Maastricht, Netherlands
[2]African Population & Health Research Center, P.O. Box 10787, Nairobi 00100, Kenya
Full list of author information is available at the end of the article

country's quest to attain food security. This would call for an enhancement of the potato sector, which is currently classified under 'orphan crops' by the Kenyan Ministry of Agriculture due to its relatively low level of development.

Kenya's potato yields have remained low even as more land is devoted to the crop. This contrasts with the experience of other regions that have experienced the green revolution. Between 1966 and 1980, the acreage under potatoes in North America and Western Europe decreased annually by more than 2% but yields increased by almost 1% [5]. In Asia, the acreage increased by about 7% in the same period and was accompanied by annual yield growth of 2%. In Africa, however, despite a 4% rise in potato-farmed land, yields remained constant in the same period.

Indeed, yields in the continent are remarkably low at less than 20 tons per hectare for Africa compared to over 40 tons for developed regions like North America (Table 1). This suggests that there is an immense potential for improvement of potato yields in Africa. A set of 'green revolution' innovations that can lead to increased potato yields - clean seeds, fertilizers, chemicals and irrigation - are well known and considerable debate has focused on approaches to enhancing their rate of adoption. However, there has been virtually no formal evaluation of the outcomes of their use. This paper assesses the relative impacts of the adoption of clean seeds, fertilizers, fungicides and irrigation on potato yields in Kenya.

Theoretical framework
Increased productivity for sub-Saharan Africa and Kenya in particular is only going to be obtained through the

adoption of innovative approaches. An innovation has been defined as a new product, new technique, new practice or a new idea [6].

We look at key demand and supply factors contributing to the low level of adoption of conventional input innovations as well as their impact once used, as we seek to examine to what extent a greater adoption of such measures would significantly improve potato yields in Kenya. Existing literature on demand factors typically focuses on individual characteristics. Studies have shown, for example, that a younger age and better education of a household head, as well as a larger farm size are associated with a higher likelihood of adopting innovations [7,8]. Rogers [7], moreover, suggests that farmers with contacts outside the local community are more likely to embrace new techniques than those embedded in a traditional lifestyle, while Unwin [8] finds people who have been farmers all their lives to be more likely to adopt innovations.

Available research on the supply-side factors shaping the uptake of innovations considers the extent to, and ways in which an innovation is made available to potential users. The innovation is usually availed by institutions and groups outside the farming community. Constraints supply or supply constraints are viewed as being established and controlled mainly by government and private institutions [6].

Ruttan [9] has drawn several generalizations from literature on green revolution. Notwithstanding exceptions due to environmental differences, these include: (1) that new high yielding varieties (HYVs) were adopted at exceptionally rapid rates in those areas where they were

Table 1 Potato production, by region, 2009 and 2010

Country/region	2009			2010		
	Production (tons)	Area (Ha)	Yield/Ha	Production (tons)	Area (Ha)	Yield/Ha
Kenya*	2,299,090	120,246	19.12	2,725,940	121,542	22.43
Uganda	689,000	101,000	6.82	695,000	102,000	6.81
Tanzania	860,980	195,690	4.40	1,472,560	172,970	8.51
Eastern Africa	9,414,400	866,491	10.86	11,403,695	912,817	12.49
Malawi	3,427,660	161,923	21.17	3,673,540	160,600	22.87
South Africa	1,866,580	55,000	33.94	2,090,210	62,200	33.60
North Africa	8,446,031	376,750	22.42	9,522,840	386,350	24.65
Africa	22,047,635	1,691,139	13.04	25,378,948	1,827,282	13.89
Asia	145,841,189	9,031,772	16.15	159,055,291	9,193,486	17.30
European Union	62,694,978	2,087,172	30.04	57,485,831	2,018,117	28.48
South America	13,880,831	874,378	15.88	14,475,788	942,423	15.36
USA	19,622,500	422,492	46.44	18,337,500	407,923	44.95
North America	24,151,275	569,249	42.43	22,760,270	547,868	41.54
WORLD	329,581,307	18,651,838	17.67	324,420,782	18,653,007	17.39

*FAO yield estimates from 1999 to 2004 were all less than 10 tons per hectare and it is not clear what changed from 2005 onwards when estimates were more than doubled. Source: Food and Agriculture Organization (FAO), accessed April 2013.

technically and economically superior to local varieties. This illustrates that technical and economical evaluations have an impact on adoption; (2) neither farm size nor tenure has been a serious constraint to the adoption of HYVs of grain. While smaller farmers and tenants tended to lag behind larger farmers in the early years following their introduction, these lags typically disappeared within a few years; (3) the introduction of HYVs has resulted in an increase in the demand for labor; and (4) land owners have gained relative to tenants.

The adoption of an innovation is primarily the outcome of a learning and communication process. This implies that there are factors related to the effective flow of information and the characteristics of information flows, information reception and resistance to adoption. Adoption will depend on an individual's general propensity to adopt innovation or his innovativeness [7]. It will also depend on the congruence between the innovation and the social, economic and psychological characteristics of the potential adopter.

Most studies on innovation diffusion end with a discussion of its resultant adoption but do not look at outcomes of the adoption process. Until the 1960s, an underlying assumption of diffusion theory was that a new product or practice offered an indisputable benefit. Innovations were viewed as pure gains - a replacement of the outdated and inefficient with something better [10]. However, more recent research has drawn attention to negative social and environmental effects of innovations. Additionally, new technologies may not always result in expected improvements in outcomes such as yields. Given the poor potato yields in Kenya, we sought to establish whether the adoption of a package of the well-established innovation inputs would lead to significant improvements in yields. To this end we examined the extent and drivers of adoption rates and analyzed their impacts on yields.

Methods

A survey was conducted in 2010 and the first quarter of 2011 in three counties of Nakuru (Njoro and Kuresoi), Nyandarua (Nyandarua South, Nyandarua West and Nyandarua Central) and Meru (Meru Central and Buuri). The three counties are located in in the Rift Valley, Central and Eastern regions of Kenya, respectively. Central region is the leading producer of potatoes in Kenya followed by Rift Valley and Eastern Region. The study counties are the main potato growing areas in their respective regions and together account for approximately 95% of total potato production in Kenya [11].

The areas studied are all in high altitude- (between 1,400 and 2,700 meters above sea level) and high-rainfall zones, experiencing mean annual rainfalls of 1,000 mm or greater. Nyandarua County has temperatures ranging

from a minimum of 2°C to a maximum of 25°C. The rainfall ranges between 700 and 1,500 mm per annum [12]. In Meru County, annual temperatures range from a minimum of 16°C to a maximum of 23°C and rainfall from 500 to 2,600 mm. Temperatures in Nakuru County range from a minimum of 12°C to a maximum of 26°C per year with rainfall ranging from 1,800 to 2,000 mm. Maximum temperatures across all study counties, therefore, are sufficiently temperate, as are minimum temperatures - with the exception of Nyandaura. The highest variability in rainfall is recorded in Meru, where some areas receive less than 1,000 mm per year, which may explain the high usage of irrigation in the county. The predominant soil type is volcanic in Nyandarua and Meru but some parts of Nyandarua have red clay soil. Nakuru mainly has loamy soils.

Since no complete household survey has been carried out in in the last 5 years, we used data from the Kenya Integrated Household Budget Survey (KIHBS) 2005/ 2006 [13] to estimate the number of households producing potatoes. The total number of such households was 790,752, of which virtually all (97%) were located in the main Central, Rift Valley and Eastern producing-regions.

KIHBS data also provided estimates of the share of potato-growing households in each target county. In Nyandarua, 97% of farmers grew potatoes compared to 34% in Nakuru and 31% in Meru. Together, the three counties accounted for about 33% of all potato growing households in Kenya.

Relevant KIHBS data is aggregated at the level of households. Equally, the respondents targeted in our study were the heads of households. Interviews captured the demographic characteristics of the household head. The household is defined as a place where members 'eat from the same pot'. In the regions studied, this was also synonymous with housing units since independent households in these rural areas do not share the same house.

To be able to generate a random sample from the three regions, we used administrative district-level information gathered through a 2009/2010 enumeration of potato farmers by the Ministry of Agriculture. For some parts of Nakuru (Njoro and Kuresoi), data were incomplete, requiring us to employ a stage-wise stratified sampling approach, estimating the number of farmers in a village and selecting one at a constant interval.

The required sample size (n) was 381 as per the formula below. We however, targeted 419 farmers, assuming a 10% non-response rate and ended up with 402 completed questionnaires.

Formula:

$$n = \frac{t^2 \times p(1-p)}{m^2}$$

Description:

n = required sample size

t = confidence level at 95% (standard value of 1.96)

p = estimated proportion of farmers growing potatoes - used 55% average as per occurrence in KIHBS [8]

m = margin of error at 5% (standard value of 0.05)

Using the KIHBS [8] data for the farmers involved in potato production, the average occurrence was 54% as below:

$$0.97 * 104637/264729 + 0.31 * 40660/264729 + 0.34$$
$$* 90381/264729 = 54.7\%$$

Calculation:

$$n = \frac{1.96^2 \times .55(1-0.55)}{0.05^2}$$

$$n = \frac{3.8416 \times 0.2475}{0.0025}$$

$$n = \frac{0.9508}{0.0025}$$

$$n = 380.32 \text{ Approximately } 381$$

The survey questionnaire was designed to collect data that could be used to generate additional variables. To increase the reliability of self-reported data, the questions asked were simple and information sought easy to recall. For instance, on yields, farmers were asked about the portion of their land they had dedicated to potatoes in the last season and the production thereof. Total production was divided by the area to generate yield data. Since the study sought to examine production in general, data on the varieties grown were not collected. The specific fertilizers and fungicides used were recorded but the ranges of fungicides were too wide to be analysed meaningfully. Several types of fertilizers were reportedly used but most farmers were unable to recall the specific kind used. As they put it, they simply follow sellers' advice on the type to purchase. The analysis, therefore, ignores fertilizer distinctions. For irrigation, the data collected were on installed irrigation facilities as opposed to actual use. It was assumed that those who had installed facilities actually used them.

The analysis used the Chi-square and Fisher's test, regression and logistic regression, where the dependent variable was dichotomous. Stata/SE 10.1 was used for the analysis.

Study results and discussions

The study found clean seeds, fungicides and fertilizers to be the most important production inputs that could easily be identified by the farmers and impacted on yields. Irrigation, the other key innovation was captured alongside other household characteristics as opposed to a farm input.

We used logistic regression to investigate the relationship between household characteristics, communication variables and adoption of the three focal input innovations. Table 2 presents results of estimating the probability of adopting the three innovations. The Chi-square statistic indicates strong significance ($P < 0.01$) of two of the models (seeds and fertilizer) over the simple model that includes only a constant. The goodness of fit for each of the models can be assessed through the pseudo R-square measure, which in our models ranges from 7.2 to 20.2%. Our models thus have good predictive ability for adoption. We will look at each of the adoption outcomes separately.

We conducted both linear and nonlinear regression analyses to determine the contribution of farm inputs to yields. In both cases, we controlled for household characteristics and communication variables. A step-wise regression process was introduced given a very low R squared value in the linear regression. This led us to a nonlinear equation for regression analysis.

Adoption of clean seeds

A very high proportion of farmers (79%) are aware that they should use clean seeds but only 4.48% actually did so. This figure, moreover, is likely to be higher than in other less dominant potato growing areas. In the study, we only considered seeds bought from certified seed producers as clean. The price of clean seeds is more than double that of recycled uncertified seeds and readily obtainable. Our estimation based on data from the Kenya Plant Health Inspectorate Services [11], the sole certifier of seeds in Kenya, indicates that available stocks of certified seeds are only about 2% of the country's seed potato requirements. FAO estimates for 2009 put the acreage on which potatoes are grown in Kenya at 120,246 ha, suggesting a seed requirement of 240,492,000 kg (120,246 × 2,000). Kenya Plant Health Inspectorate Services (KEPHIS) data, therefore, indicates that the certified seeds quantities were only 0.21% of the seed potato requirements in 2009 (Table 3). Assuming that certified seeds are multiplied at least once to give clean but not certified seeds, that is, presuming a multiplication ratio of 1:10, we estimate that clean seeds available to farmers in 2009 may only have ended up meeting 2.1% of the seed potato requirement. With such a low percentage of supply in the market, the odds are that many farmers who may wish to adopt clean seeds have not been able to do so.

As mentioned above, existing research suggests that the younger the head of household, the better the education and the larger the farm, the more likely is a household's adoption of an innovation [7,8]. Though Obare et al. [15] had contrary findings that education has no effect on adoption, our results indicated no statistically significant relationship between the level of education

Table 2 Descriptive statistics for the full dataset

Variable	Mean value/percent	SD	Min	Max
Adoption of clean seeds	0.0447761	0.20707	0	1
Proper fertilizer use	0.181592	0.385988	0	1
Use of fungicides	0.9179104	0.274843	0	1
Age (young = less than 40 yrs, old = above 40 yrs)	0.2835821	0.451298	0	1
Gender (18.41% female)	0.1840796	0.388032	0	1
Education (None & Primary =0, Secondary, Post-secondary=1)	0.4825871	0.500319	0	1
Size of land	4.457916	5.873537	0	60
Employment status (Employeed=1)	0.0895522	0.285895	0	1
Irrigation	0.2338308	0.423793	0	1
Number of cows	2.375622	2.413739	0	20
Region (Eastern=1)	0.2985075	0.458174	0	1
Visited (Visited=1)	0.5472637	0.498381	0	1
Ownership of media equipment	0.920398	0.271013	0	1
Member of farmers group	0.3109453	0.463457	0	1

and the adoption of clean seeds, In contrast, land size and number of cows owned by the household head, indicated no significant results. Having been visited by propagation agents, had a significant positive association with adoption, the latter pointing to the importance of communication about clean seeds. The gender, age, employment status, use of irrigation and region of residence of the household head did not significantly predict adoption.

Some contextual considerations can help in interpreting these findings. Land is in many cases a sign of wealth, as is the number of cows owned. Given the high cost of clean seeds, it would have been expected that higher adoption would be found among those with more land and cows is not surprising and, indeed, is to be expected. Similarly, the scarcity of seeds likely renders efforts to procure them more worthwhile for farmers who will use them on a larger area. Clean seeds are not distributed but instead have to be sourced directly from the producers. Because procurement in most cases involves transportation, there are economies of scale when procuring for a bigger farm. The number of cows owned by a potato farmer (typically used for dairy production)

Table 3 Seed potatoes certified by Kenya Plant Health Inspectorate Services (KEPHIS) in kilograms

Year	Quantity in kg
2005	64,800.00
2006	418,300.00
2007	230,600.00
2008	369,091.00
2009	496,100.00

Source: KEPHIS [14].

is an important determinant of the income available to be invested in purchasing expensive seeds. However, only education and visits by innovation propagation agents influenced adoption meaning that failure to adopt clean seeds can largely be as a result of lack of knowledge. Visits by innovation propagation agents influenced the level of awareness of clean seeds - a prerequisite for their adoption. Taken together our findings suggest that although there is high awareness on the existence of clean seeds, it takes a higher level of education to appreciate the need to navigate the highly inefficient clean seeds supply chain. Interviews with seed producers indicated that in addition to a highly insufficient quantity of clean seeds in Kenya, the few existing seed multipliers lack effective marketing and distribution systems. As a result, farmers have to incur transport costs to collect seeds. The experience of one of the authors illustrates the long distances and attendant costs that can be involved: to obtain a large enough quantity of seeds for a field experiment, trips to two seed producers located more than 200 km apart were necessary. In addition to incurring transport costs, farmers pay high prices for certified seeds, which cost on average Ksh 2,000 per 50-kg bag compared to less than Ksh 1,000 for recycled or unclean seeds. Unless one fully appreciates the benefits of clean seeds, he or she is unlikely to take the trouble to source them.

Fertilizer usage

The proportion of farmers using fertilizer in the potato growing areas is higher than the national average of 69% [13]. Potatoes respond better to chemical fertilizers than do other crops that thrive with animal manure. Use of animal manure, which may be contaminated with bacteria wilt, carries the risk of serious disease for potato plants. Farmers are, therefore, discouraged from using animal manure unless they are sure it is clean. A large majority of interviewed farmers (96%), therefore, reported using fertilizers. This raises a key question as to how their usage compares to recommended practice.

We asked each farmer about the size of land on which they grew potatoes and the amount of fertilizer they used. Using a standard recommended rate of four 50-kg bags per acre, we were able to examine the extent of adoption of fertilizers. Our findings show that only 18% of those using fertilizers are using the required quantities, 72% use less than the requirement and 8% use excess quantities. The under-use of fertilizer is certain to affect yields, given that the land on which potatoes are grown is typically over farmed - making fertilizer-use paramount for obtaining good harvests. It would appear that the recommended quantity of fertilizer per area is not commonly known.

Among the characteristics of the household head, only–ownership of irrigation equipment and the region the farmer come from were found to have a significant relationship with fertilizer usage. Age, education level, employment status and size of land were found to be insignificant. Irrigation is mainly applied in only one of the regions studied. The fact that proper usage of fertilizer is associated with irrigation use and the region of residence from indicates that proper usage is more prevalent among commercial farmers. These are farmers who are investing more on farming including the use of irrigation equipment. They therefore go an extra mile to ensure that they use the appropriate quantity of fertilizer in order to obtain optimum yields.

Among the communication variables only membership of farmer group was significant predictors of correct fertilizer use. It was surprising that visits by innovation propagation agents did not influence proper usage of fertilizer. Those who have been visited by innovation propagation agents in the last three years are more likely to apply the right quantity of fertilizer (24.54% compared to 10.44% for those not visited). The visits are highest in Eastern Region (73%), second highest in Central Region (61%) and lowest in the Rift Valley Region at 30%. This difference arises from the region of residence which emerged as a significant factor for proper application of fertilizers. The Ministry of Agriculture was identified as the leading innovation propagation agent that had visited farmers in the last three years, accounting for 80% of total visits. Though about half (55%) of the farmers were visited, the visits were very high in the Eastern region at 73% as compared to 61% for Central and 30% for the Rift Valley region. The Ministry of Agriculture has a strategy of only seeing farmers who are involved in particular projects or specifically requests a visit.

There are significantly better results for fertilizer usage for those who also use clean seeds (Table 4). The high cost of clean seeds may encourage efforts to properly apply fertilizer in order to ensure that investment in such seeds is properly recovered through good yields.

Use of fungicides

Besides bacteria wilt, another serious disease for potatoes in the tropics is late blight. The disease is controlled by spraying fungicides. Virtually all respondents (92%) reported using fungicides. Computing the extent of usage was not possible, however, given the different brands in the market, the different application regimes and the fact that application is in most cases dependent on the weather conditions. We were therefore unable to work out the extent of usage.

A correct, early timing of the first fungicide application is as important as its extent for forestalling disease. However, clearly not all farmers are cognizant of this. Responses to the question 'what prompts first application of fungicides' included, 'after germination for control purposes', 'as a preventive measure', 'when symptoms of disease are identified', and 'when weather changes' (rain and cold temperatures). About a quarter of the farmers did not answer the question and some gave multiple answers. We therefore considered it inappropriate to analyze the quality of usage. We therefore focused only on application and non-application. The number of cows, size of land owned and the region of residence emerged as a significant predictors. The first two are proxies for wealth indicating that there is more adoption of fungicides by well to do farmers. The log likelihood was only significant at 10%.

Fungicide manufacturers and retailers in Kenya engage in aggressive marketing and distribution, among some including through use of dedicated field officers to work with farmers. This likely explains the many different brands being used and the high level of uptake. Knowledge of fungicides is well-spread in the farming communities. Semi-structured interviews with the farmers revealed that most of them routinely shop for fungicides. Their application regime, however, appears to be rather haphazard.

Clean seeds, fertilizer use, fungicides, irrigation and yields

As a next step in the analysis we introduced irrigation into a regression model comprising all key variables, with a view to investigating their impacts on yields. Though irrigation has mainly been considered a household characteristic, experiences of other potato-producing countries, like South Africa and Egypt, suggest that it is an important determinant of productivity. South Africa and Egypt have average yields of 33 and 25 tons per hectare respectively, with 75 and 100% of the crop, respectively, grown under

Table 4 Adoption of clean seeds, proper use of fertilizers and use of fungicides

Variable	Seeds			Fertilizer			Fungicides	
	Coefficient	Robust SE	P	Coefficient	Robust SE	P	Coefficient	Robust SE
Age	-1.480892	1.045255	0.157	-0.3782178	0.3851948	0.326	-0.1278965	0.4698768
Gender	-0.321191	0.7842926	0.682	-0.1184211	0.4007041	0.768	-0.4856607	0.4495826
Education	0.9821727	0.5520274	0.075*	0.2845236	0.3259048	0.383	0.4638435	0.3788872
Size of land	0.0258104	0.0465198	0.579	-0.0391673	0.0398019	0.325	-0.0846073	0.0458748
Employment status	0.0255295	0.8811416	0.977	0.8092282	0.5437259	0.137	-0.0704882	0.8370973
Irrigation	0.4484032	0.799934	0.575	-1.18545	0.392375	0.003**	0.4234932	0.5356169
Number of cows	0.1624452	0.1127645	0.15	0.016982	0.0854269	0.842	0.2514047	0.1170051
Region	-0.2813659	0.8075803	0.728	2.297463	0.3866888	0***	-1.164979	0.5331393
Visited	1.692598	0.6676563	0.011**	0.4122967	0.3488554	0.237	0.343017	0.4321116
Ownership of media equipment	-0.8326009	0.8493516	0.327	0.6678926	0.8132582	0.412	0.193444	0.6770312
Member of farmers group	0.242039	0.5296039	0.648	0.7640476	0.3192164	0.017**	0.2863781	0.5342112
Use clean seeds				1.280824	0.5803187	0.027**	-0.1089398	1.212624
Proper fertilizer use	1.293441	0.6205842	0.033**				0.6277023	0.7238252
Use fungicides	0.2355052	1.908216	0.879	0.5306858	0.7029454	0.45		
Constant	−5.995303	3.703948	0.036	-3.941271	1.307758	0.003	2.065371	0.761221
Log likelihood	-56.639018		0.0***	-143.30835		0.0***	-97.736995	
Chi-square statistic	50.79			63			18.55	
Pseudo-R2	0.2179			0.2097			0.068*	
Number	379			379			379	

Results as presented are for estimation of the probability of adopting the three innovations. *P <0.1, **P <0.05, ***P <0.01 (two tailed test).

irrigation [16,17]. The regions studied in this research had an irrigation rate of 4, 59 and 13% for Central, Eastern and Rift Valley, respectively, with an overall average of 23%.

Our findings presented in Table 5 show an average yield of 14.48 tons per hectare for those using fertilizers in the right proportion as compared to 11.68 tons per hectare for those using lower than the recommended quantity of fertilizer. Farming households using in excess of the recommended quantity produce only marginally more (12.91) than those using less than the suggested amount. Those using clean seeds have an average yield of 15.75 tons per hectare as compared to 12.00 tons per hectare for those not using clean seeds. There is less variation in yields for those who use fungicides and those who do not, with 12.42 and 9.45 tons per hectare, respectively. Similarly, households applying irrigation had an average yield of 13.81 as compared to 11.66 tons per hectare for those without irrigation facility. Those both using clean seeds and applying the appropriate quantity of fertilizer were obtaining an average yield of 18.35 tons per hectare as compared to 16.39 tons per hectare for households combining the use of clean seeds and fungicides. The overall average yields for the whole sample studied was 12.17 tons per hectare. The apparent differential impact of innovations is underscored in Table 6, which shows linear and nonlinear regression results for clean seeds, proper fertilizer use, use of fungicides and irrigation as independent variables and yield as the

dependent variable. The results show a considerably higher coefficient for clean seeds than for fungicides or fertilizers, suggesting a greater impact of clean seeds on enhancing yields. Regressing the three variables on yields (seeds, fertilizer and fungicides), a low R^2 value of 5% was obtained, which only rose to 12.3% when introducing socioeconomic characteristics[a] to the model. The low R^2 value may be attributable to nonlinearity where a number of factors contribute to random or unpredictable behavior. Instances of heteroskedasticity or even non-normality may also be contributing factors. We therefore tested whether or not linear regression assumptions were violated by examining the residuals for normality. The Smirnov-Kolmogorov test gave a probability of less than 0.05 indicating that the residuals are non-normally distributed. However, the values of the standardized residuals did not exceed 3.5 or fall below −3.5 suggesting that there were no outliers. We used the Cook-Weisberg test to check for heteroscedasticity. We obtained an insignificant result, indicating lack of heteroscedasticity or homoscedasticity (presence of equal variance of the residuals along the predicted line). We tested for multi-co-linearity but obtained variance inflation factors of less than 5, indicating that the multiple models did not include two or more highly correlated predictor variables.

The histogram of the independent variable (yield) indicates that it is not normally distributed. Statistical and

Table 5 Fertilizer use per acre, clean seeds, irrigation and average yields

	Average yield (Tons/Ha)	SD	Min	Max
Fertilizer use				
Proper use	14.47705	5.640542	2.7181	25.00652
Underuse	11.68091	6.224888	1.019287	33.16082
Overuse	12.91097	6.671999	2.17448	26.09376
Seeds				
Clean seeds	15.74899	4.371697	5.4362	21.7448
Non-clean seeds	12.00645	6.340899	0.81543	33.16082
Fungicides				
Use fungicides	12.41601	6.268137	0.81543	33.16082
Do not use	9.454046	6.229853	2.71810	27.18100
Irrigation				
Has irrigation	13.81068	5.936193	1.08724	33.16082
Does not have irrigation	11.66037	6.345868	0.81543	28.26824
Combined				
Clean seeds and proper fertilizers	18.34717	2.639298	13.04688	21.74480
Clean seeds and fungicides	16.39354	3.58500	9.78516	21.74480
Proper fertilizers and fungicides	14.57256	5.603104	2.71810	25.00652
Clean seeds and irrigation	13.81068	5.936193	1.08724	33.16082
Clean seeds, proper fertilizers and fungicides, irrigation	18.34717	1.545576	16.03679	19.29851
Overall (general)	12.16793	6.310752	0.81543	33.16082

graphical tests on Stata software indicate nonlinearity for fertilizer use per acre and yields. Yields increase with additional fertilizer application up to a certain point after which they decrease. The graph of yields on fertilizer approximates a parabola with a vertex points (279.50, 14.15) representing the highest point for which fertilizer usage per acre gives the highest yield on the fitted curve. The function can be expressed as:

$$Y = 0.0402097 * F - 0.0000678 * F * F$$

Where F is the fertilizer use per acre. The peak for the x axis is 279.5, which is the optimal quantity of fertilizer recommended. The peak the y axis is 14.15 tons per hectare (Figure 1).

We then conducted a nonlinear regression analysis of yields on the four independent variables, where b1, b2, b3 and b4 are the coefficients for the independent variables, as follows:

$$nl\,(yield = (bo + \{b1\} * (fertuseacre$$
$$+ \, b2\,fertuseacre * feruseacre)$$
$$+ \, \{b3\} * seeds_{clean} + \{b4\} * usefungicides.$$

The coefficients in the nonlinear regression still show that clean seeds have the highest impact (also in Table 6). Removing the linearity assumption improves the R^2 value to 8.54%

leaving a whole 93% unexplained variance. The inclusion of social economic characteristics in the regression equation above improves R^2 value to 13.82%. We speculate that the unexplained variance could be due to other ecological factors in the value chain that have not been incorporated in the regression model. These include a high micro-variability in land quality - that is a relative instability of soils and their different responses to application of inputs [18], as well as soil conditions prior to use of clean seeds, fertilizer, or fungicides. Soils that already contain bacteria wilt will limit yields even when all required inputs are used correctly. Overmined soil may require additional organic materials besides fertilizers to obtain optimum results for a fertilizer. Climate variations are also important, given that potato diseases such as late blight are made worse by temperatures variations. These other factors not included in the model may explain the large confidence intervals reported in the model.

Conclusions
Determinants of innovation adoption
In the three models of clean seeds, proper fertilizer use and use of fungicides, age, size of land, education level, number of cows, region, visits of innovation agents, possession of irrigation equipment and membership of farmers' groups are shown to have a significant positive association with

Table 6 Nonlinear and linear regression on yields

Yield	Linear regression		Nonlinear regression	
	Coefficient	P value	Coefficient	P Value
Fertilizer use per acre	0.01168	0.001**	0.0362579	0.00***
Fertilizer use per acre squared			-0.0000605	0.001***
Clean seeds	3.11066	0.041*	2.450674	0.109
Use fungicides	1.84136	0.138	2.322216	0.037**
Constant	9.06671	0	7.172435	***
R^2	4.78		8.54	

Yield	Controlling for social-economic characteristics			
	Linear regression		Nonlinear regression	
	Coefficient	P Value	Coefficient	P Value
Fertilizer use per acre squared			-0.0000483	0.009**
Fertuse per acre	0.0070554	0.056*	0.0273886	0.001***
Use clean seeds	2.190173	0.165	1.805383	0.25
Use fungicides	2.667031	0.018**	2.352725	0.037*
Has irrigation equipment	0.5149959	0.56	0.6793034	0.44
Size of land owned	0.0548732	0.427	0.0420069	0.541
Education - Secondary Sch. & above	0.3303141	0.624	0.387795	0.563
Gender (Female)	-1.812674	0.027**	-1.794205	0.027**
Not employed	-0.3628529	0.751	-0.1604294	0.888
Retired	-1.093352	0.415	-1.040537	0.434
Age (40 -49 years)	-1.060726	0.225	-1.005677	0.246
Age (50-59 years)	0.1877138	0.833	0.4085131	0.646
Age (Above 60 years)	-1.264216	0.204	-0.9387634	0.345
Region - Central	-2.018409	0.033**	-1.752407	0.064*
Region - Rift Valley	-2.733439	0.006**	-2.257026	0.025**
Visited by Agr. Officer	0.4646729	0.507	0.2337566	0.738
Own Radio or TV	-1.090301	0.365	-1.092433	0.36
Member of farmer group	0.1880047	0.794	0.1577523	0.826
Number of cows	0.0860794	0.601	0.0973959	0.551
Constant	11.76595	0.000***	10.34928	0.000***
R squared	12.3		13.82	

* $P <0.1$,** $P <0.05$, *** $P <0.01$.

adoption in at least one model, with the region of residence being significant in two models (Table 4).

Usage of clean seeds is influenced by awareness levels as indicated by the education level (secondary school and above) and visits by innovation propagation agents. The low supply of clean seeds and the complex procurement issues make only those who fully appreciate the impact of using clean adopt them. Proper fertilizer usage is evident in the region that have also adopted irrigation indicating commercialization of agriculture including the usage of irrigation is a stronger factor in the usage of appropriate quantities of fertilizers. Use of fungicides appears to be influenced by wealth levels as represented by the size of land and the number of cows that a farmer has. The region

of residence which is also significant for use of fungicides could also be pointing to the different levels of wealth in the three regions studied.

The importance of communication variables is clearly shown in the case of usage of clean seeds. Visits by innovation propagation agents were significantly associated with the use of clean seeds. Communication variables influence adoption and when lacking in supply they constrain adoption. Visits by innovation propagation agents positively influenced the adoption of clean seeds and proper usage of fertilizers. The Ministry of Agriculture may need to review its visits strategy to accommodate routine visits to farmers at least once a year. This used to be the practice in the 1970s but was abandoned due to resource constraints.

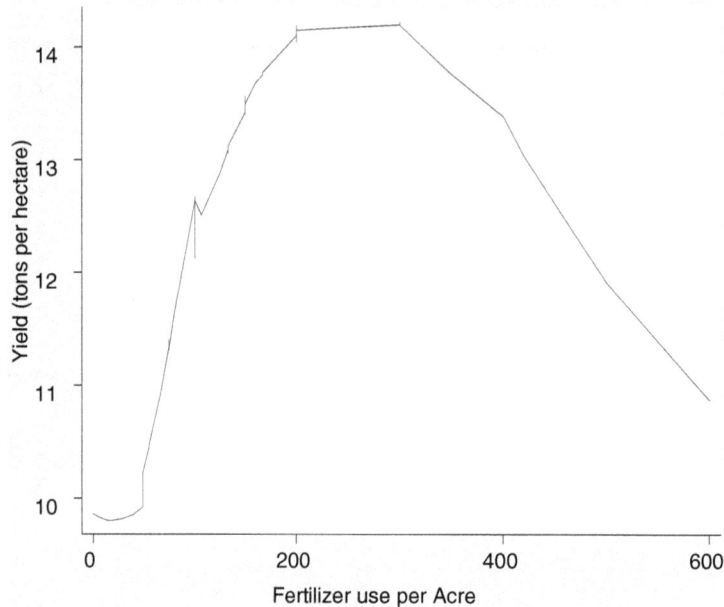

Figure 1 Graph on yields and fertilizer use.

More investments in agriculture to allow increased visits to farmers, combined with strategies to successfully and systematically convey information on 'best' agronomic practices to farming communities may help enhance the currently poor performance of potatoes in Kenya. Issues like improper use of fertilizer could likely be resolved to a great extent if farmers were adequately trained.

Adoption of innovations and yields

The adoption of clean seeds, proper usage of fertilizers, fungicides and irrigation enhances potato yields, with clean seed use giving the greatest enhancement. The existence of nonlinear relationships suggests that precision in fertilizer usage is critical since there is an optimum application point. This further underscores a need for effective training or raising awareness among potato farmers about key recommended practices for optimum yields of their crop. Given the importance of precision, the low average potato yields in the study area, reaching a maximum of 18.35 tons are unsurprising. To achieve harvests of up to 40 tons per hectare as in some developed regions, more precise application is required. Additionally, Innovations such as soil testing to determine the input requirements will help determine the appropriate application regime. Better control of adverse factors like bacteria wilt, which manifest when clean seeds are used on diseased farms, would also contribute to enhancing yields. Other factors may be uncontrollable, such as climate.

Though a wider adoption of clean seeds, proper usage of fertilizers, fungicides and irrigation as they are applied currently is unlikely to contribute to drastic improvements in potato yields for Kenya, it is still a worthwhile goal to pursue, as the present low national outputs do not do justice to the huge tracts of land devoted to the crop. An adoption of these key inputs by a higher percentage of farmers could lead to more than a doubling of Kenya's output. If this is done alongside a promotion of more potato consumption, Kenya would reduce its maize imports and enhance its food security.

Endnote

[a]Social economic characteristics used were education, employment status, gender, region, whether visited by agricultural officials, possession of radio, whether they keep records, distance from a paved road and whether a member of a farmer's group.

Abbreviations
FAO: Food and Agriculture Organization; HYV: High yielding varieties; KEPHIS: Kenya plant health inspectorate services; KIHBS: Kenya integrated household budget survey.

Competing interests
Both authors declare that they have no competing interests.

Authors' contributions
JGW carried out the field work and did the analysis. Professor MPvD was involved in the design of the study and supervised the work, which is part of a Doctorate of Business Administration at Maastricht School of Management, Netherlands. Both authors read and approved the final manuscript.

Authors' information
JGW is a doctoral candidate at Maastricht School of Management, Netherlands. He works with the African Population and Health Research Center where he holds the position of Director of Operations. MPVP is JGW's supervisor at Maastricht School of Management (MSM). Besides being a professor at MSM, he is a Professor of Water Services Management at the UNESCO-IHE Institute for Water Education.

Acknowledgements
We acknowledge the immense help we obtained from the Ministry of
Agriculture officials both at the head office and the counties studied in
the sampling and collection of data. We also acknowledge the Round Table
Africa program of the Maastricht School of Management that funded part of
the study. We also acknowledge Mr Maurice Mutisya, Dr Moses Oketch and
Dr Djesika Amendah, all of the African Population and Health Research
Center for support in data analysis. We thank Dr Isabella Aboderin' for
painstakingly editing the paper.

Author details
[1]Maastricht School of Management, Maastricht, Netherlands. [2]African
Population & Health Research Center, P.O. Box 10787, Nairobi 00100, Kenya.
[3]UNESCO-IHE Institute for Water education, Delft, Netherlands.

References
1. FAO: *FAO Statistics Division*. 2011.
2. FAO: *FAO Statistics Division*. 2009. http://faostat.fao.org/site/567/default.
    aspx#ancor, accessed in April 2012, November 2012, and April 2013.
3. FAO: *FAO Statistics Division*. 2012. http://faostat.fao.org/site/567/default.
    aspx#ancor, accessed in April 2012, November 2012, and April 2013.
4. Quin NN: **The potato contribution to population and urbanization:
    evidence from a historic experiment.** *Q J Econ* 2011, **126**(2):593–650.
5. Van der Zaag D, Horton D: **Potato production and utilization in world
    perspective with special reference to the tropics and sub-tropics.**
    Eur Assoc Potato Res 1983, **26**(4):323–362.
6. Brown L: *Innovation Diffusion: A New Perspective*. New York: Methuen &
    Company, NY; 1981. 1(X)17.
7. Rogers E: *Diffusion of Innovations*. 5th edition. New York: The Free Press; 2003.
8. Unwin T: **Household characteristics and agrarian innovation adoption in
    North-West Portugal.** *Trans Institute British Geographers, New Series* 1987,
    12:131–146.
9. Ruttan V: **The green revolution: seven generalizations.** *Int Dev Rev* 1977,
    19:16–23. December 1977.
10. Redmond W: **Innovation, diffusion, and institutional change.** *J Econ Issues*
    2003, **37**(3):665–679.
11. Ministry of Agriculture: *Annual report for 2008*. Nairobi, Kenya; 2008.
12. *Counties Profiles*. 2013. http://www.softkenya.com accessed April 2013.
13. Republic of Kenya: *Kenya Integrated Household Budget Survey 2005/6. Central
    Bureau of Statistics*. Nairobi, Kenya: Ministry of Planning and National
    Development; 2007.
14. KEPHIS: *Data provided by the organization officers*. Nairobi, Kenya; 2009. April 2012.
15. Obare G, Nyagaka D, Ngoyo W, Mwakubo S: **Are Kenya farmers allocatively
    efficient? evidence from Irish Potato producers in Nyandarua North
    District.** *J Dev Agric Econ* 2010, **2**(3):078–085.
16. Black V: *Hot Potato. GM Potatoes in South Africa, a critical analysis*. South
    Africa: The African Center for Biosafety; 2008.
17. FAO: *FAO Statistics Division*. 2005. http://faostat.fao.org/site/567/default.
    aspx#ancor, accessed in April 2012, November 2012, and April 2013.
18. Matlon PJ, Spencer DS: **Increasing food production in sub-Saharan Africa:
    environmental problems and inadequate technological solutions.**
    Am J Agric Econ 1984, **66**(5):671–676.

9

Enhancing crop shelf life with pollination

Björn Kristian Klatt[1,2*], Felix Klaus[1], Catrin Westphal[1] and Teja Tscharntke[1]

Abstract

Background: Globally, high amounts of food are wasted due to insufficient quality and decay. Although pollination has been shown to increase crop quality, a possible impact on shelf life has not been quantitatively studied.

Results: We tested how shelf life, represented by fruit decay, firmness and weight, changes as a function of pollination limitation in two European, commercially important strawberry varieties. Pollination limitation resulted in lower amounts of deformed fruits. Whereas 65% of wind-pollinated fruits were deformed, open pollination resulted in only 20% deformed fruits. During storage, the proportion of decayed fruits increased in relation to the degree of deformation. In the variety Yamaska, 80% of the fruits with high degrees of deformation decayed after four days, whereas in the variety Sonata, all highly deformed fruits had already decayed after three days. Fruit weight decreased independent from the degree of deformation. However, strongest deformations resulted in a generally lower fruit weight in Sonata, whereas in Yamaska, also medium deformed fruits had a lower weight than highly deformed fruits. Effects of deformation on firmness declines were mostly variety dependent. Whereas firmness declined similarly for all degrees of deformation for Yamaska, highly deformed fruits lost firmness fastest in Sonata.

Conclusions: Our results suggest that crop pollination has the potential to reduce food loss and waste in pollinated crops and thus to contribute to global food security. However, this relationship between pollination and food waste has so far been almost completely ignored. Future pollination research should therefore focus not only on yield effects but also on crop quality. A more comprehensive understanding of how pollination can benefit global food security should lead to a more efficient crop production to help meeting future food demands.

Keywords: Decay, Deformation, Food loss, Food waste, Fruit quality, Pollination limitation

Background

The global population is predicted to increase to up to 9 billion people by the year 2050 [1]. The main consequence will be a rising demand for food, which highlights the importance of global food security [2]. Fruits and vegetables form a substantial proportion of human food with a global consumption of more than 1.5 million tons in 2011 [3]. They contribute to a healthy human diet by providing essential nutrients such as vitamins, antioxidants and fibre [4]. Many people are lacking a sufficient nutrient supply even today [1]. Nevertheless, large portions of fruits and vegetables are being lost due to degradation during handling, transport and storage directly after harvest or are wasted at retail and consumer levels [5]. Thus, nutrient

* Correspondence: klattbk@googlemail.com
[1]Agroecology, Department of Crop Sciences, University of Göttingen, Grisebachstraße 6, D-37077 Göttingen, Germany
[2]Centre for Environmental and Climate Research, University of Lund, Sölvegatan 37, SE-22362 Lund, Sweden

supply is not only a matter of production quantities, but further depends on the quality of agricultural products, which has become a major problem with increasing attention in policy and scientific research [6].

An important factor determining the quality of fruits and vegetables is their shelf life [7,8]. In particular, fruits have a relatively short shelf life leading to declining quality during storage due to degradation of the fruit through softening, weight loss and decay [8]. Several studies have focused on the potential extension of fruit shelf life [8] by using postharvest treatments like modified storage procedures with specific coatings [9] or heat treatments [10]. In addition, quality manipulations in transgenic plants have been considered [11]. There is evidence from a few recent studies that insect pollination may not only benefit crop yield but also influence the shelf life of agricultural products. Greater firmness of insect-pollinated tomato [12], oriental melon [13], cucumber [14] and strawberry [15] only indirectly indicates possible effects of insect pollination on shelf life. Pollinator-enhanced shelf life could be an important solution to reduce postharvest losses, but data proving a direct relationship between insect pollination and shelf life are still lacking. Furthermore, firmness has been used as a proxy for shelf life, whereas it has not yet been tested whether increased firmness results in pollinator-enhanced shelf life under storage conditions.

The aim of this study was to test the direct relationship between insect pollination and crop shelf life, using strawberry as a model crop. The economic importance of strawberry is increasing globally [3], and insect pollination can improve yield as well as quality. Strawberries have a short shelf life because of fast quality loss during storage, which is due to high metabolic activity and sensitivity to fungal decay [16]. Almost 90% of fruits are lost after only four days in storage [16]. Thus, shelf life is an important determinant of postharvest quality in strawberries [17]. We analysed the impact of pollination on the shelf life of strawberries based on the degree of fruit deformation, which is another important reason for quality loss in strawberries. Deformations are caused by pollination limitation, which leads to achenes, the true nut-fruits of the strawberry, being unfertilized and thereby unable to build tissue [17]. Firmness, fruit weight and decay were used as fruit quality parameters determining shelf life [16]. We expected fruit quality to decline during storage due to decreasing firmness and fruit weight and increasing decay of the fruits. The degradation was expected to vary in relation to the degree of deformation, which is directly related to pollination limitation [17-19].

Methods

The study was conducted on a conventionally managed strawberry field near the city of Göttingen in 2012, focusing on the simultaneously yielding varieties Yamaska and Sonata. For the variety Sonata, 15 pairs of adjacent strawberry plants were randomly selected to assess whether fruit deformations were a result of pollination limitation. From each pair, one plant was covered with gauze to prevent insect pollination (wind pollination treatment), whereas the other plant was left open and thus accessible for insect pollinators. Three flowers from each plant were selected for analysis before pollination. For those flowers, fruit set was recorded and the fruits were harvested at maturity, when the entire fruit showed an intense red colour.

To assess the relationship between pollination limitation and shelf life, we focused on fruits showing different degrees of deformation, when these were obviously caused by pollination limitation. As development of all achenes depends on pollination and deformations are the result of missing achenes, pollination limitation is visible by aggregations of small unfertilized achenes at the deformation. Fruits from both varieties, Sonata and Yamaska, were harvested at maturity and then grouped in three categories based on their degree of deformation (Figure 1) following the official trade guidelines [20]. Fruits without deformations were assigned to the group 'None', fruits with slight to medium deformations were assigned to the group 'Medium' and heavily deformed as well as overall misshapen fruits were assigned to the group 'High'. All selected fruits did not show any physical damage or fungal infection. Strawberries flower in consecutive flowering periods [21]; only data collected from the second flowering period were analysed because of low numbers of fruits from other flowering periods.

The fruits were stored at 20°C for four days to simulate retail conditions [16,22]. To prevent fruits from infecting each other with fungi or being mechanically damaged during storage, fruits were carefully laid in egg boxes, eliminating direct contact. On each consecutive day, a random set of 7 to 13 fruits was selected from each group of deformation ('None', 'Medium', 'High'), and shelf life was assessed by analysing firmness, fruit weight and the proportion of decayed fruits. First, each fruit was visually inspected for surface damage and fungal decay and then weighed (BA2001 S, Sartorius). Firmness was than analysed using a texture analyser (TA-XT2 Texture Analyzer, Stable Micro Systems) following Sanz et al. [22]. The

Figure 1 Strawberry fruits with different degrees of deformation. (A) Fruit without deformations (*None*). **(B)** Fruits with slight up to medium deformations (*Medium*). **(C)** Fruits with high deformations (*High*).

peduncle and calyx were removed and fruits were bisected. Firmness was measured at the centre of each half. The texture analyser was fitted with a 5-mm-diameter probe and a 25-kg compression cell with the following adjustments: pre-test speed 6.00 mm/s; test speed 1.0 mm/s; post-test speed 8.0 mm/s; penetration distance 4 mm; trigger force 1.0 N. The maximum force in Newtons reached during tissue breakage was recorded as a measure of firmness [22], and mean values of both halves for each fruit were used for statistical analysis.

We used generalized linear models 'glm'-function in package 'stats'; [23] in R 3.1.1 [24], to test whether the amount of deformed fruits differed between open and wind-pollinated plants, using quasi-poisson distribution to account for overdispersion. The influence of fruit deformation on shelf life was analysed using generalized linear mixed effects models 'glmer' function in package 'lme4'; [25] by testing whether degrees of deformation in interaction with storage time had an effect on decay, firmness and fruit weight. According to our study design, degrees of deformation and storage time were also used as random effects. First, the model of each shelf life parameter was simplified until reaching the best fit by stepwise deleting interactions and fixed effects, using second order Akaike's Information Criterion 'AICc'-function in package 'MuMIn' [26]. In all models, the interaction between storage time and degrees of deformation had to be deleted, whilst storage time and degrees of deformation stayed. Second, we tested whether the different degrees of deformation equally contributed to explain changes in the response variable by comparing a model with degrees of deformation kept separately (full model), models with successively pooled degrees of deformation and a model without fixed effects (see Additional file 1) [27]. Again, AICc was used for model comparisons and the results were listed in Table 1. The lowest AICc for models with pooled levels indicated that these levels did not differ, whereas the lowest AICc for the full model indicated that degrees of deformation generally differed. If there was no difference between any

degree of deformation, the model with just time as a fixed effect had the lowest AICc value. Decay was modelled using binomial distribution, firmness and fruit weight assuming normal distribution. Residuals were inspected to meet model assumptions of variance homogeneity and specific distributions and data were transformed where necessary. There were several obvious measurement failures from the last day in storage for highly deformed fruits from the variety Sonata. Few mistakes have happened during harvest due to possibly harvesting the wrong variety, in the identification of the degree of deformation or during measurements, but could not be post-experimentally evaluated and thus these values were excluded from analysis.

Results

The amount of deformed fruits differed significantly between open- and wind-pollinated plants ($F_{1,24}$ = 11.088; P = 0.003; Figure 2). On average, less than 20% of the open-pollinated fruits showed deformations, whilst almost 65% of the wind-pollinated fruits were deformed. The shelf life of strawberries in both varieties was strongly determined by fruit deformation and thus pollination limitation. In Yamaska, decay differed according to all degrees of deformation, indicated by the lowest AICc for the model with unpooled fixed effects (Figure 3A; Table 1). Medium and highly deformed fruits decayed faster compared with the non-deformed fruits. After four days in storage, almost 80% of the fruits with medium and high degrees of deformation were decayed, but only 30% of the fruits without deformations were decayed. In Sonata, decay was similar for non- and medium deformed fruits, indicated by the lowest AICc for the model where these effects were pooled (Figure 3B; Table 1). However, highly deformed fruits decayed fastest. Almost 60% of non- and medium deformed fruits were decayed after four days in storage, whilst 100% of the highly deformed fruits were already decayed after the third day in storage.

Table 1 Delta AICc values resulting from model comparisons

Variety	Fruit parameter	Pooled levels				
		None	None and medium	Medium and high	None and high	Sans
Sonata	Decay (n = 123)	2.026	*0*	9.755	10.193	8.671
	Firmness (n = 123)	5.349	*0*	4.808	4.651	3.052
	Fruit weight (n = 123)	0.974	*0*	2.433	2.952	3.480
Yamaska	Decay (n = 157)	*0*	2.613	0.706	5.677	3.572
	Firmness (n = 154)	3.311	1.738	0.036	2.448	*0*
	Fruit weight (n = 157)	*0*	3.448	3.005	4.625	7.611

AICc = 0 indicates the model with the highest explanatory power. Lower delta AICc indicates better explanatory power of a model. The most explanatory models are highlighted in *italics*. Sample sizes are given in *brackets* behind fruit parameters. None = all treatment levels kept separately; Sans = model without treatment as fixed effect.

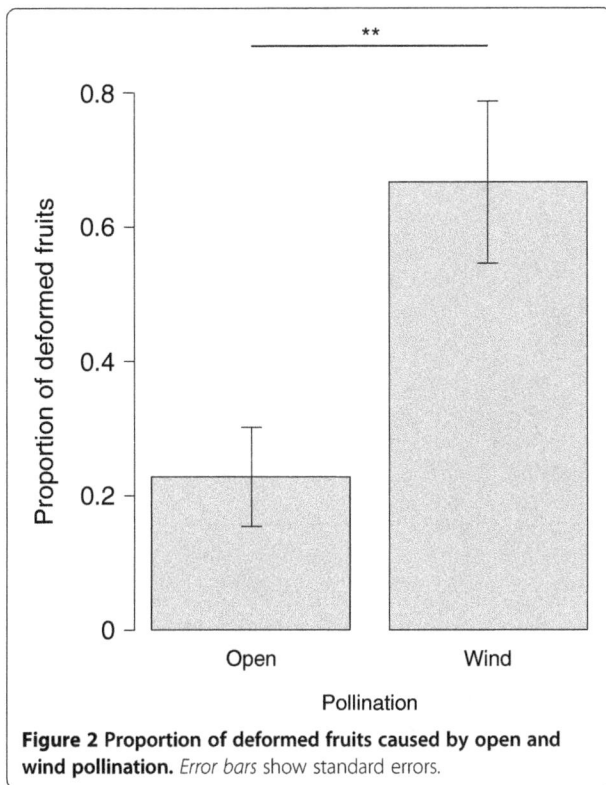

Figure 2 Proportion of deformed fruits caused by open and wind pollination. *Error bars* show standard errors.

The declines in firmness and fruit weight were also related to the degree of deformation, except for the variety Yamaska. Here, firmness similarly declined in all degrees of deformation (Figure 3C; Table 1). In Sonata, firmness decreased equally in non- and medium deformed fruits, but much faster in highly deformed fruits (Figure 3D; Table 1). In both varieties, fruit weight decreased to a similar extent in all degrees of deformation (Figure 3E,F; Table 1). However, fruit weight generally differed between all degrees of deformation in the variety Yamaska, with a highest weight in non-deformed fruits and lowest weight in highly deformed fruits. In Sonata, non- and medium deformed fruits had similar weight, whilst highly deformed fruits were much lighter.

In both varieties, fruit decay was negatively correlated to firmness (Sonata: *Spearman's correlation* = −0.91; $P <$ 0.001; Yamaska: *Spearman's correlation* = −0.91; $P < 0.001$) and fruit weight (Sonata: *Spearman's correlation* = −0.70; $P = 0.005$; Yamaska: *Spearman's correlation* = −0.79; $P <$ 0.001)), whilst firmness and fruit weight were less strong, but positively correlated (Sonata: *Spearman's correlation* = 0.65; $P = 0.015$; Yamaska: *Spearman's correlation* = 0.63; $P = 0.015$).

Discussion

Our results verify the relationship between pollination limitation and shelf life, with experimental data. The decay of strawberry fruits was strongly dependent on the degree of deformation, which was caused by pollination limitation.

Pollination limitation and fruit deformation

Open pollination produced almost solely non-deformed fruits, whilst wind pollination resulted in high amounts of deformed fruits. Deformations in strawberry fruits are a result of pollination limitation, mainly due to the absence of insect pollinators [28]. The mechanism is based on the amount of fertilized achenes [17], the true 'nut'-fruits of strawberry being an aggregated fruit [18]. Unfertilized achenes are a result of pollination limitation [28] and have no physiological functionality [29]. Aggregations of unfertilized achenes usually lead to deformations in strawberry fruits [17,30] and thus deformations can be directly linked to insufficient pollination.

Shelf life as a function of fruit deformation due to pollination limitation

We used deformations resulting from pollination limitation to test the relationship between pollination and shelf life. In both varieties, highly deformed and thus strongly pollination-limited fruits had a shorter shelf life, due to faster decay as well as lower firmness and fruit weight during the entire storage time. However, differences between medium and undeformed fruits were not the same across varieties. Fertilized achenes produce hormonal growth regulators which enhance cell progeny and size and thereby increase fruit weight [31]. Also, the firmness of strawberry fruits is functionally based on fertilized achenes. During fruit ripening, the fruit produces cell wall-degrading proteins [32], which lead to decreasing firmness. The expression of several of these proteins is limited by the growth regulators [33], which thereby decelerate fruit softening and lead to higher firmness. Cell wall-degrading proteins lead to the loss of water and fruits become softer and lighter [16] and thereby also more sensitive to mechanical damage as well as fungal decay [16]. This explains the strong correlation of the loss of firmness and weight with the decay of strawberry fruits. However, although decay and fruit weight of Yamaska was more strongly affected by both, medium and non-deformed fruits, there was no difference between undeformed and medium deformed fruits in Sonata. Thus, effects of fruit deformation and therefore pollination seem to change with variety. Although strawberries are generally dependent on insect pollination, this varies between varieties due to differences in the dependence on insect pollination. This has been shown for fruit weight and the amount of deformations caused by pollinator exclusion [19] and also for various quality aspects [15]. Reasons could be morphological differences, e.g. when anthers are located above the receptacle allowing for better self pollination [34] or differences in

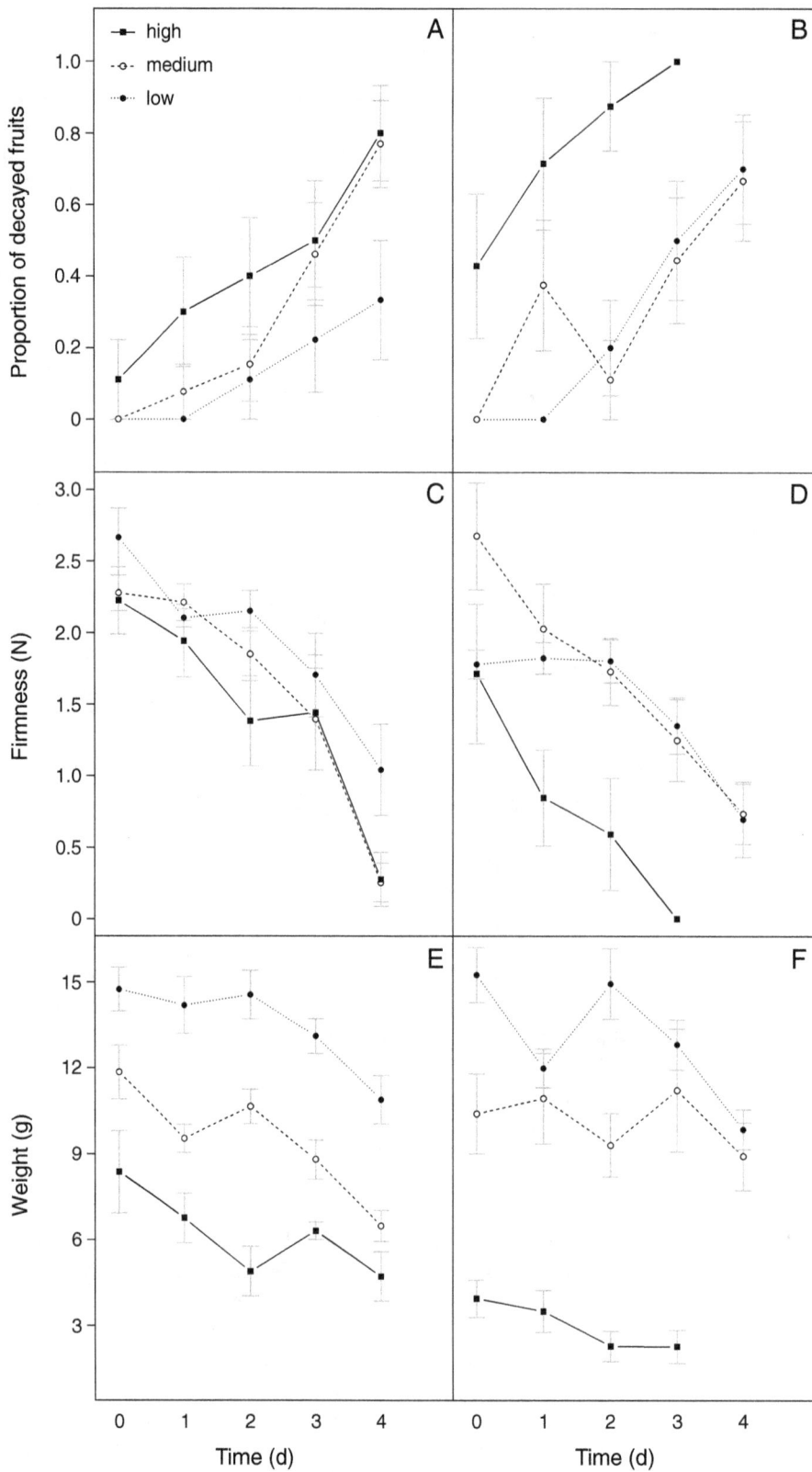

Figure 3 The dependence of fruit degradation during storage on the degree of deformation. Proportion of decayed fruits: **(A)** Yamaska, **(B)** Sonata. Firmness: **(C)** Yamaska, **(D)** Sonata. Fruit weight: **(E)** Yamaska, **(F)** Sonata. Standard errors are displayed in *grey* for better visualization.

attraction of pollinators by varieties varying in the amount of floral volatiles emitted [35].

General significance: application to other crops

In general, plant hormones play a key role for developmental processes and also determine fruit development [36]. Deformation and decreased weight are common problems triggered by pollination limitation in fruit crops such as strawberries [18]. For a few crop species, an influence of pollination limitation on fruit firmness has also been demonstrated [18]. Hence, it is likely that pollination limitation generally results in reduced shelf life in fruits and vegetables by impeding the production of plant hormones [36]. However, those effects so far have only been shown for strawberry and more research is required to confirm more general effects for other pollinated crops.

Although most pollinated crops do not belong to the major food crops, their products are of essential importance for a healthy human diet because they contain large amounts of essential nutrients [4]. Even today, nutrient supply is limited for large parts of the human population. In the following decades, an increasing world population will lead to further rising demands for food [1], especially given that the importance of pollination-dependent crops is largely increasing [2]. Our findings suggest that declining pollination services in agricultural landscapes [37] will likely increase the economic loss and waste of products from pollinated crops in the food chain as a consequence of decreasing shelf life. Thus, these products could become scarcer in the near future, leading to a general depletion in the supply of essential nutrients and further limiting availability to people in the developing world due to increasing prices. Already today, shelf life for which deformations and decay are important values appears to be of tremendous importance due to the increasing loss and waste of food [5]. Deformed fruits and vegetables have a lower market value [38] but can still be found at retail and consumer levels.

Conclusions

In conclusion, pollination is a key driver for both the appearance and the shelf life of strawberries. Similar effects of pollination limitation on pollinated crops suggest this pattern could be generally applicable, but empirical evidence for such effects is largely missing. Nevertheless, our results highlight the need to stabilize pollination services, as the importance of pollination-dependent crops is rapidly increasing [2], whilst pollination services are in danger of various anthropogenic threats [37]. Our study provides a new perspective on the relationship between food shelf life and pollination, emphasizing the need to protect and enhance pollination

services through international policies and conservation strategies.

Competing interests
The authors declare that they have no competing interests.

Authors' contributions
BKK evolved the design of the study, performed statistical analysis and interpretations of results and wrote the manuscript. FK performed lab analysis and participated in drafting the manuscript. CW and TT contributed to the design of the study, gave intellectual input to data interpretation and helped to draught the manuscript. All authors have read and approved the final version of the manuscript.

Acknowledgements
We thank K. Mecke for the permission to collect strawberry fruits, I. Smit and E. Pawelzik for the provision of lab space and instruments as well as U. Kierbaum for assistance during fieldwork. We further thank C. Brittain for language corrections and two anonymous reviewers for their helpful comments. This work has been funded by the DFG and was supported by FORMAS through the SAPES research environment.

References
1. Godfray HJC, Beddington JR, Crute IR, Haddad L, Lawrence D, Muir JF, Pretty J: **Food security: the challenge of feeding 9 billion people.** *Science* 2010, **327**:812–818.
2. Lautenbach S, Seppelt R, Liebscher J, Dormann CF: **Spatial and temporal trends of global pollination benefit.** *PLoS One* 2012, **7**:e35954.
3. FAOSTAT: *Food and Agricultural Organization of the United Nations.* Food and Agricultural Organization of the United Nations. [http://faostat3.fao.org/faostat-gateway/go/to/home/E]. Accessed 14 December 2013.
4. Eilers EJ, Kremen C, Greenleaf SS, Garber AK, Klein A-M: **Contribution of pollinator-mediated crops to nutrients in the human food supply.** *PLoS One* 2011, **6**:e21363.
5. Gustavsson J, Cederberg C, Sonesson U, van Otterdijk R, Meybeck A: *Global food losses and food waste: extent, causes and prevention.* Rome, Italy: FAO; 2011.
6. Tscharntke T, Clough Y, Wanger TC, Jackson L, Motzke I, Perfecto I, Vandermeer J, Whitbread A: **Global food security, biodiversity conservation and the future of agricultural intensification.** *Biol Conserva* 2012, **151**:53–59.
7. Manning K: *Genes for fruit quality in strawberry.* Wallingford: Cab International; 1998.
8. Soliva-Fortuny RC, Martin-Belloso O: **New advances in extending the shelf-life of fresh-cut fruits: a review.** *Trends Food Sci Tech* 2003, **14**:341–353.
9. Dhall RK: **Advances in edible coatings for fresh fruits and vegetables: a review.** *Crit Rev Food Sci Nutr* 2013, **53**:435–450.
10. Civello PM, Martinez GA, Chaves AR, Anon MC: **Heat treatments delay ripening and postharvest decay of strawberry fruit.** *J Agr Food Chem* 1997, **45**:4589–4594.
11. Brummell DA, Harpster MH: **Cell wall metabolism in fruit softening and quality and its manipulation in transgenic plants.** *Plant Mol Biol* 2001, **47**:311–340.
12. Al-Attal YZ, Kasrawi MA, Nazer IK: **Influence of pollination technique on greenhouse tomato production.** *J Agric Mar Sci* 2003, **8**:21–26.
13. Shin YS, Park SD, Kim JH: **Influence of pollination methods on fruit development and sugar contents of oriental melon (*Cucumis melo* L. cv. Sagyejeol-Ggul).** *Sci Hort* 2007, **112**:88–392.
14. Gajc-Wolska J, Kowalczyk K, Mikas J, Drajski R: **Efficiency of cucumber (*Cucumis sativus* L.) pollination by bumblebees (*Bombus terrestris*).** *Acta Sci Pol* 2011, **10**:159–169.

15. Klatt BK, Holzschuh A, Westphal C, Clough Y, Smit I, Pawelzik E, Tscharntke T: **Bee pollination improves crop quality, shelf life and commercial value.** *Proc R Soc B* 2014, **281:**20132440.

16. Hernandez-Munoz P, Almenar E, Ocio MJ, Gavara R: **Effect of calcium dips and chitosan coatings on postharvest life of strawberries (***Fragaria x ananassa***).** *Postharvest Biol Tec* 2006, **39:**247–253.

17. Ariza MT, Soria C, Medina JJ, Martinez-Ferri E: **Fruit misshapen in strawberry cultivars (***Fragaria x ananassa***) is related to achenes functionality.** *Ann Appl Biol* 2010, **158:**130–138.

18. Free JB: *Insect Pollination of Crops.* London: Academic Press; 1993.

19. Zebrowska J: **Influence of pollination modes on yield components in strawberry (***Fragaria x ananassa* **Duch.).** *Plant Breed* 1998, **117:**255–260.

20. European commission: Commission Regulation (EC) No 1580/2007; 2007 [http://ec.europa.eu/index_en.htm]

21. Darrow GM: **Sterility and fertility in the strawberry.** *J Agronom Res* 1927, **34:**393–411.

22. Sanz C, Perez AG, Olias R, Olias JM: **Quality of strawberries packed with perforated polypropylene.** *J Food Sci* 1999, **64:**748–752.

23. Venables WN, Ripley BD: *Modern applied statistics with S.* New York: Springer; 2002.

24. R Development Core Team: R: A language and environment for statistical computin; 2013 [http://www.R-project.org]

25. Bates D, Maechler M, Bolker B, Walker S, Christensen RHB, Singmann H: Linear mixed-effects models using Eigen and S4; 2014 [http://cran.r-project.org/web/packages/lme4/index.html]

26. Burnham KP, Anderson DR: *Model selection and multimodel inference: a practical information-theoretic approach.* New York: Springer; 2002.

27. Bolker BM, Brooks ME, Clark CJ, Geange SW, Poulsen JR, Stevens MHH, White J-SS: **Generalized linear mixed models: a practical guide for ecology and evolution.** *Trends Ecol Evol* 2009, **24:**127–135.

28. Connor LJ: *The role of cultivar in insect pollination of strawberries.* Proceedings of the 3rd International Symposium on Pollination; 1974:149–154.

29. Pipattanawong R, Yamane K, Fujishige N, Bang S-W, Yamaki Y: **Effects of high temperature on pollen quality, ovule fertilization and development of embryo and achene in 'Tochiotome' strawberry.** *J Japa Soc Hort Sci* 2009, **78:**300–306.

30. Nitsch JP: **Growh and morphogenesis of the strawberry as related to auxin.** *Am J Bot* 1950, **37:**211–215.

31. Roussos PA, Denaxa N-K, Damvakaris T: **Strawberry fruit quality attributes after application of plant growth stimulating compounds.** *Sci Hort* 2009, **119:**138–146.

32. Given NK, Venis MA, Grierson D: **Hormonal regulation of ripening in the strawberry, a nonclimacteric fruit.** *Planta* 1988, **174:**402–406.

33. Villareal NM, Martinez GA, Civello PM: **Influence of plant growth regulators on polygalacturonase expression in strawberry fruit.** *Plant Sci* 2009, **176:**740–757.

34. Darrow GM: **Sterility and fertility in the strawberry.** *J Agron R* 1927, **34:**393–411.

35. Klatt BK, Burmeister C, Westphal C, Tscharntke T, von Fragstein M: **Flower volatiles, crop varieties and bee responses.** *PLoS One* 2013, **8:**e72724.

36. McAtee P, Karim S, Schaffer R, David K: **A dynamic interplay between phytohormones is required for fruit development, maturation, and ripening.** *Front Plant Sci* 2013, **4:**

37. Potts SG, Biesmeijer JC, Kremen C, Neumann P, Schweiger O, Kunin WE: **Global pollinator declines: trends, impacts and drivers.** *Trends Ecol Evol* 2010, **25:**345–353.

38. European commission: *Commission Regulation (EC) No 2200/96;* 2007 [http://ec.europa.eu/index_en.htm]

10

Harmonized biosafety regulations are key to trust building in regional agbiotech partnerships

Obidimma C Ezezika[1,2,3*], Justin Mabeya[1], Abdallah S Daar[1,4,5]

Abstract

Background: The *Bacillus thuringiensis* (Bt) cotton public-private partnership (PPP) project in East Africa was designed to gather baseline data on the effect of Bt cotton on biodiversity and the possibility of gene flow to wild cotton varieties. The results of the project are intended to be useful for Kenya, Uganda, and Tanzania when applying for biosafety approvals. Using the backdrop of the different biosafety regulations in the three countries, we investigate the role of trust in the Bt cotton partnership in East Africa.

Methods: Data were collected by reviewing relevant project documents and peer-reviewed articles on Bt cotton in Tanzania, Kenya and Uganda; conducting face-to-face interviews with key informants of the project; and conducting direct observations of the project. Data were analyzed based on recurring and emergent themes to create a comprehensive narrative on how trust is understood and built among the partners and with the community.

Results: We identified three factors that posed challenges to building trust in the Bt cotton project in East Africa: different regulatory regimes among the three countries; structural and management differences among the three partner institutions; and poor public awareness of GM crops and negative perceptions of the private sector. The structural and management differences were said to be addressed through joint planning, harmonization of research protocols, and management practices, while poor public awareness of GM crops and negative perceptions of the private sector were said to be addressed through open communication, sharing of resources, direct stakeholder engagement and awareness creation. The regulatory differences remained outside the scope of the project.

Conclusions: To improve the effectiveness of agbiotech PPPs, there is first a need for a regulatory regime that is acceptable to both the public and private sector partners. Second, early and continuous joint planning; sharing of information; and transparency encourages accountability and fosters trust building. Third, direct stakeholder engagement and awareness creation builds trust between project partners and the community. A concern raised by the interviewees was the absence of a regulatory framework in Tanzania, which deterred active private sector participation in the project.

Background

The *Bacillus thuringiensis* (Bt) cotton project in East Africa was designed to gather baseline data on the effect of Bt cotton on above- and below-ground biodiversity and the possibility of gene flow to wild cotton varieties. The potential consumers of the project's results are the regulators in three East African countries—Tanzania, Kenya, and Uganda—as well as others interested in agbiotech regulation.

In Tanzania, cotton comprises the third largest agricultural export after coffee and tobacco [1]. However, the average production per hectare of 750 kg is far below the world average of 2000 kg/ha [2]. Cotton in Tanzania is cultivated by approximately 500,000 smallholder farmers [2]. One of the major challenges to cotton production in Tanzania is the cotton bollworm (*Helicoverpa armigera*), which is currently managed by use of pesticides (which

* Correspondence: obidimma.ezezika@srcglobal.org
[1]Sandra Rotman Centre, University Health Network and University of Toronto, Toronto, Ontario, Canada

pose health and environmental risks to the users), among other methods.

Cotton in Kenya is primarily grown by approximately 30,000 small-scale farmers [3]. Over the last decade, cotton production in Kenya has steadily declined [4], partly because of pests and diseases, key among them the cotton bollworm [5]. Management of the pest is primarily done, as in Tanzania, by use of pesticides (6 to 8 sprays per season) that, while effective, pose health and environmental risks to the users.

In some parts of Uganda, the current yield of cotton is approximately 400 kg/ha, which is far below the production potential [6]. As is the case in the other East African countries, a major threat to production in Uganda is the cotton bollworm, which has been reported to cause losses in yield as high as 40% in parts of the country [6].

To address the challenge posed by the cotton bollworm and other insect pests, the three East African countries have considered the use of genetically modified (GM) cotton. In Kenya, trials for Bt cotton were approved by the National Biosafety Committee (NBC) in 2003. The project, which is currently running, is being undertaken jointly by the Kenya Agricultural Research Institute (KARI) and the multinational seed company, Monsanto. The GM cotton varieties being tested in Kenya are intended to control the cotton bollworm. This flagship GM technology is expected to be commercialized in 2014 [7] under the new biosafety law that was passed in 2009, which allows for the commercialization of GM crops [8].

In Uganda, the National Agricultural Research Organization (NARO) has been conducting field trials of Bt cotton since 2009 for both insect and herbicide resistance in collaboration with Monsanto under the National Biosafety regulations, while a comprehensive biosafety law is being developed [9].

In Tanzania, the idea of using GM cotton has attracted support from top government policy makers in the Ministry of Agriculture and Food Security who have stated that the government is committed to adopting improved technologies including genetic engineering to increase cotton yields, reduce farm costs, and increase profits [10]. This will be subject to a review of the environmental laws governing GM products to enable the commercial introduction of such technologies [11]. Yet, the National Biosafety Framework (NBF), the system in place governing biosafety in Tanzania, contains a *strict liability* clause, which has been considered punitive to those wishing to develop and commercialize GM technologies in Tanzania. The clause states: "A person who imports, arranges transit, makes contained use of, releases or places on the market a genetically modified organism (GMO) or product of a GMO shall be strictly liable for any harm caused by such a GMO or product of a GMO. The harm shall be fully compensated" [12]. The response to this

clause has been limited interest by private companies to partner with Tanzanian institutions in developing such technologies, including Bt cotton. The existing legal framework in Tanzania does provide for commercialization, but efforts to ensure that there is progress in agbiotech research have been stalled by this clause. In this article we report on how the differences in biosafety regulations among the three East African countries affected trust building between the public and private sector players in the project.

The development of the Bt cotton project

The Bt cotton project in East Africa was initiated in 2006 following a successful grant application by a joint team of scientists from Tanzania, Kenya and Uganda in collaboration with agbiotech experts from the United States [13]. The goal of the three-year project was twofold: to study the effect of Bt cotton on arthropod biodiversity in East Africa and to study the possibility of gene flow from Bt cotton to wild cotton varieties. The results of the study are expected to serve as baseline data that may be useful to the regulatory organizations in the region when considering biosafety approvals for Bt cotton [13].

Funding for the project was provided by the United States Agency for International Development (USAID) through the Biotechnology and Biodiversity Interface (BBI) of the Program for Biosafety Systems (PBS) of the International Food Policy Research Institute (IFPRI). The lead partner in the project is the Tropical Pesticide Research Institute (TPRI) in Tanzania. Others include KARI in Kenya; Makerere University and NARO in Uganda; and the Agricultural Innovation Research Foundation (AIRF) in Tanzania [13]. Table 1 summarizes the roles and responsibilities of the project partners (refer to Additional file 1 for brief profiles of the implementing partner institutions).

Trust in agbiotech PPPs

The success of agbiotech projects depends on the ability of partners to engage in long-term collaborations, among other factors. Since the involvement of multinational private companies has been found to breed distrust, the presence of trust is critical in agbiotech PPP projects [14,15]. The negative consequences of such distrust are compounded in the context of GM food. For example, there is fear, among public partners, pertaining to corporate control of the seed market by genetic engineering companies [13]. This case study investigates the role of trust in the Bt cotton in East Africa partnership from the project's conception to its end. The three specific objectives of this study are to: 1) describe trust-building practices in the development of the project; 2) describe the challenges associated with trust-building in the partnership; and 3) determine what makes these practices effective or ineffective.

Table 1 Bt cotton project partner and collaborator responsibilities

Partners	Role
Tropical Pesticide Research Institute, Tanzania	Conducting gene flow studies from Bt cotton to wild indigenous cotton varieties in the region
Kenya Agricultural Research Institute, Kenya	Conducting studies on the effect of Bt cotton on both the above-ground and under-ground non-target arthropod species
Makerere University, Uganda	Management and disbursement of the project funds
National Agricultural Research Organization, Uganda	Conducting studies on the effect of Bt cotton on both the above-ground and under-ground non-target arthropod species
Agricultural Innovation Research Foundation, Tanzania	Conducting risk assessment of the Bt cotton technologies, training of stakeholders and community engagement and awareness
Biotechnology and Biodiversity Interface	Provided funding for the project
Collaborator	**Role**
Monsanto	Allowed the trials meant to be conducted by TPRI in Tanzania to run next to the Bt cotton trials at KARI, Thika Research Station in Kenya

By identifying barriers to trust and trust-enhancing practices, this study provides insight for potential funders, researchers, farmers and others involved in agbiotech initiatives. The results of this study will be useful in informing PPPs involved in agricultural biotechnology on the need for building and maintaining trust among the partners and with the community.

Methods

Data were collected by conducting interviews with key informants knowledgeable about the project; reviewing publicly available project documents and research articles; and conducting direct observations. We received Research Ethics Board (REB) approval from the University Health Network, University of Toronto for conducting the case study. Interviewees were identified by first making a list of key individuals associated with the project based on the stakeholder groups identified in the case study research protocol. This list was further populated through snowball sampling using stakeholder informants familiar with the Sandra Rotman Centre's Social Audit Project [16]. Social auditing is a process whereby an audit team collects, analyses, and interprets descriptive, quantitative and qualitative information from stakeholders to produce an account of a project's ethical, social, cultural and commercialization performance and impact. The social audit helps foster improved management practices, accountability and transparency, which in turn help to build trust both among the partners in a project as well as between the project and the public. A total of 16 individuals were identified through this process, of which ten, who were most relevant to this case study, were invited for interview. Eight individuals in total were interviewed as two of those invited were not available. Potential interviewees were sent an invitation, which included an explanation of the case study series, to participate in the interview. Informed consent was sought from the potential interviewees by

obtaining duplicate signed copies of written consent. Those who consented to participate were informed that the interview would be digitally recorded, transcribed verbatim and then analyzed.

The interviews took place in Mozambique, Uganda, Kenya and Tanzania – as per the convenience of the interviewees. The interviews were conducted using a semistructured interview guide and each lasted approximately one hour and a half. The interviewees included representatives from the Monsanto Company, TPRI, KARI, AIRF, PBS, Ministry of Agriculture in Tanzania, Vice President's Office and Ministry of Environment in Tanzania, and Makerere University. The interviews explored the partners' and the public's general perceptions of trust within the agbiotech PPP, apparent challenges to building trust, and trust-building practices in the PPP. Interviewees were also asked for their suggestions on how to improve trust in agbiotech PPPs (refer to Additional file 2 for sample questions from the interview guide).

The data were analyzed by identifying recurring and emergent themes from the transcripts, relevant literature and observations. All the data were triangulated to create a comprehensive narrative on how trust is understood and built among the partners and with the community. The case study methodology used in this study was adapted from Yin 2003 [17].

Results and discussion

Our analysis of the data reveals that the trust-building issues faced in this project revolve around challenges pertaining to the biosafety regulations in the region. Key among them is the absence of a regulatory framework in Tanzania, which is not conducive to the development and commercialization of GM technology and the evolving nature of the regulatory process in the three countries. While there have been no studies showing that the absence of a favorable biosafety law has been a hindrance to commercialization of the GM technologies in East Africa, it has been observed that the absence of clear biosafety regulations can lead to distrust among the various stakeholders [18].

Stakeholders' understanding of trust

In order to put into context the interviewees' understanding of trust-building challenges and practices, the interviewees were asked to define the word *trust* and describe its elements in the context of the partnership. Trust was described by the interviewees as the ability to have confidence in people and institutions and as a two-way positive relationship. Trust could be built upon an agreement that does not necessarily have to be formal, and delivery of results is evidence that trust is present. Overall, the interviewees understood trust to be the ability of the parties in a mutual agreement to have confidence in each other and be committed to work in a transparent and accountable manner in order to achieve the objectives of the agreement to their mutual satisfaction.

Based on the results of this study, we have derived three key lessons that would provide insight to partners in other agbiotech PPPs on fostering trust among partners and with the community and on enhancing regulatory processes.

1. Regulatory regimes acceptable to both public and private sector parties are a foundation for building trust in agbiotech PPP projects

Inconsistencies and barriers in regulatory frameworks

The respective regulatory arrangements in the three countries help explain the degree of willingness with which the private sector engaged with the partners from each of the three countries. Although the Bt cotton project in East Africa began with the BBI grant in 2006, the project activities in each country began at different times due to differences in the biosafety regulations and the management of biosafety matters among the three East African countries. Studies in Kenya began in earnest in 2007 while those in Uganda began in 2008, both having been delayed by the biosafety approval processes—which, in some instances, were perceived by the project management as a deliberate move by the regulators to slow down the process because of suspicion of incompetence on the part of the former. To forestall these suspicions the project management made efforts to address any foreseeable regulatory concerns before applying for regulatory approvals. In Tanzania, the delay was due to what the private partner considered to be unfavorable legislation in the country, which prevented Monsanto from providing the Bt cotton technology for trials in the country.

We observed that the regulatory differences and their impact on project activities did not threaten trust among the public institutions implementing the project. The interviewees from the public institutions did not point to any factor that could compromise trust among them, but they reported of a mutual working relationship among them. However, in order for the private sector to be fully engaged as a partner and active in the project activities, there was a need not only for a consistent regulatory regime in the three countries but one that is acceptable to all stakeholders. The interviewee from Monsanto emphasized that trust develops when parties deliver on commitments that are based on an acceptable biosafety legal framework. The good relationship between KARI and Monsanto, for example, shed light on two factors that contributed to trust building between the two. First, there is a favorable biosafety legal framework in Kenya [8], which enabled Monsanto to enter into an agreement with KARI. Second, the technical staff were not only competent but also committed to, and passionate about, delivering on the agreed upon milestones. Subsequently, as the objectives of the agreement were achieved, trust between the partners grew. Similarly, the interviewee from Monsanto acknowledged that, in Uganda, the National Biotechnology and Biosafety Policy 2008 [19] provided a suitable environment for private sector engagement which served as a foundation on which trust building practices would be anchored.

Contrastingly, lack of a favorable regulatory environment in Tanzania resulted in a case in which the field trials of Bt cotton scheduled to be conducted in Tanzania were transferred to Kenya due to the latter's more favorable regulatory regime. There was, as a result, no direct contact between TPRI and Monsanto in this project, according to an interviewee from the TPRI, and therefore no possibility of trust building between the two institutions.

We observe that the aforementioned inconsistencies in, and barriers to, the regulatory process posed significant hurdles to trust building among partners and with the community. According to the interviewees, there is a need for a legal framework conducive to private sector participation, which will serve as a foundation for trust building. Individual country regulatory frameworks are therefore necessary for the progress and success of GM crop development projects. For example, even if the harmonized research protocols render the results of the Tanzanian trials in Kenya acceptable to regulators in Tanzania, access to the private sector technologies for broader adoption of the GM technologies will require a regulatory regime that is more acceptable than it currently is and encourages the partners' confidence in the process.

Baseline trust and legislation as foundations for trust building

There were divergent views from the public and private sector regarding the basis on which trust among the partners in the project grew. We deduce that this may be due to differences in the working cultures of the two sectors. The interviewees from the public institutions suggested that their initial engagement was based on the presence of what they described as 'baseline trust,' on which further trust building could take place, while the private sector emphasized that trust grew based on a negotiated agreement among partners working within an acceptable legislative framework.

Immediately upon the project's inception, the public institutions decided to collaborate on an understanding that they will be able to work together to build trust. According to a scientist from PBS (the funding institution that also initiated the development of the joint proposal), this early engagement was based on some baseline trust on which the parties further built trust. 'Baseline trust' was understood by the scientist from PBS to be a particular level of trust between parties which is partly based on knowledge of past collaborations in similar projects and determined by the willingness of the parties to participate in the joint application of the grant (i.e., the initial engagement). Later on, the public institutions put in place structural management practices, including harmonization of research protocols and organizational management processes, which contributed to enhancing trust among themselves and progress in the project.

However, an interviewee from Monsanto underscored the need for a legal framework that would facilitate their active participation, emphasizing the fact that, in agbiotech PPPs, the existing legislation in a country defines the nature of the relationship among partners. The interviewee from Monsanto stated that the relationship among the partners is influenced less by trust than by legality; specifically, it is within a relationship bound by clear legislation that parties engage, negotiate and sign agreements stipulating partner roles and obligations. The presence or lack of trust, then, is determined by what ensues from such agreements (i.e., whether or not the partners deliver on agreements). For example, in the aforementioned relationship between KARI and Monsanto, trust between the two institutions was built partly as a result of the parties delivering on agreements founded on an acceptable biosafety law.

2. Early and continuous joint planning, sharing of information, and transparency encourages accountability and fosters trust building

Joint and transparent planning sessions

Despite the lack of formal documentation outlining the responsibilities of each partner institution at the inception of the project—as reported by the interviewees from the public institutions—their initial joint participation in developing the project proposal had established a strong foundation for ensuring accountability. Though the project was initially brought together by a need to unify the proposals of the three different countries for funding purposes, subsequent progress in the project reflected the value of transparent and joint planning sessions among the partners. During such meetings, financial matters were said to have been discussed openly among project partners in order to reinforce an environment of mutual accountability. The implementing partners also engaged in regular communication and information sharing, which enhanced trust. This pattern was seen through the project lifespan and was evidenced by commitment, mutual support and resource sharing – all of which contributed to building trust among the implementing partners. However, in retrospect, a TPRI partner interviewee felt that the planning could have been enhanced for better performance and trust among the partners in the project. Because of this apparent dissatisfaction with the planning, resources reserved for research were later used for non-core project activities such as holding meetings to make changes to project activities to accommodate unforeseen changes. As a result, interviewees acknowledged that there were delays in starting certain project activities, therefore necessitating a request to extend the project life.

Sharing information and resources

The sharing of resources and information is one outstanding feature of the Bt cotton project in East Africa that contributed to trust building. Information generated by any of the core partner institutions was shared among all the partners in the project through phone calls and exchange visits – all of which helped build trust. By relaying experiences including challenges and sharing data, interviewees described a mutual, collaborative relationship that enhanced trust and progress in the project. A free flow of information prevented overlaps in research and permitted scientists sufficient independence and a sense of confidence in each other.

On the topic of resource sharing, a scientist we interviewed from KARI alluded to the TPRI–KARI relationship, whereby KARI hosted Tanzania's trials in Kenya. Specifically, KARI provided land, transportation and other forms of facilitation at a subsidized rate, which helped cement trust between the two institutions. An academic from Makerere University described the practice of sharing information and resources as an accountable and collaborative approach which contributed to enhancing trust. Sharing of resources and information helped the parties synergize their strengths to improve project results. Spielman et al. [20] pointed out that synergies in agbiotech research may ideally lead to outcomes of greater quantity and with greater chances of success. We posit that the sharing of information and resources across the three countries served to build trust among the project partners and the data generated for regulatory approval may also contribute to a more unified regulatory regime in the region.

3. Direct stakeholder engagement and awareness creation builds trust between project partners and the community

Lack of public awareness and negative perception of GM crops

One factor that appeared to have contributed to slow regulatory approvals was limited awareness of GM technology among the public and specifically among government technical staff. Interviewees hypothesized that the suspicion surrounding the technology may stem from a lack of understanding of GM crops.

According to some interviewees, there is ignorance among frontline agricultural extension workers in East Africa, which is likely to influence the farmers' perceptions about, and trust in, GM crops. A recent study has shown that awareness of GM crops among members of the public and technical staff in Tanzania is poor [21]. An example of this stems from our interview with a biosafety regulator in Tanzania who stated that the public in Tanzania is not in favor of GM crops. Such skepticism, which leads to distrust in the technology, has been partly attributed to ideas propagated by misinformed media and anti-GM groups, which foster distrust between the project partners and the public. A stakeholder from KARI and another from AIRF identified the presence of anti-GM groups as contributing to the misrepresentation of GM crops. Negative and inaccurate perceptions fuelled by such groups were said to often reach farmers before they receive accurate information from the project partners, leading to misconceptions about GM crops.

We observed that lack of awareness about GM crops—even among public research institutions, regulatory institutions and ministries of agriculture—coupled with the delivery of inaccurate information to the public is likely to negatively influence public perception, and, in turn, impact regulatory decision-making processes.

Negative views about private sector involvement

The involvement of private multinational seed companies in the partnership also contributed to the community's distrust in the Bt cotton project. It has been reported that in sub-Saharan Africa the public holds unfavorable views about the involvement of the private sector in agbiotech projects [22]. Farmers, stakeholders from the seed sector, and non-governmental organizations (NGOs) focused on agriculture all view the involvement of private companies, particularly multinational ones, as being driven by a desire to monopolize the seed industry and therefore see the private sector as a potential threat to the food sovereignty of African countries [22]. The interviewees therefore perceived that public distrust about private sector involvement in the partnership stemmed from fear of corporate control of the seed technology and the view that the private sector seeks to make profits at the cost of the community. Part of this distrust was also said to be related to fear among members of the public that the Bt cotton

would have the "terminator gene", which prevents the seed from germinating if re-planted (which is a common practice among farmers). A scientist from KARI emphasized the need to continuously assure the farmers that there is no terminator gene. Negative perceptions about the private sector (including Monsanto)—the source of the Bt cotton technology—only serves to strengthen skepticism about the technology, which may affect regulatory approval processes for the technology.

Community engagement and awareness-building initiatives

In view of the poor public awareness and negative perceptions about the private sector, there was an expressed urgency for enhanced public awareness of GM technology, as public perceptions can have negative repercussions not only on the regulatory process but also on the commercialization and adoption of the Bt cotton. Pre-conceived ideas about GM crops among members of the community, irrespective of their level of education, were said to likely render the process difficult. It has been reported that substantial public information about GM technology is useful both for regulatory processes and assuring the public of ownership of the project [23].

To gain the public's trust, the core partners engaged stakeholders at every level. KARI, for example, made use of their Bt cotton field trials to create awareness through a program called "Seeing-is-Believing," where journalists, politicians, farmers and government officials were taken for visits to the trial sites to allow them to make their own comparison of Bt cotton and conventional cotton. A scientist from KARI recognized the need for farmer engagement in GM technology development in order to boost adoption of this controversial technology. The visits help demystify the technology—and the processes of developing it—to members of the public.

An interviewee from AIRF in Tanzania noted that education and training of farmers were effective for enhancing public trust in Bt cotton, especially since seminars were held in local languages. At the same time, engaging the media was seen as a strategy to preventing alarmist reporting, and as such could facilitate trust building. These impacts were expected to translate, gradually, into the development of a regulatory framework that would enable broader exploitation of the GM crops in the region. In Uganda, a similar initiative was spearheaded by representatives from the Makerere University. An interviewee from the university reported the positive remarks made by farmers who had participated in the Seeing-is-Believing tours. An interviewee from the national regulatory authority in Uganda stated, "trust is known by what you do." These initiatives reflected well on the partners and helped build the public's trust in them.

Awareness creation through public education and the innovative "Seeing-is-Believing" tours provided an opportunity for multiple stakeholders to engage directly with researchers, compare the Bt cotton against the conventional varieties, and form their own opinions about the Bt cotton technology. Interviewees noted that the awareness creation measures led to a significant decline in negative perceptions of the Bt cotton technology and was helpful in building trust.

Conclusion

The international aspect of the Bt cotton project in East Africa presented distinct challenges to trust building. These challenges include: differences in the regulatory frameworks in the three countries; structural and management differences among the three partner institutions; and poor public awareness of GM crops and negative perceptions of the private sector – all of which contributed to delayed project implementation. The project partners tackled the structural and management differences among the three partner institutions through joint planning, harmonization of research protocols and management practices. Poor public awareness of GM crops and negative perceptions of the private sector were addressed through open communication and sharing of resources (as was the case between Kenya and Tanzania), and direct stakeholder engagement and awareness creation. These efforts contributed to enhancing trust among the public partners, leading to significant achievement of the project objectives.

Notwithstanding these positive outcomes, closer engagement of the private sector remained a challenge because of the differences in the regulatory frameworks in the three countries. Whereas the regulatory frameworks in Kenya and Uganda were favorable to private sector involvement, the regulatory framework in Tanzania was not. The failure to conduct field trials of Bt cotton in Tanzania emphasizes the need for regulatory regimes that can serve as a foundation for trust building between the public and private sector. The Bt cotton project in East Africa failed to attract a private sector player into its partnership due to the regulatory differences. Effective private sector involvement relies on acceptable biosafety regulations that will boost the private sector's confidence in sharing their proprietary technologies.

Acknowledgements
The authors are grateful to each of the participants who contributed substantial time and effort to this study. Special thanks to Jessica Oh, Nadira Saleh and Jocalyn Clark for comments on earlier drafts of the manuscript. This project was funded by the Bill & Melinda Gates Foundation and supported by the Sandra Rotman Centre, an academic centre at the University Health Network and University of Toronto. The findings and conclusions contained within are those of the authors and do not necessarily reflect official positions or policies of the foundation.
This article has been published as part of *Agriculture & Food Security* Volume 1 Supplement 1, 2012: Fostering innovation through building trust: lessons from agricultural biotechnology partnerships in Africa. The full contents of the supplement are available online at http://www.agricultureandfoodsecurity.com/supplements/1/S1. Publication of this supplement was funded by the Sandra Rotman Centre at the University Health Network and the University of Toronto. The supplement was devised by the Sandra Rotman Centre.

Author details
[1]Sandra Rotman Centre, University Health Network and University of Toronto, Toronto, Ontario, Canada. [2]African Centre for Innovation and Leadership Development, Federal Capital Territory, Abuja, Nigeria. [3]Dalla Lana School of Public Health, University of Toronto, Toronto, Canada. [4]Grand Challenges Canada. [5]Dalla Lana School of Public Health and Department of Surgery, University of Toronto, Toronto, Canada.

Authors' contributions
Study conception and design: OCE, JM, ASD. Data collection: JM and OCE. Analysis and interpretation of data: JM and OCE. Draft of the manuscript: JM and OCE. Critical revision of the manuscript for important intellectual content: OCE, JM and ASD. All authors read and approved the final manuscript.

Competing interests
The authors declare that they have no competing interests.

References
1. FAOSTAT. [http://faostat.fao.org/site/342/default.aspx;].
2. Tanzania Cotton Board: **Annual report and accounts for the year ending on 30th June 2010,** Pamba House, Garden Avenue P.O. Box 9161, Dar es Salaam. 2010.
3. **Cotton Development Authority (CODA).** [http://www.cottondevelopment.co.ke/pages/The_Cotton_Industry.vrt].
4. FAOSTAT. [http://faostat.fao.org/site/567/DesktopDefault.aspx?PageID=567#ancor].
5. Wakhungu WJ, Wafula DK: **Introducing Bt Cotton Policy Lessons for Smallholder Farmers in Kenya.** 2004, 84.
6. The Republic of Uganda, Ministry Of Agriculture, Animal Industry and Fisheries: **Cotton value chain study in Lango and Acholi sub-regions.** Plan for modernisation of Agriculture (PMA). 2009.
7. Clive J: **Biotech and organic agriculture proponents have to work together to boost Africa's food security.** *Crop Biotech Update* 2012.
8. Republic of Kenya: **The Biosafety Act, 2009.** In *Kenya Gazette Supplement. Volume 10.* Nairobi, Kenya: Government Printer; 2009(Act No. 2).
9. African Agricultural Technology Foundation: **Biotech cotton in Uganda: Potential benefits, challenges & way forward.** 2010, 1-17.
10. **Space Daily.** [http://www.spacedaily.com/news/food-05c.html].
11. **Daily News Online Edition.** [http://dailynews.co.tz/index.php/parliament-news/1682-from-the-parliament21].
12. Republic of Tanzania: **The National Biosafety Framework for Tanzania.** 2004.
13. **Biovision East Africa.** [http://www.biovisioneastafrica.com/publications/BBI%20Article.pdf].
14. Friedberg SE, Horowitz L: **Converging networks and clashing stories: South Africa's agricultural biotechnology debate.** *Africa Today* 2004, **51**(1):3-25.
15. Stone GD: **Both Sides Now. Fallacies in the genetic-modification wars, implications for developing countries and anthropological perspectives.** *Current Anthropology* 2002, **43**(4):611-630.
16. Ezezika OC, Thomas F, Lavery JV, Daar AS, Singer PA: **A social audit model for agro-biotechnology initiatives in developing countries: accounting for ethical, social, cultural and commercialization issues.** *Journal of Technology Management and Innovation* 2009, **4**(3):24-33.
17. Yin RK: **Case study Research: Design and Methods.** California: Sage Publications;, 4 2003.

18. Mabeya J, Singer PA, Ezezika OC: **The role of trust building in the development of biosafety regulations in Kenya.** *Law, Environment and Development Journal* 2010, **6/2**:218-227.

19. Republic of Uganda: **National Biotechnology and Biosafety Policy.** 2008.

20. Spielman DJ, Cohen JI, Zambrano P: **Policy, investment, and partnerships for agricultural biotechnology research in Africa: Emerging evidence.** *ATDF Journal* 2006, **3(4)**:3-9.

21. Lewis CP, Newell JN, Herron CM, Nawabu H: **Tanzanian farmers' knowledge and attitudes to GM biotechnology and the potential use of GM crops to provide improved levels of food security. A Qualitative Study.** *BMC Public Health* 2010, **10**:407.

22. Ezezika OC, Daar AS, Barber K, Mabeya J, Thomas F, Deadman J, Wang D, Singer PA: **Factors influencing agbiotech adoption and development in sub-Saharan Africa.** *Nature Biotechnology* 2012, **30**:38-40.

23. **Proceedings of the Harnessing the Potential of Biotechnology for Food Security and Socioeconomic Development in Africa: 22-26 September 2008; Nairobi, Kenya.** African Biotechnology Stakeholders Forum; Nzuma JM 2008.

11

Reducing subsistence farmers' vulnerability to climate change: evaluating the potential contributions of agroforestry in western Kenya

Tannis Thorlakson[1] and Henry Neufeldt[2,3]*

Abstract

Subsistence farmers are among the people most vulnerable to current climate variability. Climate models predict that climate change will lead to warmer temperatures, increasing rainfall variability, and increasing severity and frequency of extreme weather events. Agroforestry, or the intentional use of trees in the cropping system, has been proposed by many development practitioners as a potential strategy to help farmers reduce their vulnerability to climate change. This study explores whether and, if so, how agroforestry techniques can help subsistence farmers reduce their vulnerability to climate change. From field research conducted in western Kenya, we find that households are not currently coping with climate-related hazards in a sustainable way. Farmers are aware of this, and believe that the most effective way to adapt to climate-related shocks is through improving their general standard of living. We evaluated agroforestry as one possible means of improving farmers' well-being. By comparing farmers engaged in an agroforestry project with a control group of neighboring farmers, we find that involvement in agroforestry improves household's general standard of living via improvements in farm productivity, off-farm incomes, wealth and the environmental conditions of their farm. We conclude that agroforestry techniques can be used as an effective part of a broader development strategy to help subsistence farmers reduce their vulnerability to climate-related hazards.

Keywords: Africa, Agroforestry, Climate change adaptation, Food security, Smallholder development, Vulnerability

Introduction

Climate models predict that climate change will lead to, among other things, an increase in unpredictability of rainfall, warmer temperatures, and an increase in the severity and frequency of extreme weather events [1]. These changes are expected to decrease agricultural productivity in the developing world by 10% to 20% over the next 40 years [2]. Subsistence farmers in the developing world find it particularly difficult to cope with such climate-related hazards, as they do not have the capital to invest in new adaptive practices with which to protect their homes and families [3]. Especially sensitive to climatic changes are those households that rely almost entirely on rain-fed agriculture for their livelihoods. There has been a recent focus in the international development community and literature on strategies to help subsistence farmers reduce their vulnerability to climate change [4,5].

How communities cope with exposure to current climate-related shocks and stresses can give us insight into their ability to deal with future variability brought on by climate change [6]. Scholars are calling for a more interdisciplinary combination of academic fields and farmer perceptions to understand the effects of climate-related hazards on the complex systems of rural farmers [3,5,7].

Agroforestry has been proposed as one potential strategy for helping subsistence farmers reduce their vulnerability to climate change [8-10]. Research suggests that agroforestry improves farmer well-being through improving farm productivity and incomes [9,11,12]. Yet there are few studies that explicitly examine how agroforestry techniques can reduce vulnerability to climate change [13].

In addition, many agroforestry analyses assess the impacts of scientist-managed agroforestry plots, while

* Correspondence: H.Neufeldt@cgiar.org
[2]World Agroforestry Centre (ICRAF), Nairobi, Kenya
[3]CGIAR Research Program on Climate Change, Agriculture and Food Security (CCAFS)

relatively few studies analyze existing farmer-managed agroforestry development projects [14]. Farmer-managed agroforestry projects allow farmers to choose the type of agroforestry techniques to employ and rely on farmers to modify the techniques to match their needs. Farmer-managed projects therefore more accurately represent how agroforestry techniques are used under normal circumstances. There is a need for more extensive analyses of these types of projects [14].

To address the knowledge gaps outlined above, we set out to evaluate whether, and, if so how, farmer-managed agroforestry projects reduce farmers' vulnerability to climate change. We use Turner *et al.*'s vulnerability framework to understand farmer vulnerability [15]. Turner *et al.* divide a system's vulnerability into three major components: exposure, sensitivity and resilience [15]. Exposure considers the frequency, magnitude and duration to which a system is subject to hazards. We use the term 'climate-related hazards' to cover both climate-related shocks, such as floods and droughts, and longer-term climate stresses, such as increasing rainfall variability. The sensitivity of a system is determined by both the environmental and human characteristics that contribute to how a system responds to exposures. Finally, the resilience of a system refers to actions that can improve a system's ability to cope with outside hazards.

We began our study by assessing farmers' sensitivity to climate-related hazards through examining how farmers are currently coping with floods, droughts and rainfall variability. We then sought to understand what farmers believe to be the most effective way to become more resilient in the face of these outside stresses. From our findings, we established criteria to assess if agroforestry can be an effective technique to help reduce vulnerability to climate-related hazards. In short, this study sets out to investigate three major questions: How are farmers currently coping with exposure to climate-related hazards? What do farmers believe to be the most effective method to improving their resilience? And finally, how do agroforestry practices help farmers adapt to exposure to climate-related hazards?

To address these questions, we undertook a field study of a farmer-managed agroforestry development project in the Nyando District in western Kenya. We compare farmers who have been involved in World Agroforestry Centre (ICRAF) agroforestry development projects for 2 to 4 years to neighboring farmers with no agroforestry training. We used household surveys, in-depth interviews and focus group discussions to provide both a qualitative and a quantitative dimension to our analysis.

Methods

We used a mixed methods approach that combined household surveys, in-depth interviews, focus group discussions and field observations to investigate our research questions. We used interviews and participatory activities to understand how climate-related hazards are currently impacting farmers' well-being and how farmers conceptualize their well-being. These discussions generated a list of key indicators that farmers feel are most important to the well-being of their households. Finally, the household surveys allowed us to quantify how farmer-managed agroforestry interventions impacted farmers' well-being in the face of climate-related hazards. This type of mixed method approach has been strongly encouraged in the literature to better capture the reality on the ground [16].

Site

We conducted research in two sublocations of the Nyando District (Nyanza Province) of western Kenya. The Lower Nyando sublocation is characterized by low productivity, erratic rainfall and severe soil erosion. Elevation is 1,200 m with an average annual rainfall of 1,000 mm [17]. This area is predominantly of the Luo tribe. The Middle Nyando sublocation has higher productivity, cooler temperatures and more equitably distributed rainfall. Average elevation is 1,600 meters with average rainfall of 1,500 mm per year [17]. Middle Nyando has a mix of Luo and Kalinjin tribes. Maize is the staple crop in both sublocations, with sugar cane and coffee also grown as cash crops in Middle Nyando.

Project background

Seven community groups in Lower and Middle Nyando were provided tree seedlings and agroforestry training by ICRAF in 2006 and 2008. All members of these groups were included in the treatment group. Households within the treatment group received: 5 agroforestry and agriculture training sessions, 200 to 300 seedlings, training in tree nursery management, tools and seedlings for tree nursery establishment, a small amount of food each week for involvement in community projects, and ICRAF staff support for 1 year, roughly a US$300 investment per household. ICRAF provided a mix of tree species to farmers, including: *Acacia mellifera* and *Acacia polyacantha*, *Albizia coriaria*, *Calliandra calothyrsus*, *Casuarina equisetifolia*, *Cordia abyssinica*, *Faidherbia albida*, *Gliricidia sepium*, *Grevillea robusta*, *Markhamia lutea*, *Senna siamea*, and *Warburgia ugandensis*. Two additional community groups were selected as the control group based on their proximity to the treatment groups. None of the farmers in the control group had participated in agroforestry training in the past.

Due to the distinct climatic differences between Lower and Middle Nyando, it is difficult to compare groups across sublocations. In addition, farmers in the two sublocations differ in ethnicity, market access, land size and other key characteristics. Treatment and control households within sublocations are fairly similar, as can been seen across the basic household variables presented in Table 1. A t-test for differences of means was carried out to assess differences between all treatment and control groups and unless noted in the table, no significant differences were observed. Despite similarities in basic household characteristics, it is important to note that the control and treated households were located 1 to 2 km apart.

Data collection

We surveyed 119 households in June and July 2010 to capture basic household characteristics, agroforestry practices used and biophysical farm observations. Three households were removed from the dataset due to their extreme differences across key household parameters listed in Table 1. We conducted 20 in-depth interviews with 13 farmers (6 women), 4 village elders (1 woman) and 3 community leaders (all men). In addition, we held seven interactive focus group discussions with agriculture-oriented community groups. Questions focused on observed changes in climate, farming practices productivity constraints, agroforestry practices, future goals and how households had coped with the most recent floods and droughts. Men and women were split into subgroups for a part of each focus group discussion. Detailed methodology is presented in Thorlakson [18].

Key variables used for statistical analysis were measured in a way consistent with the literature in the field. We used an estimate of total livestock value as a proxy for household wealth, as livestock is the most frequently cited indicator of farmer wealth among the Luo and Kalinjin tribes of western Kenya [19]. We collected other indicators of wealth (housing material, type of roof, and so on) but found little variation among these indicators across households in our sample. Livestock holdings were converted into an economic value using current local market prices. We measured farm productivity by converting current seasonal crop production to economic units using average 2010 crop prices in the region. Soil erosion intensity was measured on a nine-

point scale using two on-farm observations, type of erosion present and intensity of observed erosion [20]. We estimated total above ground tree biomass using Kuyah et al.'s allometric equation 1 and corresponding coefficients, which were derived from a neighboring area in western Kenya with similar growing conditions [21].

Statistical data analysis

To assess the impacts of the agroforestry development project on farmer well-being we used household wealth and farm productivity as dependent variables. We used matching techniques to increase the similarities between the treatment and control groups [22,23]. Matching gives additional weight to households across treatment groups that are most similar on selected parameters. Parameters used for matching included: household size, land tenure, household head educational level, soil type and gender of household head, as these measures were all noted in the literature to affect subsistence farmer well-being [24].

Using the matched data, linear regressions were used to evaluate the treatment's impact on the outcome variables, accounting for potential regional differences and treatment effects across the two sublocations [25]. This analysis method was validated after incorporating the basic household parameters into the regression as well, achieving similar results.

Qualitative data analysis

We transcribed all field observations, interview notes and focus group discussion notes, and tagged major topics and keywords. This allowed us to compare farmers' views on key themes across different treatment groups, locations and household characteristics [26]. Common themes that emerged included agroforestry use, climate change, drought, farm constraints, farming techniques, floods, labor, erosion, rainfall change and well being.

Results

Climate change

In 2010, the Nyando District provided a unique opportunity to study the impact of climate-related hazards on farmers, as both a drought and a flood had recently

Table 1 Mean (SD) for key household parameters

	Lower Nyando		Middle Nyando		All data
	Treated	Control	Treated	Control	
Household size	6.7 (2.7)	6.4 (2.7)	6.2 (3.3)	7.2 (2.0)	6.7 (2.7)
Household head sex (1 = male)	0.61 (0.49)	0.6 (0.51)	0.93 (0.25)	0.93 (0.26)	0.77 (0.43)
Education of household head (form, 0 to 16)	5.4 (4.6)	4.2 (4.9)	7.2 (3.5)	6.1 (4.2)	5.9 (4.3)
Land size (hectares)	0.97* (0.38)	0.69 (0.32)	1.46 (1.38)	1.42 (1.3)	1.2 (1.1)
Holds title to Land? (1 = yes)	0.91* (0.23)	1 (0)	0.63 (0.49)	0.60 (0.50)	0.80 (0.42)
N	46	15	30	28	119

*Mean values are significantly different from control group households at the 5% level.

affected the area. Lower and Middle Nyando experienced drought-like conditions in September and October 2009 when the short-rains season failed. Specific data on the intensity of the drought in this region is sparse, but food shortage was widespread in the region due to water shortages [27]. In addition, the Lower Nyando region was also hit with a significant flood in March and April of 2010 that displaced 180 people and destroyed 7 homes across the Nyando District [28].

Farmers interviewed grouped climate-related hazards into three major topics: increased variability of the timing of rains, droughts, and floods. We focus on these three major types of hazards, though we understand that other changes are also expected to occur due to climate change.

Impacts due to exposure to climate-related hazards

Results from household surveys show that farmers' farm productivity decreased by 60% and 39% in Lower and Middle Nyando respectively during the 2009 to 2010 growing season in comparison to a typical growing season as experienced in 2008 to 2009. Farmers attribute this decrease to a combination of the drought, flood and rain variability experienced in the previous 12 months. Maize, the staple crop in the region, followed similar production trends (Table 2).

Due to the climate-related hazards experienced in the 2009 to 2010 season, households reported experiencing intense periods of hunger. 100% and 70% of households in the Lower and Middle Nyando region, respectively, experienced at least 1 additional month of hunger as compared to a typical year. Average duration of hunger periods for households were 4.5 and 2.3 months in Lower and Middle Nyando, respectively. A hunger period, as defined by farmers, is a time when the household had severe difficulties obtaining enough food to feed all household members.

Coping strategies during exposure to climate-related hazards

During periods of hunger, the most common coping strategy reported was to reduce food consumption through restricting the size, diversity and number of meals taken each day. Households that were involved in off-farm activities intensified their work in these areas

and others engaged in casual farm labor. Selling of livestock during drought and flood periods was a popular coping strategy, with 55% of farmers selling livestock in 2009 to 2010 to deal with food shortage. Consequently, livestock prices dropped 25% to 50% during this period. See Table 3 for a list of common coping strategies.

Farmers were also forced to use more detrimental coping strategies to cope with the reduced productivity in 2009 to 2010. From discussions with farmers, we defined detrimental coping strategies as those that have harmful long-term impacts on household productivity. Farmers reported selling oxen reserved for plowing during periods of drought, leading to lower farm productivity the following season as people then had to plow by hand. In all, 66% of farmers reported consuming seeds reserved for planting. This consumption had negative repercussions, as many farmers were forced to plant fewer seeds the following season due to the depletion of their personal seed stores and constraints in capital. In Middle Nyando, limited capital following the drought also restricted farmers' ability to purchase fertilizer and other chemical inputs regularly used. Some Middle Nyando farmers reported being forced to lease part of their farms for 2 years to wealthy farmers in the area in order to feed their families. This coping mechanism is especially detrimental as it prevents farmers from accessing their main source of livelihood, their land, for 2 years. According to some farmers, engaging in casual labor during periods of hunger also represents a detrimental coping strategy as it delays the planting in their own farms.

Farmers involved in an agroforestry development project typically used fewer detrimental coping strategies during hunger periods. Farmers with mature trees on their land were able to sell seedlings, timber and firewood and consume fruit from their trees during periods of hunger. Farmers reported that this diversification of coping strategies allowed them to rely less on other traditional coping strategies. See Table 3 for a comparison across groups.

Adaptation to climate-related hazards

The most effective way farmers found to reduce their vulnerability to these climate-related hazards was to

Table 2 Farm productivity and maize production during the 2009 to 2010 flood and drought year compared to an average year (2008 to 2009)

	2009 to 2010 farm productivity (Ksh)	Normal year farm productivity (Ksh)	Percentage difference	Maize production 2009 to 2010 (kg)	Normal year maize production (kg)	Percentage difference
Lower Nyando	8,200	20,500	−60%	85	220	−61%
Middle Nyando*	43,000	70,500	−39%	900	1,300	−30%

Farm productivity measured in Kenyan shillings (Ksh).
*Middle Nyando residents did not experience a flood in 2010.

Table 3 Proportion of farmers using coping strategies to deal with flood and drought in 2009 to 2010

	Lower Nyando		Middle Nyando	
	Treated (%)	Control (%)	Treated (%)	Control (%)
Reduce quantity, quality or no. of meals	82	66	54	86
Help from government, NGO, church	40	47	11	25
Borrow money	31	40	29	46
Casual labor	24	40	32	18
Sell possessions or livestock	73	66	36	43
Consume seeds	67	80	50	71
Consume or sell fruit from trees	40	25	68	38
N	45	15	28	28

NGO = non-governmental organization.

diversify income to include off-farm activities. Farmers who engaged in off-farm activities, such as wage-earning jobs or owning small shops, reported being better able to cope with climate-related hazards than their farming neighbors. Farmers with higher average farm productivity also reported fairing better during rainfall variations as they had more stores to draw on when current production was low.

Well-being

We discussed with farmers how they believed they could improve their overall standard of living when exposed to the hazards described above.

Importance of food security

Farmers interviewed were most interested in ways to improve their household's food security, especially during periods of outside shocks. Food security, as defined by farmers, is the ability to obtain an adequate diet for all household members throughout the year, without being forced to use long-term savings to purchase food. To achieve food security, farmers reported being interested in opportunities to start small business ventures or obtaining credit to purchase farm implements to improve their farm productivity. Farmers also expressed interest in opportunities to improve their agricultural knowledge and to learn about alternative income opportunities as other indirect pathways to improve food security.

Our quantitative analyses support farmers' assertion that farm productivity is tied to food security. Controlling for other key variables, our findings show that a household in the 75th percentile of farm productivity is on average 11% more food secure than a similar household in the 25th percentile of farm productivity ($P <0.0001$).

Other components of well-being

It is only after a household reaches relatively food security that they begin investing in long-term processes for improving other components of their well-being. This

stepwise process to improving well-being became clear when contrasting Lower and Middle Nyando farmers' goals. Lower Nyando farmers are still very food insecure and thus rarely report focusing on any goals not directly related to improving their household's food supply. However, some Middle Nyando farmers report feeling food secure throughout the year and discuss goals related to expanding landholdings, improving their children's education and investing in long-term projects to ensure financial security.

Farmers also reported that they have begun to put more emphasis on the environmental conservation of their land. Both treatment and control focus groups concluded that their well-being had significantly declined due to soil erosion on their farms. As one farmer explained, 'Soil is our livelihood'. A number of community groups had recently been formed to focus on environmental issues and soil conservation practices in the area, suggesting that environmental conservation is perceived as a key way through which communities believe they can improve their well-being.

Constraints to achieving well-being are summarized in Table 4. Most farmers agreed that unpredictable weather and lack of access to capital are the two largest constraints to improving their lives, but environmental degradation was also cited as a major concern.

Vulnerability reduction

In addition to food security, most farmers also cited an ability to cope with shocks and stresses as a key characteristic of a successful household. During times of stress, successful households are food secure for 2 to 3 months longer, often giving support to their neighbors and family. Successful households are not forced to sell livestock or belongings, take their children out of school, or significantly reduce meal portions during exposure to outside shocks.

Farmer concerns about their household's vulnerability to outside shocks was evident in almost all interviews. As one farmer explained, 'We are reliant on the rains

Table 4 Major constraints farmers identified to achieving well-being

	Lower Nyando		Middle Nyando		All data (%)
	Treated (%)	Control (%)	Treated (%)	Control (%)	
Weather	66	80	86	92	73
Capital	59	73	64	86	68
Farm inputs and implements	39	46	40	43	41
Environmental health of land	52	66	24	80	53
Health of household	25	26	43	57	36
N	45	15	28	28	116

from God, and there is nothing we can do to change these patterns'. Currently, households feel unable to deal with the unexpected problems that arise from extreme weather events, sicknesses, job loss, low cash crop market prices, and so on. Farmers continuously reiterated their need to find better ways in which to deal with exposure to outside shocks, particularly rainfall variability and drought, which frequently disrupt their lives. Farmers were most interested in improving their off-farm incomes, diversifying income sources and improving general farm productivity to reduce their sensitivity to climate-related hazards.

During interviews, farmers also emphasized their desire to remain autonomous in deciding what type of specific adaptation measures they choose to employ. Many farmers complained that some specific climate-change adaptation measures suggested by agricultural extension workers or non-profit organizations, such as planting drought-resistant maize, were actually detrimental to their farm yields during normal or heavy rains. With the uncertainties farmers face in weather patterns from year to year, they were unwilling to invest in strategies that were less productive under certain weather conditions. Farmers reported being interested in receiving information and advice on potential adaptation strategies, as long as outside constituents did not decide what activities would take place in their communities without their consent.

Agroforestry

Using our results that farmers were most interested in general well-being improvements to adapt to climate-related hazards, we assessed agroforestry's potential in providing these general well-being improvements in the face of climate-related hazards. When interpreting these results, it is important to note that the agroforestry project assessed has only been in operation for 2 to 4 years and thus the long-term effects of agroforestry involvement are not captured in our analysis.

Improvements in farm productivity

Our results suggest that agroforestry improves farm productivity and household wealth. 43% of farmers

noticed an improvement in farm productivity after planting trees on their land. Farmers found trees improved their farm productivity by decreasing soil erosion and increasing soil fertility. Farmers who reported no change in farm productivity explained either they had not planted nitrogen-fixing trees in their fields or the trees were not yet mature enough to assess the effects.

Overall, only 12% of farmers in the agroforestry program chose to intercrop nitrogen-fixing trees in their fields. Many farmers expressed concern that planting trees in their fields would reduce productivity of their crops and were unwilling to take such a risk. All farmers who have begun intercropping trees reported significant improvements to their productivity after incorporating nitrogen rich leaves into their soil.

For farms using agroforestry techniques, our quantitative data suggests a slight improvement in farm yields in both Lower and Middle Nyando when compared with the control group. Our linear regression model estimates that Lower Nyando farmers involved in an agroforestry project improved their farm productivity, on average, by about 1,500 Ksh (US$19) per year when compared to the control group. (For all currency conversion, the July 2010 current rate of 80Ksh = US$1 was used.) A 1,500 Ksh increase is the equivalent of increasing an average Lower Nyando household's maize yields by 35%. However, the standard errors in this analysis are quite large ($P = 0.678$) (Table 5). In Middle Nyando, the results from statistical analysis show farm productivity increase by 2,100 Ksh (US$26) for treated units but again with a high standard error ($P = 0.549$). This increase is equivalent to improving an average Middle Nyando household's maize yields by 20%. The high uncertainty in the quantitative results of the study was likely in part due to the small sample size, short duration of agroforestry participation and the non-randomized selection of households.

Improvements in household wealth and income diversity

Overall, farmers were most interested in trees' ability to provide them with additional farm income. During focus group discussions, farmers ranked the potential income

Table 5 Agroforestry's effect on farm productivity and household wealth

	Lower Nyando			Middle Nyando		
	Treated	Control	Difference	Treated	Control	Difference
Farm Productivity (Ksh)	4,600	3,100	1,500	16,200	14,100	2,100
SD			3,700			3,400
P value			0.68			0.55
Household Wealth (Ksh)	62,200	38,400	23,800	58,900	67,300	−8,400
SD			13,900			12,900
P value			0.09			0.516
N	45	15		28	28	

Values from linear regression models run on matched data.

benefits from trees as the most helpful aspect of the trees on their land. The excitement farmers expressed in the income benefits from tree products stemmed in part from the limited opportunities for income generation in the area. As one elder farmer explained, 'There is just no way to earn an income here... No one has money to buy anything from anyone else'.

Among the farmers in Lower Nyando who have had trees for 4 years, 87.5% of farmers reported income improvements. Benefits were reported from the sale of fuel wood, timber, fruit and seedlings and through savings in food purchases due to an increase in farm productivity. For those farmers with mature fruit trees, average seasonal profits were 3,250 Ksh (US$40). Farmers who had not seen improvement in their income after planting trees explained that their trees were still too young to provide any benefits.

In our quantitative analysis we used livestock holding as a surrogate for household income as it is difficult to measure wealth in real terms among small-scale farmers. Household wealth, as measured by current livestock holdings, improved for Lower Nyando participants involved in an agroforestry project. Treated units in Lower Nyando had livestock holdings worth 24,000 Ksh (US$300) more than control units in the region, on average ($P = 0.092$, Table 5). The Lower Nyando statistical findings agree with our qualitative observations about agroforestry's ability to improve household wealth.

For Middle Nyando, project involvement decreased average value of livestock holdings by 8,000 Ksh (US $100), ($P = 0.516$, Table 5). It is not surprising that Middle Nyando farmers have not improved their wealth through agroforestry involvement, as these farmers planted their trees only 2 years ago and do not yet have mature trees that can provide timber, fruit or fuel wood for sale. Due to their remote location, Middle Nyando farmers have also had less success selling tree seedlings to neighboring communities so have been unable to receive substantial income benefits from this source. Lower Nyando farmers, however, reported selling fruit,

timber, fuel wood and seedlings on a regular basis to local markets. From discussions with farmers in the area, it appears that the inconclusive results on wealth in Middle Nyando are in part due to a lack of infrastructure in the area, that is currently acting as a barrier to access markets for their tree crops.

Other benefits

Involvement in agroforestry practices also provides a number of other general improvements that helped farmers increase the environmental sustainability of their farms. In all, 70% of farmers involved in agroforestry projects cited soil erosion control as a key benefit. Soil erosion was particularly detrimental to people affected by the 2010 floods in Lower Nyando, with many farmers complaining of decreased soil fertility due to the intense soil erosion during the heavy rains. Farmers considered tree planting to be the most effective method of soil erosion control.

Our field observations support farmers' claims that increased tree density reduces soil erosion. As can be seen in Figure 1, the data show a downward trend when the tree biomass per hectare is plotted against a scale of soil erosion observed, showing that, on average, farmers with higher tree biomass per hectare experience less soil erosion (correlation coefficient = −0.31).

Involvement in agroforestry also provides substantial labor savings to women household members by reducing time spent on fuel wood collection. Some women reported walking over 20 km to purchase fuel wood in neighboring districts. Women in low-tree-density areas also reported being threatened by their neighbors in a struggle over fuel wood resources.

Our research shows that agroforestry involvement leads to substantial reductions in fuel wood purchased and the time that households use to collect fuel wood (Table 6). Fuel wood was the second most commonly cited use of trees on the farm. Women with mature trees on their land felt that they now have access to a safer and more stable supply of fuel wood. These women reported devoting

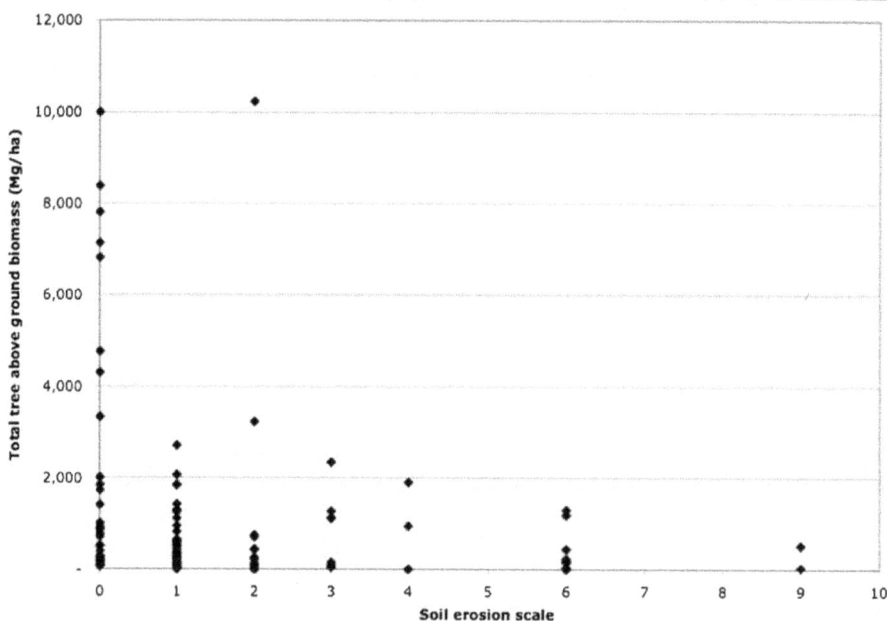

Figure 1 Scale of soil erosion plotted against total above ground tree biomass per hectare (Mg/ha). Tree biomass was estimated using Kuyah et al.'s equation number 1 [21].

more time to income-generating activities and farm care now that fuel wood stocks are nearby.

It also appears that there is an opportunity in low-tree-density areas to increase off-farm incomes of women through local fuel wood sales. There is a substantial demand for purchased fuel wood in these areas, yet only two farmers reported having enough excess fuel wood on their land to sell fuel wood at the market.

Specific coping strategies

In addition to the general benefits listed above, agroforestry also provides specific coping options to farmers during exposure to floods and droughts. Some farmers reported having fruit from their trees as their only steady source of sustenance during the floods, as other crops were underwater or had been washed away. During the drought, many farmers reported selling fuel wood and timber to produce additional income for food purchases. Tree roots prevented extensive soil erosion and ensuing soil fertility loss for farmers in Lower Nyando during the 2010 floods. The alternate coping strategies provided by trees allowed farmers additional flexibility in their management of the climate stresses they faced.

Discussion

Achieving well-being or a similarly acceptable quality of life is a fundamental goal in most development projects focused on poverty alleviation [29]. Our findings also showed that farmers expressed an interest in improving their well-being, but with a particular emphasis on the need to reduce vulnerability when exposed to outside shocks and stresses. This finding agrees with those of Place *et al.* and encourages a move away from the focus only on poverty alleviation to focus on the reduction of the vulnerability of poor populations [16].

Using these ideas expressed by farmers, we modified, for the use of our analysis, the definitions of well-being discussed in the literature [30-32]. We define well-being as the ability to improve one's household income, assets, and food security in the face of outside stresses and shocks (modified from 30, 31). This definition is unique because it takes into account the need to achieve well-being improvements while exposed to climate-related hazards and it specifically stresses the importance of food security and other tangible changes.

We found that food security is a concept very important to the farmers interviewed but it is not always

Table 6 Fuel wood acquisition

	Lower Nyando			Middle Nyando		
	Treated	Control	Difference	Treated	Control	Difference
Weekly time spent on fuel wood collection (min)	360	540	180	220	260	40
Percentage of households purchasing fuel wood	17%	66%		3%	7%	
N	45	15		28	28	

emphasized in the literature's definitions of well-being. Food insecure farmers in Lower Nyando explained that all of their long-term decisions were dependent on their food supply. These farmers were forced to stop long-term projects, including the care of their seedlings, when faced with the significant food shortage during the 2009 to 2010 season. In contrast, farmers in Middle Nyando, where average hunger periods were less severe, felt much better equipped to plan and save for their futures. This contrast between how farmers in Lower and Middle Nyando think about long-term goals supports Appadurai's conclusion that more marginalized groups have a more limited ability to aspire [33]. It also supports evidence that food insecure farmers tend to be less innovative. This has recently been shown in a representative survey of 700 farmer households in four countries in East Africa [34].

Farmers' emphasis on food security should provide a reminder to development academics of the importance of dealing with basic necessities when considering more complex development interventions. It is paramount that development projects operating in food-insecure areas show how they can help households improve their food security from the outset. This lesson is especially relevant to agroforestry development projects, as farmers are asked to invest substantial time in tree seedling care before any benefits are accrued. For households that are still struggling to ensure they have enough to eat, tree seedlings and other long-term investments are often neglected until this need is met.

The impacts of the recent floods, droughts and rainfall variability on farmers are consistent with much of the literature of similar climate-related hazards in the area [35]. The literature on coping strategies also highlights similar strategies to those used by farmers in this study [36]. However, from our analysis it is clear that the current use of these traditional coping strategies will not be able to sustain households in the face of more frequent exposure to climate-related hazards predicted by climate change models such as Nelson et al.'s [37]. Already, farmers experience intense periods of food insecurity and are forced to engage in detrimental coping strategies that threaten the long-term sustainability of their farms and households' well-being. This finding is consistent with other analyses of the ability of subsistence farmers to cope with climate variability in the region [4].

The literature exhibits some tension on whether to focus climate change adaptation on these specific adaptation strategies or on the more general well-being improvements [38,39]. Through our discussions with farmers on future adaptation options, farmers largely agree with the literature that supports general well-being

improvements to deal with future climate-related hazards [37,40-42]. Specifically, farmers are interested in improving their off-farm incomes and farm productivity. As climate change predictions continue to be highly unpredictable, it is important to focus on adaptation strategies that are robust and help ensure farmers' well-being under a variety of forms of climate-related hazards.

Farmers report, and the literature agrees, that some specific adaptation strategies, such as the current forms of drought-resistant crops, may actually increase farmers' vulnerability under certain climate-related hazards [41]. Farmers interviewed refused to use more drought-tolerant types of maize, as they could not afford the loss of productivity associated with these strains if the rains did come. These results highlight the need to do place-based studies of adaptive strategies to assess which specific projects will be most effective at reducing farmer vulnerability under a wide variety of climate hazards. Through our analysis, it appears that agroforestry can be an effective adaptation strategy, as agroforestry was shown to be beneficial under a wide range of climatic conditions. This interpretation also provides evidence for the introduction of financial safety networks, such as index insurances, as a means of reducing climate risk while allowing farmers to continue using improved seedling material for higher yields under normal climatic conditions. The index insurance could help farmers cope with greater losses during weather extremes such as droughts.

Farmers also expressed a strong desire to remain autonomous in making decisions about the implementation of climate-change adaptation strategies. However, it was also clear that households will need, and are interested in receiving, information and advice on potential climate-change adaptation strategies. This distinction between outside organizations' role as an advisor on adaptation options and an implementer of specific strategies is somewhat nuanced, but will be important to ensure farmer buy-in to future adaptation strategies. Adger et al. highlight the need to facilitate climate-change adaptation decisions among households, as individual autonomy alone does not necessarily lead to the most effective decisions [43]. Future research is needed to find effective advising strategies that can help farmers access information on and capital to invest in adaptation strategies.

Our findings provide quantitative results, supported and nuanced by qualitative descriptions, on how agroforestry techniques can help farmers mitigate their vulnerability to climate-related hazards, both through improving general well-being and through providing specific coping measures that are effective in the face of a wide range of climate-related hazards.

Despite the high standard errors within our quantitative analysis, in conjunction with our qualitative findings they suggest a positive correlation between agroforestry involvement and farm productivity and household wealth, similar to other literature in the field [44]. It is unsurprising that we observed only a small improvement in farm productivity because few farmers used intensive agroforestry techniques in their fields. Intercropping of nitrogen-fixing trees had low adoption among farmers because farmers considered planting trees in their field as risky. Planting trees along household boundaries appears to show greater acceptability among farmers and may be a good first step to integrate agroforestry practices into farms without the perceived risk of intercropping.

Improvements in household wealth were also limited due to the maturity of the trees. Most income comes from selling fruit, excess fuel wood and timber, and many farmers interviewed had only a limited number of mature trees on their land at the time of the survey. Limited improvements are therefore expected among the farmers who have only been using agroforestry techniques for 2 to 4 years. In addition, we feel that using livestock as a surrogate for household wealth did not capture the full change in farmers' wealth due to agroforestry practices. Future studies should consider a more robust measurement of wealth among subsistence farmers.

Our findings also highlight the length of time it takes for agroforestry practices to benefit farmers, as many of the impacts reported were quite small. We expect the magnitude of improvements in farm productivity and household wealth will increase as trees mature, as all of the qualitative data collected supports this trend. Future studies of this area would need to confirm such an assertion. These conclusions underline the need to provide agroforestry techniques in collaboration with other development initiatives that can deliver short-term benefits to farmers while waiting for the investments in trees to pay off.

We approached our research with an understanding that agroforestry practices may be used and perceived differently by men and women [45]. However, we did not find that women participated in agroforestry in a substantially different way, as all farmers reported planting, tending to and harvesting from trees. The one gender difference we observed was that women were primarily responsible for fuel wood collection and thus ranked fuel wood collection as a more significant benefit than men did.

Our study was particular to a specific agroforestry project and location, allowing us to deeply understand the unique context of these communities [46,47]. However, we feel that the conclusions can still provide significant guidance to future studies on reducing farmers' vulnerability to climate-related hazards. Our findings suggest that it is likely that for extremely poor households, improving general well-being will be the most effective way to reduce vulnerability to future hazards associated with climate change. This finding can be generalized to other subsistence farming communities, as the most basic problems faced by farmers during climate-related hazards are widespread. Whether this finding can be applied to more successful small-scale farmers remains to be explored.

Finally, we found that fully engaging farmers in the research process significantly improved our analysis. Farmers report benefiting from the research process because it provided them with an opportunity to discuss constraints to their well-being and think about potential solutions. Communities found that discussions of vulnerability allowed them to reflect on their current farming practices and engage in a community conversation of future adaptation options. This dialogue really flourished when we presented our initial findings to the farmers in a community gathering we organized at the end of our field research. This conclusion supports White's assertion that encouraging reflection among communities is of value in and of itself [47].

Discussions with farmers at this community gathering stressed the responsibility field researchers have to provide feedback and results to the participants in their study. As the local chief explained, 'Many scientists have come here, but you are the first to return with results'. We hope that future studies can continue to build on this approach and engage farmers more fully throughout the research process. This will allow the scientific community to further the dialogue with farmers in how we can work together to ensure environmental sustainability and well-being improvements in the face of future climate-related hazards.

Conclusions

Our findings show that farmers are interested in improvements in their food security, income, farm productivity, and environmental sustainability of their farms in the face of outside shocks and hazards. Our results agree with the literature in calling for a focus on reducing vulnerability to climate-related hazards through robust adaptation measures that can be beneficial regardless of the type of climate hazard experienced. From analysis of current coping strategies being employed, it is clear that subsistence farmers in the Nyando District are not coping with climate-related hazards in a sustainable way. As climate change increases the frequency and intensity of these events, existing coping strategies may no longer be able to support households through difficult times.

Even in a short-term analysis of a farmer-managed agroforestry study with low uptake of intensive techniques, agroforestry practices provided farmers noticeable benefits. Involvement in agroforestry improved farm productivity, household wealth, increased income diversity, reduced soil erosion and provided a number of specific coping strategies to help households during exposure to climate-related hazards.

These conclusions beg the question, what can be achieved with even more effective implementation of agroforestry practices? In order to enhance the effectiveness of agroforestry in improving farmer well-being in the face of climate-related hazards, our findings suggest the following for future agroforestry development projects:

Pair agriculture and agroforestry training

Agroforestry is a long-term process and, as we showed in this analysis, benefits can take a long time to accrue. Therefore, improving the sustainability of agroforestry projects is key. It will be important to include shorter-term benefits to farmers that are coupled with agroforestry implementations so farmers do not get discouraged during the initial project stages.

Pairing agroforestry and agricultural training is an excellent opportunity to provide short-term benefits from improving basic agriculture knowledge with long-term extension for agroforestry practices. We also found that if farmers understand the potential of their trees to enhance their well-being, they put more concentrated efforts toward tree care and management, giving further reason to focus on the training aspect of agroforestry programs.

Improve market accessibility to enhance income-generating opportunities provided by agroforestry techniques

As our analysis showed, one of the most effective ways to reduce farmers' vulnerability to climate change is through improving incomes of farmers. Because tree crops are more resistant to climatic shocks, they can provide support for farmers during these times of stress. Agroforestry techniques have the potential to provide income to farmers through the sale of fuel wood, timber, fruit and seedlings. In comparing the benefits derived from agroforestry involvement across the two sublocations, market access played a key role in improving household incomes due to agroforestry projects. Therefore, market access needs to be improved. This can be done on a governmental scale through improving infrastructure or, more locally, through establishing cooperatives that pool resources to access markets.

Couple access to farm implements and capital with agroforestry projects

Lack of access to farm implements and capital were listed as key constraints to overall farm productivity in the Nyando region, and almost no one has access to small-scale loans in this area. Although not directly related to agroforestry, any major constraint to farm productivity reduces farmers' ability to cope with climate change. In addition, by improving access to loans and to farm implements through an agroforestry development project, farmers are able to see tangible benefits in the short term from their project involvement before their trees have matured. The coupling of access to credit and agroforestry training has been found to be an effective way to reduce vulnerability to climate change in other studies [4]. Access to weather index insurance may be another support mechanism worth exploring to couple with future agroforestry development projects.

Organize educational farm visits to successful agroforestry projects to increase adoption of agroforestry techniques

There is substantial room for the expanded use of agroforestry techniques in improving farm productivity, as adoption of intensive agroforestry techniques was low among farmers. The key reason for this low adoption rate was that farmers perceive these techniques to be high risk. All farmers interviewed who engaged in intensive agroforestry techniques had seen the benefits on someone else's farm before implementing the techniques themselves. Many farmers suggested this as the most effective way to convince them that agroforestry techniques may be useful in their area.

Concluding remarks

Agroforestry, like any single coping strategy, will not prove to be the silver bullet to climate change adaptation. However, we found that agroforestry practices do have substantial potential to help farmers improve their well-being and the environmental sustainability of their farms. Through these improvements, and by providing some additional specific adaptation strategies, agroforestry practices can reduce farmer vulnerability to climate change. By adopting the recommendations outlined above, we hope that future agroforestry projects can contribute, as a part of larger development initiatives, to helping subsistence farmers better adapt to future climate change.

Competing interests
The authors declare that they have no competing interests.

Authors' contributions
TT conceived the study design, collected field data, analyzed results and drafted the manuscript. HN participated in the study design, contributed to the draft of the manuscript and performed critical reviews. Both authors read and approved the final manuscript.

Authors' information
TT was lead author.

Acknowledgements
First and foremost, we would like to thank the individual farmers for their participation throughout the research process. We also immensely grateful

to Walter Adongo, Joash Mango, Amos Odhiambo and Brian Mateche for their integral work as members of the research team. This research would not have been possible without the support of the Comart Foundation and the Harvard University Weatherhead Center. Finally, a special thanks to William Clark and Andrew Harris, for their insights and constructive feedback.

Author details

[1]Sustainability Science Program, Harvard University, Cambridge, MA, USA. [2]World Agroforestry Centre (ICRAF), Nairobi, Kenya. [3]CGIAR Research Program on Climate Change, Agriculture and Food Security (CCAFS).

References

1. Alley R, Berntsen T, Bindoff N, Chen Z, Chidthaisong A: **Solomon S, Qin D, Manning M**. In *Climate Change 2007: The Physical Science Basis. Contribution of Working Group I to the 4th Assessment Report of the Intergovernmental Panel on Climate Change*. Cambridge, UK: Cambridge University Press; 2007.

2. Nelson GC, Rosegrant MW, Koo J, Robertson R, Sulser T, Zhu T, Ringler C: *Climate Change: Impact on Agriculture and Costs of Adaptation*. Washington DC: IFPRI; 2009.

3. Thompson J, Millstone E, Scoones I, Ely A, Marshall F: *Agri-food System Dynamics: Pathways to Sustainability in an Era of Uncertainty*, STEPS Center Working Paper 4. Brighton, UK: STEPS Center; 2007.

4. Gabrielsson S, Brogaard S, Jerneck A: **Living without buffers: illustrating climate vulnerability in the Lake Victoria Basin**. In *Uncertain futures*. PhD thesis.: Lund University, Centre for Sustainability Studies; 2012.

5. Morton J: **The impact of climate change on smallholder and subsistence agriculture**. *Proc Natl Acad Sci U S A* 2007, **104**:19680–19685.

6. Kates RW: **Cautionary tales: adaptation and the global poor**. *Climate Change* 2000, **45**:5–17.

7. Smit B: **Adaptation, adaptive capacity and vulnerability**. *Glob Environ Chang* 2006, **16**:282–292.

8. Challinor A, Wheeler T, Garforth C, Craufurd P, Kassam A: **Assessing the vulnerability of food crop systems in Africa to climate change**. *Climate Change* 2007, **83**:381–399.

9. Verchot LV, Noordwijk MV, Kandji S, Tomich T, Ong C, Albrecht A, Mackensen J, Bantilan C, Anupama K, Palm C: **Climate change: linking adaptation and mitigation through agroforestry**. *Mitig Adapt Strat Glob Chang* 2007, **12**:901–918.

10. Bank W: *World Development Report 2008: Agriculture and Development*. Washington DC: World Bank Publications; 2008.

11. Garrity D: **Science based agroforestry and the achievement of the millennium development goals**. In *World Sgroforestry into the Future*. Edited by Garrity D, Okono A, Grayson M, Parrott S. Nairobi, Kenya: World Agroforestry Centre; 2006.

12. Leakey R: **Agroforestry: a delivery mechanism for multi-functional agriculture**. In *Handbook on Agroforestry: Management Practices and Environmental Impact*. Edited by Kellimore LR. New York, NY: Nova Science Publishers; 2010:461–471.

13. Scherr SJ, Franzel S: **Introduction**. In *Trees on the Farm: Assessing the Adoption Potential of Agroforestry Practices in Africa*. Edited by Franzel S, Scherr SJ. New York, NY: Cabi Publishing; 2002.

14. Scherr SJ, Franzel S: **Promoting new agroforestry technologies: policy lessons from on-farm research**. In *Trees on the Farm: Assessing the Adoption Potential of Agroforestry Practices in Africa*. Edited by Franzel S, Scherr SJ. New York, NY: Cabi Publishing; 2002.

15. Turner BL, Kasperson RE, Matson PA, McCarthy J, Corell R, Christensen L: **A framework for vulnerability analysis in sustainability science**. *Proc Natl Acad Sci U S A* 2003, **100**:8074–8079.

16. Place F, Adato M, Hebinck P: **Understanding rural poverty and investment in agriculture: an assessment of integrated quantitative and qualitative research in western Kenya**. *World Dev* 2007, **35**:312–325.

17. Kenya Food Security Steering Group: *Food security district profile, Nyando District, Nyanza Province*. http://www.nyando.org/reference/foodsecurity.pdf.

18. Thorlakson T: *Reducing Subsistence Farmers' Vulnerability to Climate Change: The Potential Contributions of Agroforestry in Western Kenya. World Agroforestry Centre Occasional Paper 16*. Nairobi, Kenya: World Agroforestry Centre; 2011.

19. Shipton PM: *Mortgaging the Ancestors: Ideologies of Attachment in Africa*. New Haven, CT: Yale University Press; 2009.

20. Okoba BO, Graaff J: **Farmers' knowledge and perceptions of soil erosion and conservation measures in the Central Highlands, Kenya**. *Land Degrad Develop* 2005, **16**:475–487.

21. Kuyah S, Dietz J, Muthuri C, Jamnadass R, Mwangi P, Coe R, Neufeldt H: **Allometric equations for estimating biomass in agricultural landscapes: I.** *Aboveground biomass. Agric Ecosyst Environ* 2012, **158**:216–224.

22. Rosenbaum PR, Rubin DB: **Constructing a control group using multivariate matched sampling methods that incorporate the propensity score**. *Am Stat* 1985, **39**:33–38.

23. Ho D, Imai K, King G, Stuart E: **Matching as non-parametric preprocessing for reducing model dependence in para-metric causal inference**. *Polit Anal* 2007, **15**:199–236.

24. Kabubo-Mariara J, Linderhof VGM, Kruseman G, Atieno R, Mwabu G: *Household Welfare, Investment in Soil and Water Conservation and Tenure Security: Evidence from Kenya*. Amsterdam, Netherlands: Poverty Reduction and Environmental Management, Working Paper No 06/06; 2006.

25. Imai K: **Toward a common framework for statistical analysis and development**. *J Comput Graph Stat* 2008, **17**:892–913.

26. Weiss RS: *Learning from Strangers: The Art and Method of Qualitative Interview Studies*. New York, NY: The Free Press; 1994.

27. Kenya Red Cross: *Drought operations update: alleviating human suffering*. No. 1/09. http://reliefweb.int/sites/reliefweb.int/files/resources/ FE8B12A7BF328277C125764000435C 92-Full_Report.pdf.

28. Office for the Coordination of Humanitarian Affairs (OCHA): *Kenya: 2010 floods and landslides situation update. Report number 3*. http://reliefweb.int/ sites/reliefweb.int/files/resources/D86F059443444C20C1257726003F27 7D-Full_Report.pdf.

29. Costanza R, Fisher B, Ali S, Beer C, Bond L, Boumans R, Danigelis NL, Dickinson J, Elliott C, Farley J, Gayer DE, MacDonald Glenn L, Hudspeth J: **Quality of life: an approach integrating opportunities, human needs, and subjective well-being**. *Ecol Econ* 2007, **61**:267–276.

30. Chambers R, Conway GR: *Sustainable Rural Livelihoods: Practical Concepts for the 21st Century*. Brighton, UK: Institute of Development Studies Discussion Paper 296; 1991.

31. Lindenberg M: **Measuring household livelihood security at the family and community level in the developing world**. *World Dev* 2002, **30**:301–318.

32. White S: **Analysing wellbeing: a framework for development practice**. *Dev Pract* 2010, **20**:158–172.

33. Appadurai A: **The capacity to aspire: culture and the terms of recognition**. In *Culture and Public Action*. Edited by Rao V, Walton M. Palo Alto, CA: Stanford University Press; 2004.

34. Kristjanson P, Neufeldt H, Gassner A, Mango J, Kyazze FB, Desta S, Sayula G, Thiede B, Förch W, Thornton PK, Coe R: **Are food insecure smallholder households making changes in their farming practices? Evidence from East Africa**. *Food Sec* 2012, **4**:381–397.

35. Mogaaka H: *Climate Variability and Water Resources Degradation in Kenya: Improving Water Resources*. Washington, DC: World Bank Working Paper No 69; 2006.

36. Eriksen SH, Brown K, Kelly PM: **The dynamics of vulnerability: locating coping strategies in Kenya and Tanzania**. *Geogr J* 2005, **171**:287–305.

37. Nelson GC: *Food Security, Farming, and climate change to 2050*. Washington DC: International Food Policy Research Institute; 2010.

38. Brooks N, Adger W, Kelly P: **The determinants of vulnerability and adaptive capacity at the national level and implications for adaptation**. *Glob Environ Chang* 2005, **15**:151–163.

39. Smit B, Pilifosava O: **Adaptation to climate change in the context of sustainable development and equity**. *Sustain Dev* 2003, **8**:879–906.

40. Ahmed S, Chaturvedi S, Saroch E, Chopde S, Sharma S, Dixit A, Gyawali D: *Singh Rathore M, Mudrakartha S, Moench M, Rehman T, Wajih SA, Upadhya M, Sharma RK: Adaptive Capacity and Livelihood Resilience: Adaptive Strategies for Responding to Floods and Droughts in South Asia*. Boulder, CO: The Institute for Social and Environmental Transition; 2004.

41. Krysanova V, Buiteveld H, Haase D, Hattermann FF, Niekerk K, Roest K, Martínez-Santos P, Schlüter M: **Practices and lessons learned in coping with climatic hazards at the river-basin scale: Floods and droughts**. *Ecol Soc* 2008, **13**:1–27.

42. Smucker TA, Wisner B: **Changing household responses to drought in Tharaka, Kenya: vulnerability, persistence and challenges.** *Disasters* 2008, **32:**190–215.

43. Adger W, Arnell N, Tompkins E: **Successful adaptation to climate change across scales.** *Glob Environ Chang* 2005, **15:**77–86.

44. Syampungani S, Chirwa PW, Akinnifesi FK, Ajayi OC: **The potential of using agroforestry as a win-win solution to climate change mitigation and adaptation and meeting food security challenges in southern Africa.** *Agric J* 2010, **5:**80–88.

45. Kiptot E, Franzel S: **Gender and agroforestry in Africa: a review of women's participation.** *Agrofor Syst* 2012, **84:**35–58.

46. Gabrielsson S: *Uncertain futures. PhD thesis.* Lund University, Centre for Sustainability Studies; 2012.

47. White S: *Bringing Wellbeing into Development Practice. Wellbeing in Developing Countries Research Group Working Paper.* Bath, UK: University of Bath; 2009.

12

Effects of surrounding crop and semi-natural vegetation on the plant diversity of paddy fields

Nur Rochmah Kumalasari[1,2][*] and Erwin Bergmeier[1]

Abstract

Background: The ecosystems around, and plant composition in, paddy fields in Java are varied, owing to differences in climate, altitude, and traditional farming practice. This study examines the effects of different types of surrounding land use and vegetation on the plant diversity in paddy fields.

Methods: We studied three upland (400–850 m asl) and three lowland areas (10–50 m asl) in the island of Java, Indonesia. Samples of vegetation were taken in fields and bunds (partition between paddy field plots) in two rice cultivation seasons from October 2011 through to June 2012, including the peak of rice cultivation in Java between October and February. We used Analysis of Variance Matrix Unbalanced to analyze the effects of area, complexity, location, and season on plant composition. Tukey's Honestly Significant Difference (HSD) test was performed to determine significant differences between groups in the sample.

Results: We recorded 14 crop species and 221 non-cultivated plant species, of which 171 species occurred in paddy fields and 190 on bunds. Species numbers in upland areas were higher than in lowland areas. In fallows, twice as many species as in cultivated rice fields were found. The presence of semi-natural vegetation within short distance had no significant effect on plant species numbers in paddy fields. Multiple cropping and intercropping around paddy fields and on bunds had a marked effect on plant diversity.

Conclusions: Differences in plant species numbers and composition between lowland and upland areas are more pronounced than the effects of local environmental complexity. To enhance high and varied plant diversity on the field and landscape scale, traditional multiple and intercropping systems should be supported.

Keywords: Bund, Cultivation season, Fallow, Java, Landscape complexity, Monoculture, Multiple cropping, Rice field, Weeds

Background

Environmental factors and management are the most important general drivers of plant species richness, abundance, and composition in agroecosystems [1-4]. Weeds, defined here as non-cultivated plants of crop fields, may hamper land management and food production but support biological diversity, diversify food webs and potential food resources, and improve soil health [5,6]. Weed communities differ in relation to environmental and landscape structural heterogeneity [2,7], and

even the seed banks in arable lands are influenced by landscape complexity [8,9].

The ecosystems around paddy fields in Java are varied, owing to differences in climate and altitude and because traditional farming varies across villages [10]. Available resources and the local natural environmental conditions, such as season, water availability, landscape, and natural vegetation, determine the type and complexity of land use. For instance, in the rainy season, water availability is increased both in irrigated and rain-fed paddy fields, a factor that is important for rice plant growth and weed management [11].

Javanese rice farmers commonly divide a year in three rain-dependent cultivation seasons. The prime cultivation (between the end of October till March) starts in the wet season, the second cultivation period is in the

* Correspondence: nurrkumala@gmail.com
[1]Department of Vegetation & Phytodiversity Analysis, Albrecht-von-Haller Institute of Plant Sciences, Georg-August University of Göttingen, D-37073 Göttingen, Germany
[2]Department of Nutrition and Feed Technology, Faculty of Animal Science, Bogor Agricultural University (IPB), Bogor, Indonesia

medium-wet season around April–July, and the third is the dry season between August and October [12]. The optimum cultivation date is indicated by the soil moisture level required for planting and in general by the water availability to be expected in the coming period [13]. Compared with 1961–1970, the length of the rainy season changed slightly and the number of wet months had a degressive trend in the period after 1991 [14], resulting in a shift in timing of rice planting and cultivation management.

To remain competitive and to provide food, feed, and industrial crops to satisfy the increasing demands of people, smallholder rice farmers in Java commonly practice intercropping or multiple cropping (polycultures). To increase their incomes, rice farmers cultivate two or more crops per year on the same land [15]. Intercropping and multiple cropping have a favorable effect on sustainable agriculture [16], as soil fertility is increased [17-19] and weed infestation prevented to some extent. Intercropping was found to influence the weed composition in adjacent fields and reduced weed densities by two-thirds [20,21]. Weed seed densities under crop rotation decreased by one-third in comparison to continuous cropping [22]. Owing to the ecology and phenology of plants, weed community composition and density are thus related to intercropping.

In this study, the effects of different types of surrounding vegetation on the plant diversity in paddy fields of smallholders in Java are examined. The objective was to understand whether the weeds in the paddy fields are influenced by customary multiple-crop systems nearby paddy fields, by the vegetation at their margins, and by surrounding semi-natural vegetation. The aim was to provide a scientific basis for developing an integrated weed management strategy for paddy fields respecting local agricultural traditions of food security and agroecosystem functioning.

We hypothesize, first, that weed species numbers and composition in paddy fields are lower than at their margins (bunds) and, second, that surrounding crops and semi-natural vegetation influence the weed diversity.

Methods
Characteristics of study areas and data sampling
We studied sites of 5 × 5 km in six areas in the island of Java, Indonesia, i.e. Cugenang, Karanganyar, Malang, Karawang, Brebes, and Gresik. The first three are representative of upland rice cropping systems (400–850 m asl). In the upland areas, the annual rainfall amounts to 3,100–4,600 mm and is concentrated in the period between September and May (Figure 1). The temperature ranges between 20°C and 27°C, depending on altitude. The other three study sites are representative of lowland

rice cropping systems (10–50 m asl). The annual rainfall of 1,300–2,550 mm is mainly during the southwest monsoon between September and April in Gresik and Brebes and from December in Karawang. The temperature ranges from 26°C to 30°C (Figure 1).

The size of the terraces under cultivation varied depending on topography, contour, and ownership; but in flat areas, terraces were usually around 50 × 100 m. In upland areas of Central Java, terraces were between 56 and 176 m^2 in size and shaped so as to descending to the river [23]. Bunds are linear structures dividing rice fields to control the water depth. Varying in style with the riverbed slope [24], they are at least 20 cm high and between 15 and 150 cm wide, with the shape depending on the paddy topography. The water depth is controlled and regulated by the spillway height (around 5 cm) (Figure 2). In Indonesia, bunds are plowed in and rebuilt each season using the mud of the plow layer to make a boundary between paddy fields, to provide a footpath, and to control water level [25].

Paddy fields were classed as either cultivated or fallow. On the basis of our survey of the sites, we classed paddy fields, bunds, and landscapes as either simple or complex. We noted the distance from the sampled paddy field to the nearest semi-natural vegetation to identify 'complex' or 'simple' conditions of paddy fields (Table 1). 'Complex landscape' means farmland where rice has been planted together with other crops in short distance. 'Complex bund' is farmland where rice has been planted together with other crops or trees on the bunds. Further, we noted the bunds' width and plant cover. 'Simple' paddy fields, bunds, and landscapes are farmlands where rice has been planted without another crop in the fields, on terraces and/or bunds, and with no or distant semi-natural vegetation around.

'Other crops' in this work are any arable or horticultural crop plants cultivated in spatial sequence and/or interplanted in the paddy fields or on bunds [26], such as corn, mungbean, soybean, chili, onion, etc. They form part of polyculture or multiple cropping systems that are commonly and extensively practiced by Indonesian farmers in the dry season. In 'semi-natural' vegetation, species composition and/or abundance has been altered through anthropogenic disturbances. While no clear natural analog may be known, they consist of a largely spontaneous and native set of plants shaped by ecological and phytogeographical processes (definition modified after Federal Geographic Data Committee [7]).

Statistical analysis
This analysis was conducted at the area level, by considering all weed species surveyed within each area. To reduce statistical noise, species present in only one plot in one area were omitted.

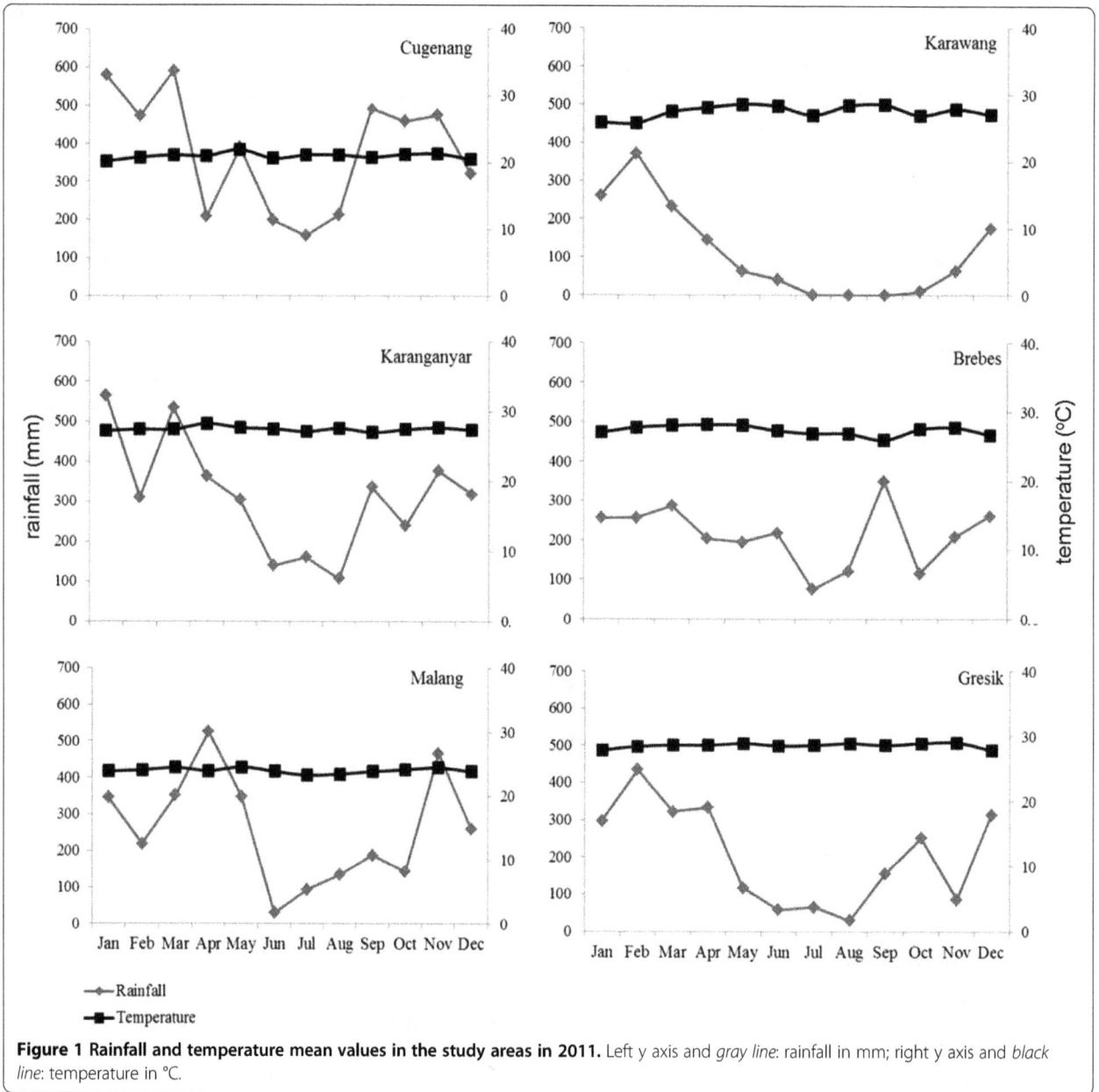

Figure 1 Rainfall and temperature mean values in the study areas in 2011. Left y axis and *gray line*: rainfall in mm; right y axis and *black line*: temperature in °C.

We used Analysis of Variance Matrix Unbalanced to analyze the effect of the sample sites, complexity, location, and season on weed composition using presence/absence data. We divided the sample plots in three categories, i.e. cultivated paddy fields, fallows, and bunds. We determined the complexity in terms of presence or absence of semi-natural vegetation and other crops surrounding the paddy fields (Table 1). The 'complexity' condition of fields and bunds was ranked 1 = simple or 2 = complex. Similarly, at the landscape level, monoculture and multiple cropping were distinguished. Cultivation seasons were 1 = October–December or 2 = May–June.

The effects of complexity, season, and area on rice weed diversity at the field level were analyzed for all weed species. The variables were log- or square-root transformed if necessary to meet the normality requirement for the residual. To determine the key factor(s) influencing weed diversity and quantitative interrelationships, backwards stepwise multiple regression analysis was applied using the criteria of probability of $p < 0.05$ to accept and $p > 0.1$ to remove the variable from the analysis. The analyses were performed in R $\times$64 3.0.2 software (R Development Core Team, Vienna, 2013) using the 'Rcmdr' and 'library CAR' packages. Furthermore, Tukey's Honestly Significant Difference (HSD) test was performed to determine significant differences between groups in the sample.

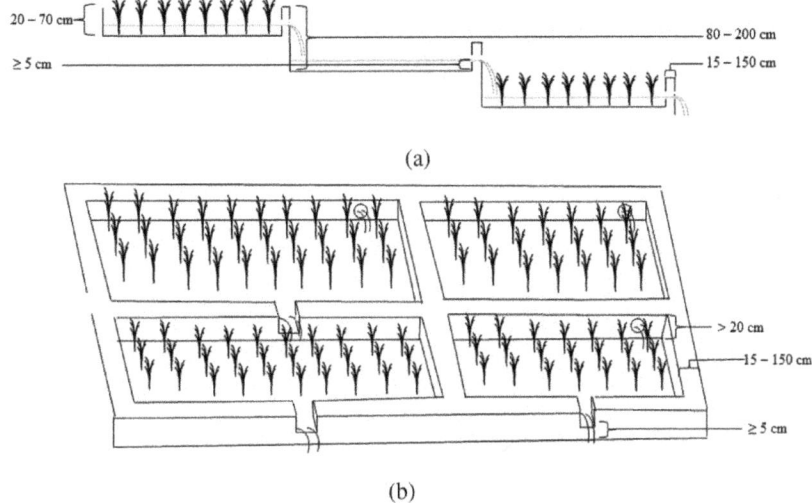

Figure 2 Scheme of bund, terrace, and field dimensions. Upland **(a)** and lowland **(b)** rice fields. Samples of vegetation were taken in a period from October 2011 through to June 2012, i.e. including two cultivation seasons October–December and May–June. The weed vegetation was surveyed in each of the six paddy field sites on a regular grid of sample plots. Each site comprised 33 plots of paddy fields, 20 m^2 in size, and 33 plots of bunds, 10 m^2 in size. Total vascular plant species composition was determined per plot.

Results

Characteristics and variation of the paddy field environment

We mapped seven types of semi-natural vegetation, i.e. conservation forest, bamboo forest, community garden, basin, grassland, fallow, and artificial lake vegetation. In terraces, farmers planted horticultural plants, viz. cassava, mungbean, soybean, chili, tomato, onion, sugarcane, and corn. On cropped bunds, the farmers planted trees and grass for animal feed and own consumption, such as banana, coconut, guava, *Sesbania grandiflora*, *Setaria spachelata* var. *splendida*, and *Pennisetum purpureum*.

In each area, we found various environment types around paddy fields (Table 2). In Gresik, most farmers build a special large-sized bund of around 1.5 m height and 1–1.5 m width, meant to prevent water overflow in the rainy season, to maintain suitable water depth for small-scale fish farming, and to keep water until the dry season.

The start of cultivation time varied across Java (Table 3). In upland areas, the first rice cultivation started around October. In lowland areas, the beginning of cultivation depended on rainfall and water availability. Lowland farmers commonly cultivate rice in the first and second cultivation season. Some farmers in Karanganyar and Brebes cultivate horticultural plants in the dry (third) season when water is insufficient for rice cultivation. Farmers in Gresik raised fish in their rice fields between January and March at the peak of the rainy season and with water overflow from paddy fields.

Plant diversity in paddy fields and bunds

Overall, we recorded 14 crop species and 221 non-cultivated plant species (weeds), of which 171 species occurred in paddy fields and 190 on bunds. Only 116 species (52%) were shared between paddy fields and bunds. Weed composition differed markedly between paddy fields and bunds (Tukey's HSD test: $t = -11.7925$, df = 383.841, p value <2.2e-16). Mean weed species numbers in paddy fields were 9.5, whereas on bunds species numbers were significantly higher (18.4; p <0.05).

Plant species numbers differed much between cultivated fields, fallows, and bunds (Figure 3). In fallows, twice as many species as in cultivated rice fields were found. The highest species numbers occurred in the three upland areas while the lowland areas and particularly the district of Karawang had much lower weed species numbers. The presence of semi-natural vegetation within short distance had no significant effect on weed species numbers in paddy fields. Species numbers in 'complex' bunds and in multiple-crop landscapes were higher than in 'simple' bunds and in monoculture landscapes, respectively. Table 4 shows the results of the Analysis of Variance Matrix Unbalanced on the transformed data. Weed species numbers differed markedly between sample categories, areas, and seasons, as did the environment complexity categories of landscapes and bunds.

Table 1 Complexity criteria adopted for paddy fields, bunds, and paddy landscape sections

	Semi-natural vegetation <100 m apart	Other crop than rice <100 m apart	Other crop in bund <100 m apart
Complex paddy field	+	+/−	+/−
Complex landscape	−	+	−
Complex bund	−	−	+

Table 2 Environment types of paddy fields in each area

	n	Cugenang	Karawang	Karanganyar	Brebes	Malang	Gresik
		33	*33*	*33*	*33*	*33*	*33*
Type of field (%)							
Simple	94	51.5	66.7	33.3	51.5	42.4	39.4
Complex	104	48.5	33.3	66.7	48.5	57.6	60.6
Mean distance of complex field to semi-natural vegetation (m)		46.3 ± 33.3	37.3 ± 22.3	39.1 ± 29.6	61.3 ± 38.4	40.3 ± 32.0	48.6 ± 37.2
Type of landscape (%)							
Monoculture	73	18.2	81.8	36.0	33.3	36.0	15.2
Multiple crop	125	81.8	18.2	64.0	66.7	64.0	84.8
Multiple crop: mean distance to other crops (m)		6.2 ± 9.3	5.7 ± 4.4	14.9 ± 11.1	19.1 ± 20.5	7.9 ± 8.0	4.2 ± 5.8
Type of bund (%)							
Simple	160	80.0	100	87.9	100	30.3	87.9
Complex	38	20.0	0	12.1	0	69.7	12.1
Mean bund width (cm)		74.5 ± 32.1	48.8 ± 22.6	48.6 ± 24.3	41.1 ± 10.1	53.3 ± 24.2	98.0 ± 62.8
Bund vegetation cover (%)		64.8 ± 27.0	26.0 ± 24.4	59.7 ± 29.4	32.4 ± 27.8	61.4 ± 29.8	65.5 ± 27.9

Inspection of all environmental effects showed that weed species numbers were significantly higher in fallow fields with complex landscape and bund, in Cugenang or Malang in the first season (Table 5). The total species number in bunds (190) was higher than in fallow fields (142) and cultivated fields (133).

At the landscape level, multiple-crop farming systems around paddy fields resulted in increased species numbers by about 10% in the paddy fields, compared to monoculture systems. Compared to simple bunds, crop cultivation on bunds had a significant effect, with about 25% higher species numbers in paddy fields (Table 5).

Discussion

Plant diversity in rice fields

The composition, total mass of plants in a stand, and population of rice weeds vary greatly under different habitat conditions and land use. In traditional agricultural systems such as in Java, farmers commonly use weeds as fodder resource, either grazed directly by ruminants or else cut, harvested, and given as fresh feed to cattle [27]. Moreover, many rice field plants are useful to farmers and villagers as medicinal or food resource, or for cultural and religious ceremonies [28,29]. Integrated weed management must therefore be biodiversity-friendly, attempting to sustain weed species of value for farmers and agroecosystems. The use of weeds as in traditional rice farming systems is moreover beneficial in order to overcome problems of pollution through herbicides. Habitat compartmentation, niches, overall diversity, and its regional variation in paddy field landscapes are as yet insufficiently studied. In particular, the effects of land use and landscape patterning on biodiversity and ecosystem functioning deserve to be further explored. In our study, we examined therefore which land use and environmental factors influence the plant diversity of rice fields and its local variation in Java.

Our results showed that the weed species numbers on the bunds were significantly different from those in cultivated paddy fields. The presence of crops on bunds was correlated with an increase in weed species numbers in paddy fields, supporting Palmer and Maurer [30] who found that high diversity in multiple-crop systems promoted the number of weed species.

Javanese farmers commonly let weeds grow or cultivate other crops on the bunds. Different crops use different soil resources and have different canopy structure, thereby

Table 3 Seasonality of paddy field cultivation in the Javanese study areas from October through September

Cultivation season (month)	10	11	12	1	2	3	4	5	6	7	8	9
Cugenang	r	r	r	r/f	r	r	r	r/f	r	r	r	r/f
Malang	r	r	r	r/f	r	r	r	r/f	r	r	r	r/f
Karanganyar	r	r	r	r/f	r	r	r	r/f	h	h	h	h
Gresik	r	r	r	s	s	s	r	r	r	r/f	f	f
Brebes	h/r	r	r	r	f	r	r	r	h	h	h	h
Karawang	f	f/r	r	r	r	r/f	f/r	r	r	r	r/f	f

r, rice plants cultivated; h, horticultural plants cultivated; f, fallow; s, small fish farming; r/f, f/r, or h/r transitional time (rice plants cultivated to fallow; fallow to rice plants cultivated, or horticultural plants cultivated to rice plants cultivated).

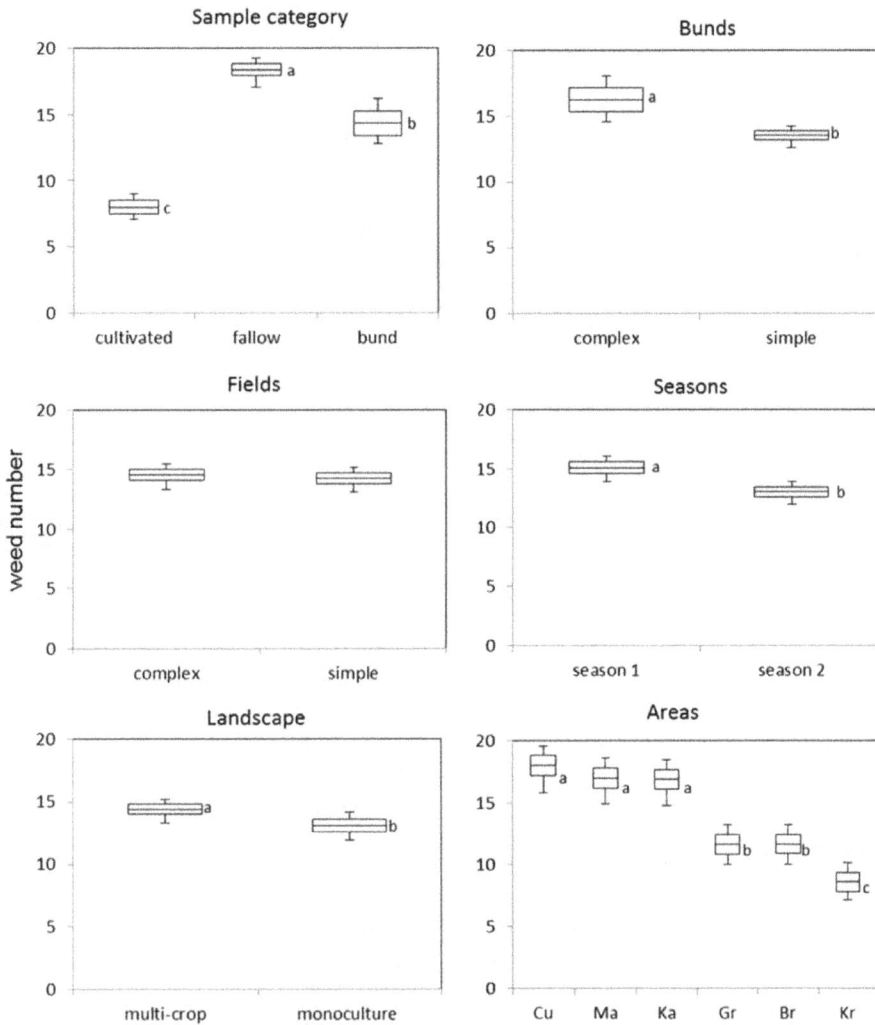

Figure 3 Mean (±SE) weed species numbers of sample categories; field, landscape, and bund types; areas; and seasons. Cu = Cugenang, Ma = Malang, Ka = Karanganyar, Br = Brebes, Kr = Karawang, Gr = Gresik.

creating microhabitat conditions that may further increase the number of weed species and enhance the differences between weed communities in fields and on bunds. Subudhi et al. [23] reported that the structure of bunds affected rice yield as well as soil and moisture content. Organic matter, pH level, and nutrients (N, P, K, Ca, and Mg) were found higher in paddy fields with trees on the bunds, especially leguminous trees [19].

As a result of weed control, overall diversity of weeds in cultivated plots was less than in the fallows and on the bunds. The higher number of weeds in fallows suggests that the soil seed bank in paddy fields is significant

Table 4 Testing for differences in the number of weed species by using Analysis of Variance Matrix Unbalanced

	Df	Sum Sq	Mean Sq	F value	Pr (>F)	
Sample category	2	9,299.7	4,649.9	133.4537	<2.2e-16	***
Field (simple vs. complex)	1	3.3	3.3	0.0794	0.778330	
Landscape (monoculture vs. multicrops)	1	216.4	216.4	5.2796	0.022114	*
Bund (simple vs. complex)	1	316.5	316.5	7.7232	0.005720	**
Area	5	4,125.9	825.2	20.1338	<2.2e-16	***
Season	1	352.9	352.9	8.6111	0.003542	***

Significance codes: *0.01; **0.001; ***0.

Table 5 Tukey's HSD test on the mean species numbers in different environments

Treatments	Mean species numbers	SE
Sample category		
Cultivated	7.9	±0.5[c]
Fallow	18.4	±0.5[a]
Bund	14.4	±0.9[b]
Landscape		
Multiple crop	14.4	±0.4[a]
Monoculture	13.1	±0.5[b]
Bund complexity		
Complex	16.3	±0.9[a]
Simple	13.6	±0.4[b]
Area		
Cugenang	17.9	±0.8[a]
Malang	16.9	±0.8[a]
Karanganyar	16.9	±0.8[a]
Gresik	11.6	±0.8[b]
Brebes	11.6	±0.8[b]
Karawang	8.6	±0.8[c]
Season		
Season 1	15.1	±0.5[a]
Season 2	13.0	±0.4[b]

[a,b,c] indicate statistically significant differences at the level $p < 0.01$.

[31] and triggers the germination of plants adapted to the relatively dry conditions of fallows. Local differences in weed composition among paddy fields are therefore chiefly due to season and cultivation stage.

The effects of rice cultivation on weed diversity vary with the production methods applied, whether fields are managed organically or with agrochemicals, and which irrigation and planting systems are used [1]. In a review, Navas [16] concluded that there are two factor groups determining weed diversity, i.e. local environmental conditions (climate, soil, land use history) and crop/treatment related factors (current or preceding crop type, land drainage, fertilization, tillage, and herbicide application). Irrigation and its frequency are to be added here [32]. We found that cultivated paddy fields in the rainy season serve as a suitable habitat for many hydrophilous plants (Kumalasari, unpublished). The hydrophilous plants have lifecycle along rice cultivation and most of them ended after harvest. This condition is related with water availability that depends on different number of rainfall in each season, water management, and rice plant stage.

The effect of complexity of fields, bunds, and landscapes

Contrary to our expectations, weed species numbers in paddy fields were unrelated to the semi-natural vegetation nearby. In contrast, in temperate arable landscapes, weed species diversity was influenced by landscape complexity [21]. There may be an effect though, not studied by us, between paddy weed diversity and semi-natural vegetation, acting at shorter distances. In any case, the influence of adjacent agricultural management, in particular of other crops nearby, on plant species numbers seems to be more pronounced than that of other vegetation formations [33], which may be explained by the fact that paddies and semi-natural formations have but few species in common.

Positive effects of intercropping on organic carbon, soil nitrogen, some micronutrients [18], total soil nutrients and thus, on nutrient uptake [17] are well-known. As those nutrients are essential for plant growth in general, both crops and weeds profit from multiple-crop systems [34]. In the present study, intercropping around paddy fields and on bunds had a considerable effect on weed species numbers (Table 5). Also, other researchers reported on the effects of agricultural seasonality, such as fallow management [35] and intercropping [36], on weed species in tropical fields.

The intervening time between harvest and the next rice cultivation in Java varies; meanwhile, the land typically lies fallow. The duration of the fallow period is determined by rainfall and may be as short as three weeks in the upland areas of Java and up to 16 weeks in the lowlands. Our research showed that season has an effect on species numbers, thus supporting [37] in that there was more weed growth in the dry season than in the wet season (but see [38]).

Conclusions

The comparison of paddy weed species numbers in different areas and environments showed that sample category, area, and season contribute most to overall local species richness. Our study underlines moreover the significance of local management and landscape patterning for plant species diversity. Whether landscapes and bunds are 'complex' or 'simple' has an effect on species numbers in paddy fields. Altogether, differences in weed species numbers and composition between areas are more pronounced than the effects of local environmental complexity. Thus, to support species diversity in lowland paddy fields, intercropping of horticultural plants should be maintained. In the upland areas, multiple cropping and the presence of fruit trees or crops on the bunds were positively related to plant species diversity and are thus recommended for an integrated agricultural management and to support agroecosystem functioning for the local smallholder communities. It is further essential to maintain, pass on, and teach the traditional knowledge of plant life in paddy fields and the local farming practice to preserve the plant diversity and to perpetuate its use.

Competing interests
The authors declare that they have no competing interests.

Authors' contributions
NRK conceived of the study, carried out the field work, performed the statistical analysis, and drafted the manuscript. EB coordinated the study, contributed to the design, and helped to draft the manuscript. Both authors read and approved the final manuscript.

Acknowledgements
This work was supported through a scholarship from Erasmus Mundus Experts Asia (mobility program) and a PhD grant from BIOTROP (2012). We would like to thank Dr. Sri S. Tjitrosoedirdjo for invaluable support of the field work and laboratory research. We are grateful to Prof. Em. Soekisman Tjitrosemito and Prof. Luki Abdullah for their substantial contributions in designing the field work. Many thanks are also due to Adnan who assisted the data analysis.

References

1. Donald PF: Issues in international conservation: Biodiversity impacts of some agricultural commodity production systems. *Conserv Biol* 2004, **18**:17–37.
2. Dornelas M, Moonen AC, Magurran AE, Bàrberi P: **Species abundance distributions reveal environmental heterogeneity in modified landscapes.** *J Appl Ecol* 2009, **46**:666–672.
3. Espinosa-García FJ, Villaseñor JL, Vibrans H: **The rich generally get richer, but there are exceptions: Correlations between species richness of native plant species and alien weeds in Mexico.** *Divers Distrib* 2004, **10**:399–407.
4. Otto S, Vasileiadis VP, Masin R, Zanin G: **Evaluating weed diversity with indices of varying complexity in north-eastern Italy.** *Weed Res* 2012, **52**:373–382.
5. Marshall EJP, Brown VK, Boatman ND, Lutman PJW, Squire GR, Ward LK: **The role of weeds in supporting biological diversity within crop fields.** *Weed Res* 2003, **43**:77–89.
6. Way MJ, Heong KL: **The role of biodiversity in the dynamics and management of insect pests of tropical irrigated rice-a review.** *Bull Entomol Res* 1994, **84**:567–587.
7. FGDC: *National Vegetation Classification Standard Version 2. FGDC-STD-005-2008.* Reston, Virginia, USA: Vegetation Subcommittee, Federal Geographic Data Committee, FGDC Secretariat, U.S. Geological Survey; 2008.
8. José-María L, Sans FX: **Weed seedbanks in arable fields: effects of management practices and surrounding landscape.** *Weed Res* 2011, **51**:631–640.
9. Roschewitz I, Gabriel D, Tscharntke T, Thies C: **The effects of landscape complexity on arable weed species diversity in organic and conventional farming.** *J Appl Ecol* 2005, **42**:873–882.
10. Sukristiyonubowo S: **Nutrient balances in terraced paddy fields under traditional irrigation in Indonesia.** Gent: Gent University, Applied Analytical and Physical Chemistry; 2007. http://buck.ugent.be/fulltxt/RUG01/001/215/679/RUG01-001215679_2010_0001_AC.pdf.
11. Hoque MZ: *Cropping systems in Asia: On-Farm Research and Management.* Los Baños (Philippines): International Rice Research Institute (IRRI); 1984.
12. Sumarno: *Periodisasi Musim Tanam Padi sebagai Landasan Manajemen Produksi Beras Nasional. (periodization of rice cultivation season as a base of national production management).* [http://203.176.181.70/bppi/lengkap/st080206-1.pdf]
13. Tazhibayeva K, Townsend RM: *The Impact of Climate Change on Rice Production: heterogeneity and uncertainty.* [http://www.robertmtownsend.net/sites/default/files/files/papers/working_papers/ImpactofClimateChangeonRiceYieldsDec2012.pdf]
14. Pramudia A: *Musim Hujan di Sentra Produksi Padi sudah Berubah (rainy season has changed in rice production centers).* [http://203.176.181.70/bppi/lengkap/kl060903.pdf]
15. Godoy R, Bennett CPA: **The economics of monocropping and intercropping by smallholders: The case of coconuts in Indonesia.** *Hum Ecol* 1991, **19**:83–97.
16. Navas ML: **Trait-based approaches to unravelling the assembly of weed communities and their impact on agro-ecosystem functioning.** *Weed Res* 2012, **52**:479–488.

17. Mohammaddoust-e-Chamanabad HR, Asghari A, Tulikov AM: **The effects of weed-crop competition on nutrient uptake as affected by crop rotation and fertilizers.** *Pak J Biol Sci* 2007, **10**:4128–4131.
18. Prasad B, Umar SM: **Effect of rice based six multiple cropping sequences under two cycles of crop rotations on yield and fertility status of soil.** *Plant Soil* 1990, **127**:251–258.
19. Sae-Lee S, Vityakon P, Prachaiyo B: **Effects of trees on paddy bund on soil fertility and rice growth in Northern Thailand.** *Agroforestry* 1992, **18**:213–223.
20. Gomez P, Gurevitch J: **Weed community responses in a corn-soybean intercrop.** *Appl Veg Sci* 1998, **1**:281–288.
21. Rahnavard A, Ashrafi ZY, Alizade HM, Sadeghi S: **Studies on the effect of fertilizer application and crop rotation on the weed infested fields in Iran.** *J Agric Technol* 2009, **5**:41–50.
22. Cardina J, Herms CP, Doohan DJ: **Crop rotation and tillage system effects on weed seedbanks.** *Weed Sci* 2002, **50**:448–460.
23. Subudhi CR, Behera B, Samantary SK: **Field bund structures for production of rice in North Eastern Ghats of Orissa, India.** *Res J Agric Sci* 2010, **1**:475–476.
24. Hoshikawa K, Kobayashi S: **Study on structure and function of an earthen bund irrigation system in Northeast Thailand.** *Paddy Water Environ* 2003, **1**:165–171.
25. Walker SH, Rushton KR: **Water losses through the bunds of irrigated rice fields interpreted through an analogue model.** *Agric Water Manag* 1986, **11**:59–73.
26. Gallaher RN: *Multiple Cropping Systems.* [http://www.eolss.net/sample-chapters/c10/E5-15-02-04.pdf]
27. Kosaka Y, Takeda S, Sithirajvongsa S, Xaydala K: **Plant diversity in paddy fields in relation to agricultural practices in Savannakhet province.** *Laos Econ Botany* 2006, **60**:49.
28. Edirisinghe JP, Bambaradeniya CNB: **Rice fields: an ecosystem rich in biodiversity.** *J Natl Sci Found Sri Lanka* 2006, **34**:57–59.
29. Kumalasari NR, Abdullah L, Bergmeier E: **Nutrient assessment of paddy weeds as ruminant feed in Java.** *Livest Res Rural Dev* 2014, **26**:59. [http://www.lrrd.org/lrrd26/4/kuma26059.html]
30. Palmer MW, Maurer TA: **Does diversity beget diversity? A case study of crops and weeds.** *J Veg Sci* 1997, **8**:235–240.
31. De Rouw A, Casagrande M, Phaynaxay K, Soulileuth B, Saito K: **Soil seedbanks in slash-and-burn rice fields of northern Laos.** *Weed Res* 2013, **54**:26–37.
32. Li RH, Qiang S: **Composition of floating weed seeds in lowland rice fields in China and the effects of irrigation frequency and previous crops.** *Weed Res* 2009, **49**:417–427.
33. Gaba S, Chauvel B, Dessaint F, Bretagnolle V, Petit S: **Weed species richness in winter wheat increases with landscape heterogeneity.** *Agric Ecosyst Environ* 2010, **138**:318–323.
34. Newton AC, Begg GS, Swanston JS: **Deployment of diversity for enhanced crop function.** *Ann Appl Biol* 2009, **154**:309–322.
35. Awanyo L: **Dealing with weedy problems in agriculture: the role of three agricultural land use management practices in the forest-savanna ecological zone of Ghana.** *Area* 2008, **40**:446–458.
36. Akobundu IO, Ekeleme F, Chikoye D: **Influence of fallow management systems and frequency of cropping on weed growth and crop yield.** *Weed Res* 1999, **39**:241–256.
37. Chauhan BS, Johnson DE: **Row spacing and weed control timing affect yield of aerobic rice.** *Field Crop Res* 2011, **121**:226–231.
38. Saito K, Azoma K, Rodenburg J: **Plant characteristics associated with weed competitiveness of rice under upland and lowland conditions in West Africa.** *Field Crop Res* 2010, **116**:308–317.

13

Landscape diversity and the resilience of agricultural returns: a portfolio analysis of land-use patterns and economic returns from lowland agriculture

David J Abson[1,2*], Evan DG Fraser[2,3] and Tim G Benton[4]

Abstract

Background: Conventional agriculture is increasingly based on highly specialized, highly productive farms. It has been suggested that 1) this specialization leads to farms that lack resilience to changing market and environmental conditions; and 2) that by decreasing agricultural diversity, the resilience of the farming system also decreases.

Methods: We used agricultural gross margin (GM) forecasts from 1966 to 2010 and remote sensing data from agricultural landscapes in the lowland UK, in conjunction with modern portfolio theory, to test the hypothesis that decreasing land-use diversity results in landscapes that provide higher, but more volatile, economic returns. We considered the role of spatial scale on the expected levels of volatility and resilience of agricultural returns.

Results: We found that: 1) there was a strong linear trade-off between expected GMs and the expected volatility of those GMs in real lowland agricultural landscapes in the UK; 2) land-use diversification was negatively correlated with expected GMs from agriculture, and positively correlated with decreasing expected volatility in GMs; 3) the resilience of agricultural returns was positively correlated with the diversity of agricultural land use, and the resilience of agricultural returns rose quickly with increased land-holding size at small spatial extents, but this effect diminished after landholdings reached 12,000 hectares.

Conclusions: Land-use diversity may have an important role in ensuring resilient agricultural returns in the face of uncertain market and environmental conditions, and land-holding size plays a pivotal role in determining the relationships between resilience and returns at a landscape scale. Creating finer-grained land-use patterns based on pre-existing local land uses may increase the resilience of individual farms, while maintaining aggregate yield across landscapes.

Keywords: Resilience, Agro-diversity index, Agro-ecology, Specialization, Landscape heterogeneity, Land use

Background

During the past 60 years, changes in the agricultural industry have led to a global agrifood system dominated by large, capital-intensive farms [1-3]. These farms are increasingly specialized in terms of the crops they produce, and hence are dependent on inputs from other sectors of the economy [4-6]. This change in agriculture has been driven by the search for increased economic efficiency, economies of scale, and reduced marginal costs of production. However, the homogenization of agriculture may have an unintended drawback, and some evidence suggests that these more specialized farms are also less resilient [7-12] and that they experience increased income volatility [13-15]. Hence, there may be trade-offs between agricultural returns and the resilience of those returns in modern farming systems.

We present the results of an empirical study that used data on forecasted annual average agricultural gross margins (GMs) between 1966 and 2010 and data on land-use diversity (derived from census data and satellite

* Correspondence: abson@uni-leuphana.de
[1]FuturES Research Center, Leuphana Universität, Lüneburg, Germany
[2]Sustainability Research Institute, University of Leeds, Leeds, UK

imagery) to examine the relationships between landscape units with different levels of agricultural diversity and the amount and volatility of the expected GMs from agriculture that each different landscape unit provided. We examined this relationship at a range of different spatial scales to address two core research questions. We investigated first, whether more specialized landscapes (that is, those with lower land-use diversity) have higher average GMs, and second, whether more specialized landscapes have more volatile returns. In addition, we examined the role of spatial scaling of land-use patterns in real landscapes on these two relationships. Together, these analyses indicate the extent to which, and at what scales, there may be a trade-off between expected GMs and the volatility and resilience of those expected GMs.

Resilience and agricultural systems

The central theoretical concept in this paper is that of 'resilience', derived from systems dynamics thinking, which the literature broadly describes as the tendency of a system to return to its original state following a disturbance. Resilience therefore has a number of properties: the ease with which a system can be disturbed (resistance), the way in which a system returns to its pre-disturbance state (that is, its speed and trajectory), and the propensity for a system to move to an alternative stable state following disturbance [16,17]. Resilience is often interpreted as a measure of either the size of the perturbation required to flip a system into a new dynamically stable state (regime shifts or system identity shifts) [18,19] or the capacity of a system to maintain its current equilibrium state in the face of perturbations [20].

Operationalizing resilience in many empirical situations is complex, thus system behavior typically either needs a systems model or experimental perturbation to assess the way in which the system responds. Both of these factors are difficult to simulate for large-scale, complex systems. In some extreme examples, a regime shift can be identified by very significant changes. Notable examples include the Dust Bowl period of the 1930s in North America, when a prolonged drought rendered millions of hectares of farmland unproductive, and displaced hundreds of thousands of people from their homes [21]; the Ethiopian Famine in the 1980s, when a relatively minor drought triggered a catastrophic famine [22-24]; or the Irish Potato Famine, when the failure of a single crop caused a permanent depopulation of western Ireland [25,26]. Although extremely important, studying such tragedies lends itself to a qualitative case study-based research approach, and are difficult to analyze quantitatively, for a sample of other case studies see [27,28].

Attempts to quantify resilience in the absence of clear regime shifts are hampered by the multi-dimensional nature of the concept, particularly given that the different properties of resilience may be quantified in incommensurable units. As most systems are continually disturbed and fluctuate around a quasi-equilibrium state [20], examining resilience as the relationship between the size of disturbance and the effect of that disturbance [29-31] is perhaps more generally useful. For most applications to agricultural systems not subject to catastrophic change, this element of resilience can be articulated as the stability of agricultural returns in the presence of different exogenous shocks [32]. Agricultural returns are inherently volatile, and change in response to a range of exogenous (for example, disease outbreaks, climate, currency exchange rates, market forces, rapidly changing subsidy systems) and endogenous (for example, crop choice) factors [33,34], with the returns from different agricultural sectors being sensitive to different exogenous drivers of change [35].

One of the key themes deriving from the resilience literature is the hypothesis that agricultural landscapes that are more heterogeneous may also be more resilient in terms of the stability of agricultural returns, as such diverse landscapes should reduce risk (defined in terms of the expected variance in returns [28,36-41]). However, there is potentially an inherent trade-off, in that a diversified strategy reduces volatility at the cost of reduced expected mean returns. The concept of 'bet-hedging' captures this dichotomy; in highly variable systems, strategies that trade off the variance against mean returns can often be superior [42-44]. Hence, in this study, we were interested in determining whether land-use diversity influences the volatility and resilience of the expected GMs in agricultural landscapes.

Land-use diversification has the potential to reduce resilience (expected volatility of GM per unit of expected GM) because the returns generated from an individual land use are dependent on a relatively narrow range of weather conditions and the vagaries of commodity price. Both weather conditions and commodity markets have become increasingly erratic [45,46], causing concerns that farm returns have become less resilient [47]. For example, between 1990 and 2007, the average annual net income of a UK farming enterprise (excluding horticulture) was approximately £23,000; however, this averaged figure hides the significant volatility in these returns over this time period, with the average return ranging from approximately £45,000 in 2002 to just £8,700 in 2000 [48].

In this research, we quantified the volatility of agricultural returns in terms of the expected standard deviation (SD) of GMs and economic resilience (or rather one important aspect of economic resilience) as the coefficient of variation (CV) in expected GM. CV is a normalized measure of dispersion of a probability distribution, which is defined as the ratio of the SD to the mean. In this case, we used the ratio of the expected (mean) GMs to the expected SD of the expected GMs as our measure of resilience. We based this on the assumption that agricultural land-use portfolios (the choice of agricultural land-use investments within a landscape) that provide a lower expected variance to returns ratio would be more resilient. It should be noted that we did not address the resilience of individual farmers, which would require detailed knowledge of the assets, capacities and access to formal and informal institutional support of individual farmers; rather, we sought to investigate the potential role of land-use diversification on the volatility and resilience of returns from agriculture.

We examined this question at a range of different spatial scales (from 25 to 3600 hectares) to investigate the degree to which spatial extent would influence the results.

Modern portfolio theory (MPT) provides analytical tools for investigating the relationships between land-use choices, expected GMs, and the expected variance in those GMs on a landscape scale. MPT was developed in the field of finance in the 1950s, to quantify the optimum level of diversification that would balance risks (the expected variance in returns) and the expected mean return of a given investment portfolio [49]. The key concept in portfolio management is that income streams are additive, whereas risks may partially cancel each other out [49,50]. The logic is that diversification in a portfolio can reduce the risk (or the expected variance) of the portfolio's returns to perturbations, as long as not all possible investments respond in the same way to the same shocks; that is, provided there is not perfect covariance over time in the returns from different agricultural activities. This concept can be applied to agricultural systems by considering the different land-use choices as the individual elements of a portfolio. Therefore, the key to reducing expected variance in returns is for a farmer to select a diversity of land uses that will respond differently to market, institutional, or environmental perturbations. For example, when this concept is applied to an agricultural system of wheat and oats, it is clear that the inputs needed to produce both of these crops are roughly the same (because the crops are of a similar type, namely cereals), and thus the costs of these inputs are likely to increase or decrease by the same amount (this is called a systematic risk). However, the market price of these crops is inversely correlated; wheat prices often increase at the same time as the price of oats decreases (this is called a unique risk) [50]. Thus, by investing in both wheat and oats, the farmer can diversify away the unique risks associated with market-price volatility.

The application of MPT to natural rather than financial assets has, to date, been limited. It has been suggested that the principles of MPT could be transferable to the field of biodiversity conservation [51], and MPT has previously been used to quantify the risk and return profiles of individual farmers in Northern Ireland [52] and to the genetic diversity within cereal crops [53,54]. It has also been suggested that MPT is an appropriate tool for assessing vulnerability of food systems through the diversification of crop production and the basket of food entitlements [38]. However, the application of MPT to agricultural landscape patterns represents a novel approach to operationalizing agricultural ecosystem resilience.

In this study, we used published data for land use and expected average agricultural GM data in conjunction with MPT to analyze the relationships between land-use diversity, expected mean returns for agriculture, and the expected variance and resilience of those returns in three UK lowland agricultural regions. This analysis differs from previous applications of MPT to agricultural land-use investments [52] in that it used real land-use patterns to assess the relationships between expected returns and expected variance of returns for actual land-use portfolios.

Methods

For the study, we first identified three representative lowland agricultural regions in southern England for which we could obtain detailed data on agricultural land cover. We then utilized 353 study sites each 1 km^2 in size, and three, regional sub-extents, each 576 km^2 in size, to explore different aspects of the relationships between land-use diversity and expected agricultural GMs at a various landscape scales within these three different agricultural regions (Figure 1).

We used published satellite-derived land-cover data and livestock estimates to quantify spatially explicit agricultural land-use patterns in each region. To assess diversity, this land-use data was used to calculate a diversity index score for each landscape unit. For assessment of expected agricultural GMs, we used published annual forecasts of expected agricultural GMs to calculate the average GMs (including income from agricultural subsidies) of the farming activities found in each region over the period 1966 to 2010. To assess the relationship between agricultural returns, resilience, and diversity, we used a number of metrics that allowed us to assess the expected mean, SD, and CV of agricultural GMs in these landscapes, using the analytic tools of MPT, and then we related these to land-use diversity.

Study sites

Three lowland regions broadly representative of lowland English agriculture were selected for investigation. Each region represents a different spatial arrangement of agricultural activities. Region 1 (south-west region) is primarily (but far from exclusively) a livestock and dairy farming region. Region 2 (south central region) represents a more

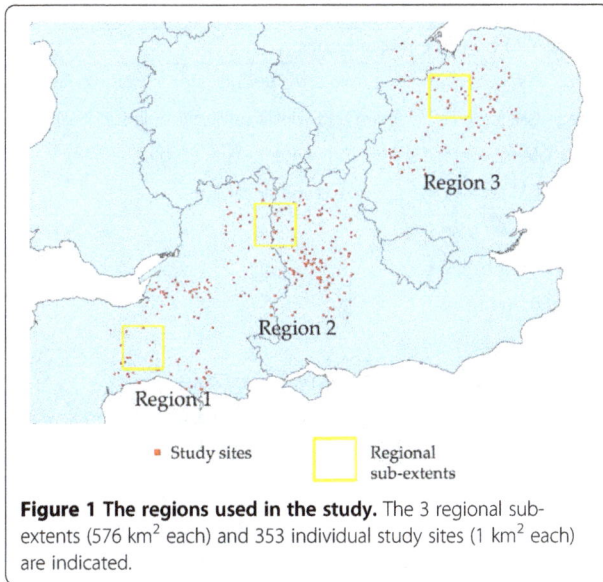

Figure 1 The regions used in the study. The 3 regional sub-extents (576 km² each) and 353 individual study sites (1 km² each) are indicated.

mixed agricultural landscape, including horticulture, arable, and dairy farming. Region 3 (in eastern England) is dominated by larger expanses of arable farming compared with the other two regions, with increasing concentrations of horticultural production in the northeast corner. Within these three regions, 353 individual sites each 1 km² in size (the small red squares in Figure 1) and three regional sub-extents of 576 km² each (the yellow squares in Figure 1) were selected for analysis. Using individual study sites allowed us to explore the relationship between landscape diversity, resilience, and agricultural returns between the study regions. Within the regional sub-extents, we analyzed nine portfolio sizes ranging from 25 to 3600 hectares, to reflect the range of farm land holdings typically found in UK lowland agricultural landscapes.

Data used to assess agricultural returns

We found only a single source of data that could provide consistent quantification of returns from UK agricultural activities over a suitable time frame, namely the *John Nix Farm Management Pocketbook*. The John Nix pocketbooks provide forecasts of annual farm GMs per hectare for different agricultural activities for the years 1966–2010 (no pocketbooks were produced for 1970, 1973, 1975, or 1982). The John Nix GM forecasts relate to the average expected margins of individual agricultural activities and not to the expected margins for individual farms. GM is defined as the difference between farm revenue (including subsides) and the associated variable costs for a given activity. Although GM does not include fixed costs and, therefore, is not a perfect measure of agricultural returns [55], it is a widely applied measure within the field of agricultural economics.

However, it is important to note that the John Nix GM forecasts represent estimated average returns for England, and thus are likely to underestimate the actual variance in returns for individual landscapes, as they cannot account for variability in yields for a given field. Nevertheless, these data do provide an indicator of the covariance in GM for different land uses over a relatively long period (44 years), and therefore provide an insight into the role of land-use diversification as a means of reducing expected variance in returns.

Data used to assess land use

To identify the land uses in the study regions, we made use of the 2000 *Land Cover Map* (LCM2000) [56], which provides a satellite-based assessment of land cover for all of the UK in the year 2000. For agricultural crops (including hay/silage) the relationships between land cover and land use are clear, and 12 agricultural crops were identified within the LCM2000 land-cover data for which GM data was available (Table 1).

Associating LCM2000 grassland types with the GM data was more difficult, as the LCM2000 data provides only information on land cover and not land use. For example, the LCM2000 reports managed grassland as a land cover; however, this may be used for raising different types of livestock, each of which will have different GMs. There are three primary land uses for lowland grassland: dairy, beef, and sheep production. Therefore, a number of assumptions had to be made in order to attribute GMs from these three land uses to grassland land covers. The LCM2000 data was reclassified as either managed grassland (intensive, managed calcareous, and grazing marsh) and unmanaged grass (rough grass, rough acid grass, unimproved/neutral grass, and calcareous unmanaged grass). Livestock estimates were drawn from the June Agricultural Census (JAC) for the year 2000 to estimate the livestock-based land uses for the LCM2000 grassland data. The JAC provided total livestock numbers for 4 km² grid squares. The JAC grid square, within which each of the centroids of the 353 study sites fell, was spatially joined to the study sites using the geographic mapping system ArgGIS [58]. This provided average per head estimates for the three livestock types in each study site. To allow for direct land-use comparisons between the three livestock types, livestock numbers were converted into livestock units (LUs). LUs represent the average land requirements for different livestock types, and we based these on LU conversion factors of 1 for dairy, 0.7 for beef and 0.12 for sheep [59]. The LU ratios for each livestock type (based on LU values and per head estimates) was then used to estimate the proportion of grassland (identified from the LCM2000 data) used by each livestock type for each study site.

Table 1 The assumptions used to select GM estimates[a]

Land use	Assumptions for GM from the pocketbooks
Dairy	Average GM for average stocking rate (two cows per hectare) of Holstein Friesians
Managed grass (beef)	Lowland average GM for average spring and winter calving (single suckling)
Unmanaged grass (beef)	Upland average GM for average spring and winter calving (single suckling)
Sheep	Lowland average GM for average spring and winter calving
Hay, silage	Silage sales minus silage costs
Barley	Average GM for winter barley
Maize	Average GM for fodder maize
Wheat	Average GM for winter-sown wheat
Cereal (spring)	Average GM for spring-sown cereals (wheat, barley, and oats)
Cereal (winter)	Average GM for winter-sown cereals (wheat, barley, and oats)
Field beans, peas	Average GM for winter-sown beans and dried peas
Horticulture	Average GM for carrots, onions, and broad beans
Linseed	Average GM for linseed
Potatoes	Average GM for maincrop potatoes
Oilseed rape	Average GM for winter-sown oilseed rape
Sugar beet	Average GM for sugar beet

Abbreviations: GM, gross margin.
[a]Sources: *John Nix Farm Management Pocketbooks* and the UK Department for the Environment, Food and Rural Affairs [57].

We assumed that managed grassland (70% of the total grassland extent) was used only for dairy and beef production, and that the unmanaged grass was used only for sheep and beef production. In practice, a small proportion of the lowland managed grassland is used for sheep production. However, the JAC data suggested that less than 2% of the grassland in the study regions are given over to sheep production and it is likely that only 10% [60] of this would be on managed grass. Therefore, we assumed that sheep would be confined only to unmanaged grasses. Through this process, four new land-use classes were created: dairy, beef (improved grass), beef (rough grazing), and sheep, for which the GM estimates from John Nix could be applied. Because of the difference in productivity between managed and unmanaged grasslands, the GM estimates for upland beef production were used to value beef on rough grazing, whereas the lowland GM estimates were applied to beef on improved grass. The lowland GM estimates were used for sheep. Table 1 details the final 16 land-use classes valued in the MPT analysis, and the assumptions used to estimate GM for each land use.

The LCM2000 land-cover map and JAC data were used in ArcGIS [58] to identify the agricultural land uses (including estimates of the livestock uses for grasslands) within each study site, and for each landscape in the regional and sub-regional analyses. Fragstats [61] was used to calculate the area covered and the percentage of the total landscape of each agricultural land use for each spatial extent. All land-use estimates were converted to per-hectare measurements when calculating annual expected returns. This allowed direct comparison between landscapes with different agricultural extents.

Evaluating diversity
Shannon's diversity index (H′) [62] was used as the indicator of agricultural landscape diversity. This widely used index was selected because it takes into account both the abundance and the evenness of agricultural land uses present in a given landscape. Moreover, given the discriminatory power of this index, it is a particularly useful measure for comparing diversity between similar landscapes [63]. Shannon's index was calculated as:

$$H' = - \sum_{1=0}^{n} (P_i * lnP_i) \qquad (1)$$

where P_i = proportion of the landscape occupied by the land-use patch type i.

Analysis: applying modern portfolio theory to explore the relationships between productivity, resilience, and diversity
We used MPT to calculate the expected GMs, expected SD in GMs, and the CV of GMs for different land-use portfolios, where these metrics were assessed based on the inter-annual covariance of the forecast GMs of each land-use type over the analysis period (1966 to 2010). The calculations, all of which are based on the work of Sharpe [50], are detailed below.

The expected rate of return (E) was averaged over all possible outcomes with weights equal to respective probabilities. E_i is the expected rate of return for the ith land cover given by

$$E_i = \sum_{t=1}^{M} P_t R_t \tag{2}$$

where P_t is the probability of the R_t where R is the return on investment for the year t and M is the total number of years over which returns from the land cover are known. As all outcomes are equally likely (one outcome per year) then

$$E_i = \sum_{t=1}^{M} \frac{R_t}{M} \tag{3}$$

under the condition that

$$\sum_{i=1}^{N} X_i = 1 \tag{4}$$

When this condition is met, any given spatial arrangement of land covers within a landscape represents an investment landscape portfolio p. If R_{pj} is the jth return on the portfolio, X_i is the fraction of the landscape invested in land cover i, and N is the number of land-cover investments, then the portfolio return for a given year is

$$P_{pj} = \sum_{i=1}^{N} X_i R_i \tag{5}$$

and the expected return of the land-cover portfolio p is the weighted average (the proportion of the landscape under each land cover) of the sum of the expected returns

$$E_p = \sum_{i=1}^{N} X_i E_i \tag{6}$$

The variance of returns for the ith land cover is, therefore,

$$\sigma_i^2 = \frac{\sum_{t=1}^{m} (R_i - E_i)^2}{m - 1} \tag{7}$$

and the variance of the land cover portfolio, p, is the expected value of the squared deviations of the return on the land-cover portfolio from the expected return on the land-cover portfolio. For a portfolio with two investments (1 and 2), and given that the expected value of the sum of a series of returns is equal to the sum of the expected value of each return, and the expected value of a constant (percentage of landscape under a land cover) multiplied by a return is equal to the constant times the expected return, we have

$$\sigma_p^2 = X_1^2 \sigma_1^2 + 2X_1 X_2 E_p[(R_1 - E_1)(R_2 - E_2)] + X_2^2 \sigma_2^2 \tag{8}$$

The covariance of returns on investments 1 and 2 is given by

$$cov_{12} = E_p[(R_1 - E_1)(R_2 - E_2)] \tag{9}$$

therefore

$$\sigma_p^2 = X_1^2 \sigma_1^2 + 2X_1 X_2 cov_{12} + X_2^2 \sigma_2^2 \tag{10}$$

For a land-cover portfolio with N land covers, the variance of the portfolio is, therefore, given by:

$$\sigma_p^2 = \sum_{i=1}^{N} \left(X_i^2 \sigma_i^2 \right) + \sum_{i=1}^{N} \sum_{j \neq i}^{j=1}{}^{N} \left(X_i X_j Cov_{ij} \right) \tag{11}$$

The value of E_p and σ_p^2 were calculated for the different landscapes to allow the investigation of the relationships between agricultural land-use diversity, average expected economic GMs, and the relative resilience (that is, the CV) of those margins over time. Here it should be noted that such measures of variance include both the 'upside' and 'downside' variances. Upside variance refers to variations above the mean, whereas where downside variance (or semi-variance) considers only deviation below the mean (in this case, the expected return Ep). There is an argument that that only downside variance should be considered, because deviations above the mean are desirable; however, semi-variance is difficult to apply to portfolios and may not be relevant. For instance, in cases where the distribution of returns is symmetric, the evaluation of portfolios based on upside and downside variance will be the same. Therefore, semi-variance was not used here. All GM values were converted to 2010 prices using the UK Treasury's gross domestic product deflator [64].

Results

Historic returns and resilience of individual land uses

The average expected GMs per hectare for each of the agricultural land uses have changed considerably since 1966 (Figure 2). There was a clear pattern of declining average GMs across all land uses, with a particular steep decline in GMs within the period 1972 to 1986. After 1986, the GMs continued to decline, but at a less rapid rate.

Across all land uses, horticulture (including sugar beet and potatoes), and dairy showed the highest expected returns over the analysis period of 1966 to 2010 (Table 2). However, these GMs may be misleading in comparison to the other GMs presented here, owing to the lack of information about the fixed cost of

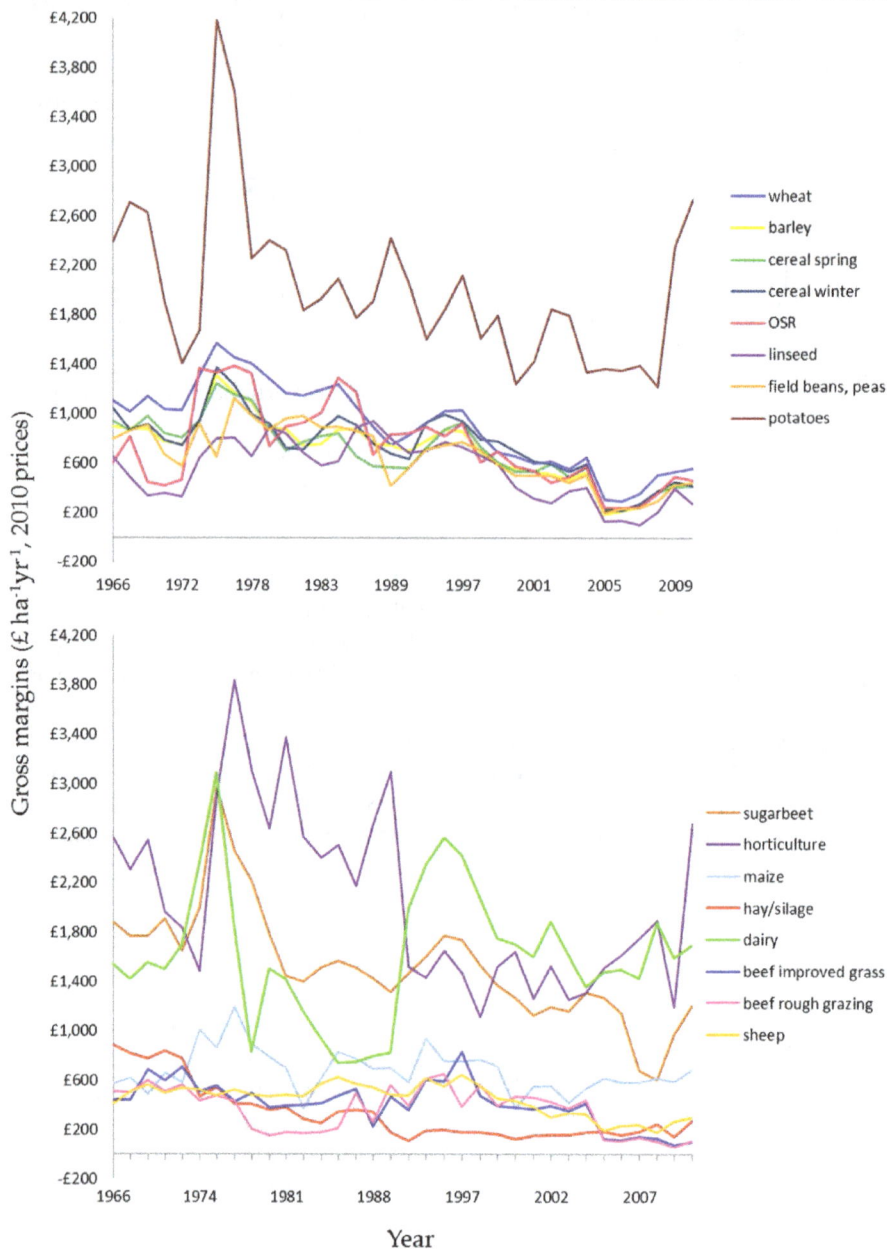

Figure 2 Mean gross margin (GM) values for selected land uses from 1966 to 2010 (in 2010 prices). Data taken from the *John Nix Farm Management Pocketbooks.*

machinery in the estimates, which are likely to be high for these particular land uses. Cereal, oilseed rape, and maize fell into the middle range of expected returns, while livestock farming and hay/silage production had lower than average expected GMs (Table 2). Linseed in particular showed poor economic resilience as measured by the expected variance:return ratios, with the expected SD in GM being 87% of the expected GMs over the 1966 to 2010 period. The data (Table 2, Figure 2) highlight the volatility in individual land-use investments in lowland agricultural landscapes in the UK.

Historic correlations and covariance of land use, gross margins, and resilience

The correlation coefficients (a normalized measure of covariance) are presented (Table 3) as they are easier to interpret than covariance. The covariance structure of these data was the basis for the MPT analysis of expected GMs and expected variance of GMs for the different agricultural land-use portfolios presented below. The forecast GMs from cereal, oilseed crops (linseed and rape), and field beans and peas were found to be closely correlated over the 44 years analyzed in this

Table 2 Descriptive statistics of UK annual average GM per hectare (in 2010 prices)[a,b]

Land use	Proportion of land base across the three study regions, %	Predicted GM, GBP/hectare/year	StDev of predicted annual GMs across analysis period), GBP/hectare/year	CV of GM, %
Wheat	19.2	£909	£338	0.37
Barley	10.3	£716	£260	0.36
Cereal spring	2.3	£703	£258	0.37
Cereal winter	4.8	£755	£268	0.36
Oilseed rape	6.7	£744	£339	0.46
Linseed	5.2	£280	£243	0.87
Field beans, peas	5.6	£668	£244	0.37
Potatoes	3.3	£2,020	£644	0.32
Sugar beet	2.7	£1,209	£462	0.38
Horticulture	1.7	£2,067	£707	0.34
Maize	1.1	£678	£173	0.26
Hay/silage	7.6	£336	£231	0.69
Dairy	10.7	£1,613	£542	0.34
Beef (improved)	9.3	£413	£182	0.44
Beef (rough)	9.2	£361	£180	0.50
Sheep	0.2	£448	£129	0.29
Mean weighted by investment proportion		£815	316	0.44

Abbreviations: CV, coefficient of variation; GM, gross margin; GBP, Great British pounds.
[a]Based on historic forecast GMs for agricultural activities from 1966–2010
[b]Source: *John Nix Farm-Management Pocketbooks.*

study with an average Pearson's correlation coefficient of 0.83 (Table 3). GMs from beef and sheep production were also closely correlated. The weakest correlations were between fodder crops (maize, hay/silage), dairy, and beef, horticulture, and cereal production. Although Table 3 does not provide complete information about the covariance structure through time, it does suggest that landscapes containing a mixture of cereal, livestock, and horticultural land uses were those most likely to have the highest resilience.

Relation between land-use diversity, expected gross margins, and expected variance in gross margins

The relationships between the expected GMs, the expected SD of those GMs, and the landscape diversity for the bundles of land uses found within each of the 1 km^2 study sites (n = 353) were significant (Figure 3). There was a strong linear relationship (r^2 =0.82, P<0.0005) between expected GMs and the expected variance in those GMs (Figure 3a); higher landscape diversity imply lower expected GMs (r^2 = 0.30, P<0.0005) (Figure 3b). Similarly, the expected variance (SD) in GM declined with increased land-use diversity (r^2 = 0.45, P<0.0005) (Figure 3c). The relationship between the CV (coefficient of variation of GMs and variance of GMs) and land-use diversity (Figure 3d) was very weak (r^2 = 0.05, P = 0.02), with high variances in expected GMs in

homogenous landscapes balanced by high expected GMs, and low variance in GMs in heterogeneous landscapes counteracted by generally low levels of expected GMs. It should be noted that no attempt to imply direct causation between landscape diversity and resilience is intended, as the resilience of GMs is entirely explained by the historic covariance structure of GM data for the land-use portfolio at each study site rather than by the land-use diversity of the sites *per se.*

Portfolio size, diversity, and the resilience of lowland agricultural landscapes in the UK

To further explore the relationships between land-use diversity, portfolio size, and GMs, we calculated land-use diversity, expected GM, expected SD of GM, and CV of GM using nine different landscape portfolio sizes (the spatial extent over which the land-use portfolio returns were calculated) within the three regional sub-extents. The portfolio sizes ranged from 25 to 3600 hectares, and were designed to capture the range of landholding size in lowland UK agricultural regions. As with the study site analysis, it can be seen that areas with lower expected GMs (Figure 4a) tended to have lower expected variance in GMs (Figure 4b). The relationship between expected GM, expected SD of GM, and CV (Figure 4c) was less clear. In the region 1 and 2 sub-extents, the lower GMs and variances in GMs tended to

Table 3 Correlation matrix of gross margins from different lowland agricultural land uses (1966 to 2010)

	Wheat	Barley	Cereal (spring)	Cereal (winter)	Oilseed rape	Linseed	Field beans	Potatoes	Sugar beet	Horticulture	Maize	Hay/silage	Dairy	Beef (improved grass)	Beef (rough grass)
Barley	0.936														
Cereal (spring)	0.9505	0.932													
Cereal (winter)	0.9185	0.9425	0.9506												
Oilseed rape	0.8242	0.8066	0.7424	0.7951											
Linseed	0.7428	0.7732	0.6379	0.7508	0.7526										
Field beans	0.8999	0.8246	0.8326	0.8088	0.7635	0.747									
Potatoes	0.6426	0.606	0.6527	0.6372	0.53	0.492	0.4904								
Sugar beet	0.8655	0.8606	0.9082	0.8792	0.7031	0.596	0.6771	0.7173							
Horticulture	0.6557	0.5542	0.5362	0.4973	0.5218	0.5333	0.6155	0.7038	0.5189						
Maize	0.5836	0.5633	0.5426	0.5924	0.688	0.5578	0.4979	0.4676	0.6092	0.4165					
Hay/silage	0.5348	0.492	0.5732	0.4644	0.1288	0.1015	0.4386	0.3919	0.5394	0.4117	0.1069				
Dairy	0.0976	0.1426	0.2317	0.2339	0.0353	0.0248	−0.089	0.205	0.3094	−0.287	0.2202	−0.05			
Beef (improved grass)	0.6509	0.7132	0.7444	0.727	0.4715	0.4544	0.5685	0.2081	0.6135	0.1202	0.2643	0.421	0.2474		
Beef (rough grass)	0.2294	0.4179	0.3959	0.4754	0.1372	0.2179	0.1739	0.0444	0.3632	−0.166	0.0633	0.3017	0.3172	0.7103	
Sheep	0.7713	0.7692	0.7384	0.798	0.6818	0.7416	0.7971	0.314	0.6103	0.3431	0.4485	0.3141	0.0147	0.8028	0.475

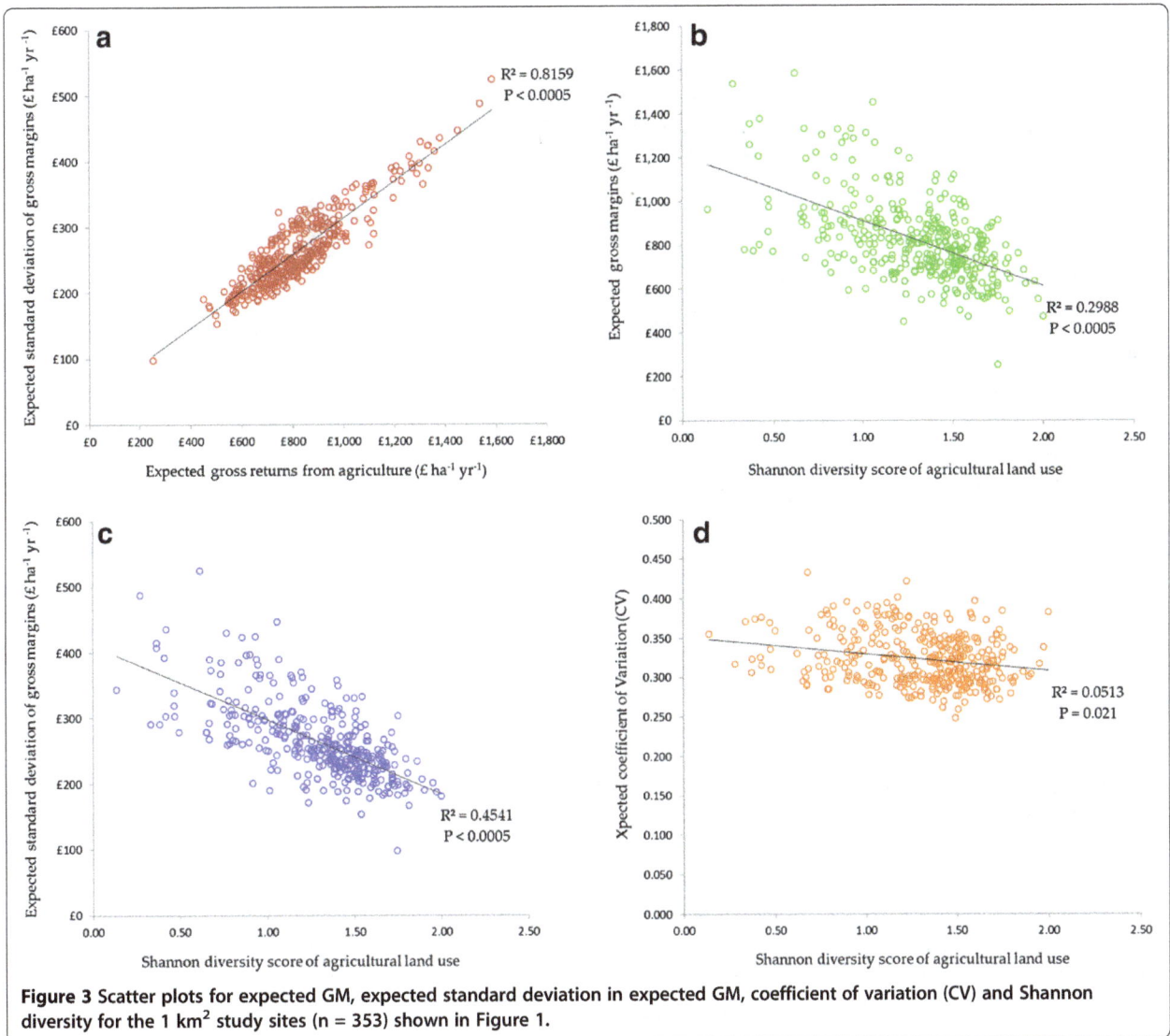

Figure 3 Scatter plots for expected GM, expected standard deviation in expected GM, coefficient of variation (CV) and Shannon diversity for the 1 km² study sites (n = 353) shown in Figure 1.

have a higher CV, with higher expected GMs unable to counteract the increased expected variance in GMs. A clear pattern of the relationships between GM, variance, and CV could not be seen in the region 3 sub-extent. However, it can also be seen that as the portfolio size over which the analysis was undertaken increased, there was a general increase in the economic resilience of the portfolios (that is, reduced CV) (Figure 4c), owing to decreasing expected variance in returns (Figure 4b).

The regional sub-extent analysis also showed that as the portfolio spatial extent increased, the diversity of the land uses within each portfolio also increased (Figure 5a), with diversity increasing most rapidly as portfolio size increased from 25 to 400 hectares, but continuing to increase at a slower rate up to the maximum portfolio size (3600 hectares). Agricultural economic resilience increased rapidly (decreasing CV) with increased portfolio size (Figure 5b). From this, we can infer that increased land-use diversity,

and, therefore, increased diversification of the investments in each portfolio, results in increased economic resilience, but that there are few gains in resilience beyond a portfolio size of 1200 hectares. By selecting larger portfolios within the fixed sub-extents, it would also be possible to increase the economic resilience for a given portfolio of investments while maintaining the same overall GMs returned across each regional sub-extent.

The expected mean GMs per hectare for the region 1 and 2 sub-extents were similar (GPB£710 and GBP£637/hectare/year, respectively; Figure 5). However, the mean CV across the sub-extents at portfolio sizes beyond 100 hectares was considerably lower in the more mixed agricultural region (sub-extent 2) than in the more heterogeneous livestock dominated region (sub-extent 1). The arable dominated regional sub-extent (region 3) had the highest mean returns per hectare (GBP£1,113/hectare/year) and the highest mean CV at all portfolio sizes

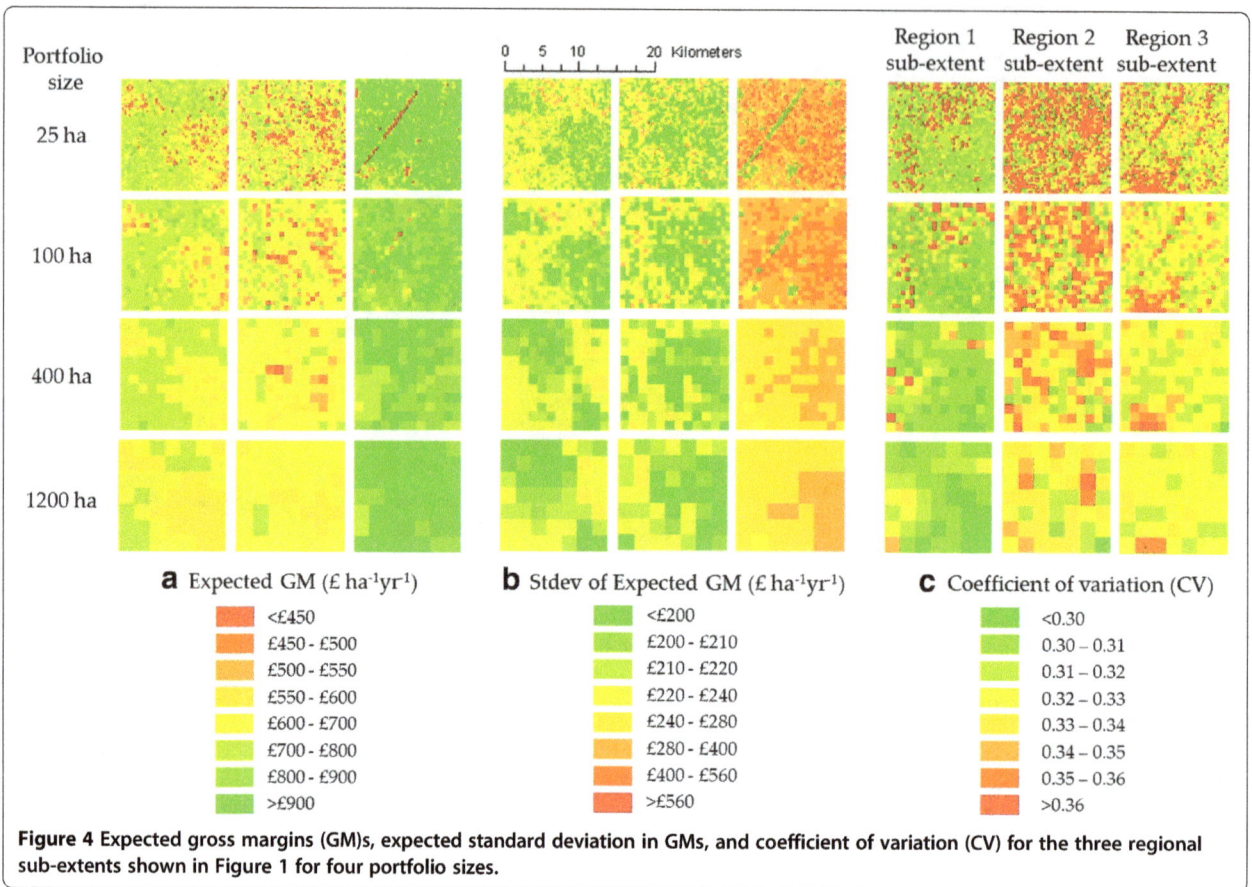

Figure 4 Expected gross margins (GM)s, expected standard deviation in GMs, and coefficient of variation (CV) for the three regional sub-extents shown in Figure 1 for four portfolio sizes.

(Figure 5b). All the regional sub-extents showed similar changes in mean CV with increased portfolio size. As the portfolios increased in size (thus taking in more land-use types), the mean economic resilience of the sub-extents increased (Figure 5b), with the CV dropping by around 5% as the landscape portfolios size increased from 25 to 400 hectares, and a reduction in CV of approximately 7% on average in the move from a portfolio size of 25 hectares to one of 1200 hectares. Beyond 1200 hectares, further increases in landscape portfolio extent made little difference to the mean resilience to returns structure, despite continued increases in portfolio diversity (Figure 5a,b).

It is notable that even small portfolios had CVs that were significantly lower than the weighted mean CV for individual land uses (Table 2), so even small amounts of land-use diversification can increase the resilience of agricultural returns, regardless of scale. A crucial finding here is that small increases in portfolio size dramatically decreased the CV of expected GMs in real lowland agricultural landscapes (suggesting increased economic resilience) and this held true in all three regions, with the majority of these gains occurring in the move from 25-hectare to 800-hectare landscape portfolios. This suggested that increases in diversity in lowland agricultural

landscapes is likely to lead to increased economic resilience of those land-use, and that the most homogenous landscapes will benefit most from increased land-use diversity.

Discussion

Agricultural GMs are often used in academic research as indicators of the economic functioning of farm enterprises [65-67] and agricultural landscapes [68,69]. However, as shown in this paper, there are clear and statistically significant trade-offs between high expected GMs and the expected variance of those margins in the face of the constant environmental, economic, and policy perturbations in the real landscape, as studied here. Considering both the expected returns and the expected variance of agricultural returns may therefore provide a more complete understanding of the economic functioning of particular agricultural activities or agricultural landscapes.

This research suggests that one way to increase agricultural resilience would be to increase the diversity of land use within a landscape. This would provide a lower, but more stable, level of expected returns compared with a single land use, which gives high expected returns, but also high expected volatility of returns in the face of exogenous perturbations. The resilience of returns from

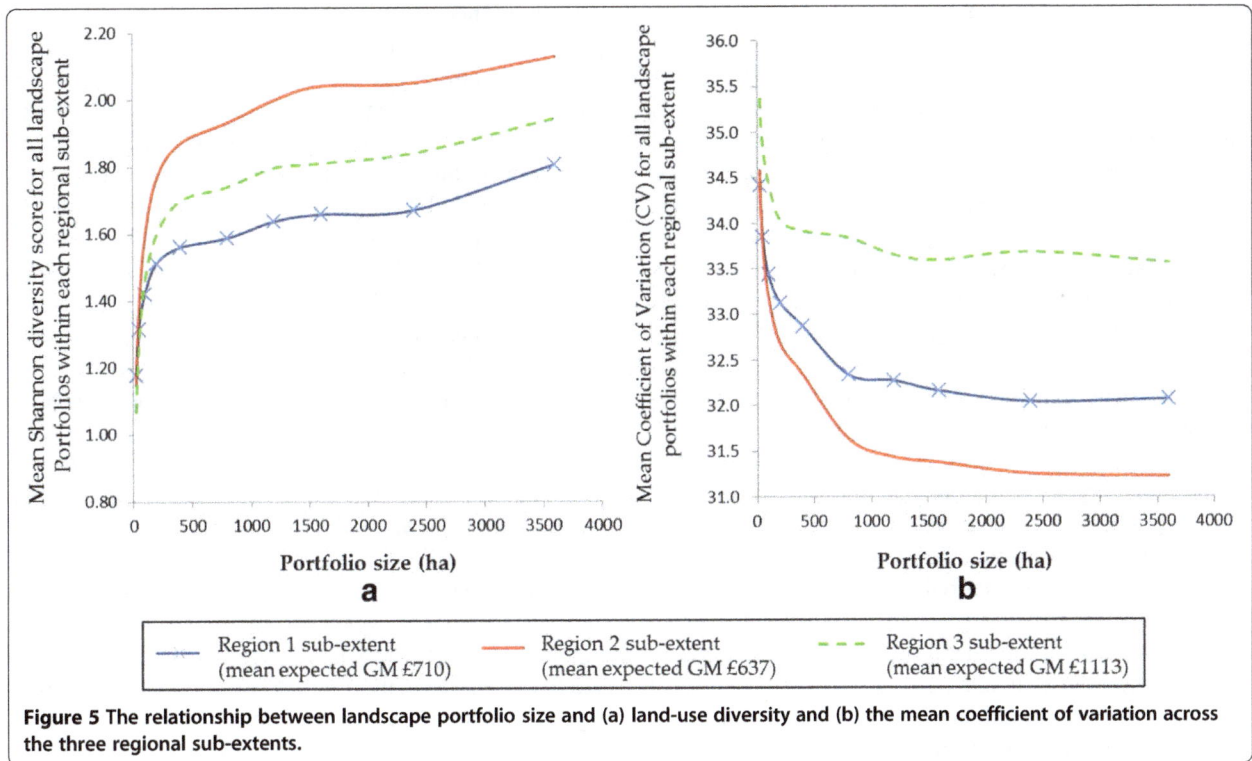

Figure 5 The relationship between landscape portfolio size and (a) land-use diversity and (b) the mean coefficient of variation across the three regional sub-extents.

agriculture is, in part, dependent not only on the agricultural economic resilience of individual land uses, but on the interactions of the expected returns between the different land uses present within any given landscape portfolio. More specifically, a mixture of cereal crops, livestock, dairy, and fodder crops creates a covariant returns structure that lowers the volatility of the aggregate returns, thereby increasing economic resilience across the landscape.

However, caution must be used in the interpretation of these results in terms of the resilience of individual farm enterprises, because the use of generalized estimates of GM is likely to have over-simplified the volatility of agricultural returns at the farm scale. Moreover, many other factors, such as capital assets, individual capacities, adaptive management, and institutional support, all play vital roles in increased agricultural economic resilience [70-72], and these were not assessed in this analysis. Moreover, given the trade-off between expected GM and the expected variance in GM, choices of land-use investments are likely to be determined in part by the risk to returns preferences of individual farmers [15,53]. Nevertheless, these preliminary findings do suggest that, all other things being equal, land-use diversification may provide a means of increasing economic resilience in lowland agricultural landscapes.

Importantly, these findings are not based on theoretical idealized mixture of land uses, but on the pre-existing land use choices in real landscapes. Thus, our findings

represent meaningful land-use strategies that are appropriate to the topographic, environmental, and economic contexts of the landscapes in which they occur. Future monitoring of the relationships between agricultural land-use diversification and economic resilience that draw on regional sample surveys of actual, rather than predicted, GMs may help further clarify the relations between land use, incomes, and economic resilience.

One key finding from this paper is the possibility of determining an optimal spatial extent over which the agricultural economic resilience is maximized for UK lowland agricultural landscapes. In real landscapes, increasing the physical size of the portfolio over which returns are estimated increases the mean expected economic resilience (that is, the CV). There were rapid increases in the mean resilience return ratio (as measured by the CV) as portfolios increased in size from 25 hectares (0.25 km^2) to around 400 hectares (4 km^2). Beyond approximately 1,200 hectares (12 km^2), increasing the size of the portfolio had little effect on the portfolio performance.

These findings suggest that if the UK lowland land-use diversity found at the sizes of 1200-hectare or even 400-hectare landscape extents were to be replicated at smaller spatial extents, this would significantly increase the resilience of UK landscapes at scales more closely associated with the average farm holdings of UK farmers (57 hectares). The resilience gains obtained from such an increase in land-use diversity are likely to be greatest in the most homogenous agricultural landscapes. Nevertheless, the benefits of increasing such relatively fine-grained land-use diversity seems to occur in grassland, arable, and mixed-agriculture landscapes. Moreover, the potential benefits of increasing agricultural land-use diversity is not limited to increasing economic resilience of farmed landscapes, but is likely to also provide co-benefits for other important functions of the agricultural landscapes. For example, it has been suggested that habitat heterogeneity is a key component in biodiversity conservation [73,74], that amenity values are positively associated with agricultural landscape diversity [75] and that the provision of cultural ecosystem services are greater in diverse agricultural landscapes [76].

It is necessary to note here that the agricultural GM data are not a perfect measure of economic returns, nor can they account for either the potential losses of economies of scale or the potential synergies that may come through increased moves towards finer-grained agricultural landscapes. Nevertheless, the relationships between land-use diversity and increased agricultural economic resilience that we found in this study are based on a sub-regional finer graining of existing land uses, rather than by the replacement of existing land uses with others that might not be suitable for local climatic or typological conditions. This, in turn, suggests that fine-scale land-use diversity managed at a sub-regional scale that takes advantage of existing local land uses can increase farming resilience without affecting the aggregate yields of agricultural goods produced by those sub-regional landscapes.

Finally, these findings suggest that there may be a role for fine-grained agricultural land-use diversification as a means of increasing the resilience of returns from agriculture. Replicating the existing the land-use diversity found within typical UK lowland agricultural extents of 400 to 1200 hectares at a farm scale would create diversified farming portfolios that might reduce the volatility of farmers' returns. Alternatively, an increase in agricultural economic resilience could potentially be achieved within existing land-use patterns through some form of portfolio sharing or other collective approaches to economic management at landscape scales. This is an area of research that warrants further investigation. In the face of increasingly volatile commodity markets and weather patterns [77], enhancing economic resilience in agricultural landscapes is no longer simply a desirable goal, but an increasingly important requirement of creating sustainable agricultural ecosystems. Article 30 of the recent Common Agricultural Policy (CAP) reform proposals are intended to increase (arable) crop diversification at the farm scale as part of the CAP 'greening' initiative [78]. However, the proposed CAP reforms

provide no incentives to diversify the wider matrix of arable, horticultural, and livestock land uses at a landscape scale. The findings presented here provide empirical evidence for the long-standing theory of links between landscape-scale diversity and economic resilience. Landscape approaches to agricultural ecosystem management are increasingly being called for in relation to achieving objectives in conservation [79] and ecosystem services management [80,81]. Although they are exploratory, these results suggest that rural-development policies that include a focus on the co-ordination of land-use management at landscape level may also be beneficial in terms of increasing the economic resilience of lowland agricultural regions.

Agricultural land-use diversification at the landscape scale might be aided by policies that facilitate the sharing of resources, thereby reducing the need for the economies of scale, and the resultant homogenization of landscapes, that are required for the use of modern agricultural machinery. Such co-operation requires the building of trust between farmers [82], and could be assisted by the formation of institutions such as environmental co-operatives that have multiple objectives, such as the maintenance of landscape character and biodiversity or ecosystem service conservation, or the use of collaborative agricultural environment schemes [83].

Conclusions

Our research has produced a number of key findings. First, there is a trade-off between expected mean returns and the volatility of those expected returns, such that specialization in farmscapes is associated with maximizing mean returns, but a higher volatility of those returns. Secondly, land-use diversity is positively correlated with the expected stability of returns, and negatively correlated with expected returns. Thirdly, there is considerable scale dependency in the relationships between land-use diversity and the resilience of agricultural returns. Small spatial extents (less than 400 hectares) in UK lowlands do not currently provide sufficient portfolio diversification to minimize the CV in expected returns of agricultural production.

Perhaps, most importantly, this research suggests that the resilience of agricultural returns within lowland agricultural landscapes could potentially be increased through fine-grain land-use diversification without affecting the aggregate returns or land-use portfolio at the landscape level. Given the current volatility of agricultural returns, it seems reasonable that land-use diversity and volatility or the resilience of agricultural returns should be given greater consideration in research and policy interventions on the socio-economic functions of agricultural ecosystems.

Abbreviations
CV: Coefficient of variation; GBP: Great British pound; GM: Gross margin; JAC: June Agricultural Census; LCM2000: Land Cover Map 2000; LU: Livestock units; MPT: Modern portfolio theory; H': Shannon's diversity index.

Competing interests
The authors declare that they have no competing interests.

Authors' contributions
DJA, EDGF, and TGB participated in the conception and design of the study and the interpretation of the results. DJA acquired the data, and carried out the GIS and MPT analysis. DJA, EDGF, and TGB were involved in the co-ordination and drafting the manuscript. All authors read and approved the final manuscript.

Acknowledgements
DJA was supported by an ESRC/NERC Interdisciplinary Award. We thank the two reviewers, whose comments greatly improved the manuscript.

Author details
[1]FuturES Research Center, Leuphana Universität, Lüneburg, Germany. [2]Sustainability Research Institute, University of Leeds, Leeds, UK. [3]Department of Geography, University of Guelph, Guelph, Ontario, Canada. [4]Institute of Integrative and Comparative Biology, University of Leeds, Leeds, UK.

References
1. Healey MJ, Ilbery BW: **The industrialization of the countryside: an overview.** In *The Industrialization of the Countryside.* Edited by Healey MJ, Ilbery BW. Norwich: Geo Books; 1985:1–28.
2. Pretty JN: *The Living Land.* London: Earthscan; 1998.
3. Horlings LG, Marsden TK: **Towards the real green revolution? Exploring the conceptual dimensions of a new ecological modernisation of agriculture that could 'feed the world'.** *Glob Environ Chang* 2011, **21:**441–452.
4. Blaxter K, Robertson N: *From Dearth to Plenty: The Second Agricultural Revolution.* Cambridge: Cambridge University Press; 1995.
5. Fraser EDG: **Crop diversification and trade liberalization: linking global trade and local management through a regional case study.** *Agr Hum Values* 2006, **23:**271–281.
6. Kim K, Chavas JP, Barham B, Foltz J: **Specialization, diversification, and productivity: a panel data analysis of rice farms in Korea.** *Agr Econ* 2012, **43:**1–14.
7. Döös BR: **Environmental degradation, global food-production, and risk for large-scale migrations.** *Ambio* 1994, **23:**124–130.
8. Giampietro M: *Multi-scale Integrated Analysis of Agroecosystems.* Boca Raton: CRC Press; 2004.
9. Trenbath B, Conway G, Craig IT: **Threats to sustainability in intensified agriculture.** In *Agroecology: Researching the Ecological Basis for Sustainable Development.* Edited by Gliessman SR. New York: Springer-Verlang; 1990:337–365.
10. Walker BH, Steffen W: **An overview of the implications of global change for natural and managed terrestrial ecosystems.** *Conservation Ecology* 1997, **1:**2.
11. Fraser EDG, Mabee W, Slaymaker O: **Mutual dependence, mutual vulnerability: the reflexive relation between society and the environment.** *Glob Environ Chang* 2003, **13:**137–144.
12. Government Office for Science: *Foresight. The Future of Food and Farming: Final Project Report. Final Project Report.* London: The Government Office for Science; 2011.
13. Blank S: **Income risk varies with what you grow, where you grow it.** *Calif Agr* 1992, **46:**14–16.
14. Hanson J, Johnson D, Peters S, Janke R: **The profitability of sustainable agriculture on a representative grain farm in the mid-Atlantic region 1981–1989.** *Northeast J Agric Resour Econ* 1990, **19:**90–98.
15. Baumgärtner S, Quaas MF: **Managing increasing environmental risks through agrobiodiversity and agrienvironmental policies.** *Agr Econ* 2010, **41:**483–496.
16. Berkes F, Colding J, Folke C: *(Eds): Navigating Social-Ecological Systems: Building Resilience for Complexity and Change.* Cambridge: Cambridge University Press; 2003.
17. Folke C: **Resilience: The emergence of a perspective for social-ecological systems analyses.** *Glob Environ Chang* 2006, **16:**253–267.
18. Scheffer M, Carpenter SR: **Catastrophic regime shifts in ecosystems: linking theory to observation.** *Trends Ecol Evol* 2003, **18:**648–656.
19. Cumming GS, Collier J: **Change and identity in complex systems.** *Ecol Soc* 2005, **10:**29.
20. In *Panarchy: Understanding Transformations in Human and Natural Systems.* Edited by Gunderson LH, Holling CS. Washington: Island Press; 2001.
21. Worster D: *Dust Bowl: the Southern Plains in the 1930s.* London: Oxford University Press; 2004.
22. Haile T: **Causes and characteristics of drought in Ethiopia.** *Ethiop J Agric Sci* 1988, **10:**85–97.
23. Comenetz J, Caviedes C: **Climate variability, political crises, and historical population displacements in Ethiopia.** *Global Environ Change B Environ Hazards* 2002, **4:**113–127.
24. Fraser EDG: **Travelling in antique lands: using past famines to develop an adaptability/resilience framework to identify food systems vulnerable to climate change.** *Clim Chang* 2007, **83:**495–514.
25. Fraser EDG: **Social vulnerability and ecological fragility: building bridges between social and natural sciences using the Irish Potato Famine as a case study.** *Conserv Ecol* 2003, **7:**9.
26. O'Grada C: *The Great Irish Famine.* London: Macmillan; 1989.
27. Fraser EDG, Rimas A: *Empires of Food: Feast Famine and the Rise and Fall of Civilizations.* New York: Free Press; 2010.
28. Fraser EDG, Stringer LC: **Explaining agricultural collapse: macro-forces, micro-crises and the emergence of land use vulnerability in southern Romania.** *Glob Environ Chang* 2009, **19:**45–53.
29. Pimm S: **The complexity and stability of ecosystems.** *Nature* 1984, **307:**321–326.
30. Simelton E, Fraser EDG, Termansen M, Forster PM, Dougill AJ: **Typologies of crop-drought vulnerability: an empirical analysis of the socio-economic factors that influence the sensitivity and resilience to drought of three major food crops in China (1961–2001).** *Enviro Sci Policy* 2009, **12:**438–452.
31. Simelton E, Fraser EDG, Termansen M, Benton TG, Gosling S, South A, Arnell N, Challinor A, Dougill AJ, Forster PM: **The socioeconomics of food crop production and climate change vulnerability: a global scale quantitative analysis of how grain crops are sensitive to drought.** *Food Secur* 2012, **4:**163–179.
32. Fraser EDG, Termansen M, Sun N, Guan D, Simelton E, Dodds P, Feng K, Yu Y: **Quantifying socioeconomic characteristics of drought-sensitive regions: evidence from Chinese provincial agricultural data.** *Comptes Rendus Geoscience* 2008, **340:**679–688.
33. Bakker MM, Govers G, Ewert F, Rounsevell M, Jones R: **Variability in regional wheat yields as a function of climate, soil and economic variables: assessing the risk of confounding.** *Agric Ecosyst Environ* 2005, **110:**195–209.
34. Di Falco S, Chavas JP: **Rainfall shocks, resilience, and the effects of crop biodiversity on agroecosystem productivity.** *Land Econ* 2008, **84:**83–96.
35. van Meijl H, van Rheenen T, Tabeau A, Eickhout B: **The impact of different policy environments on agricultural land use in Europe.** *Agric Ecosyst Environ* 2006, **114:**21–38.
36. Di Falco S, Perrings C: **Crop biodiversity, risk management and the implications of agricultural assistance.** *Ecol Econ* 2005, **55:**459–466.
37. Elmqvist T, Folke C, Nystrom M, Peterson G, Bengtsson J, Walker B, Norberg J: **Response diversity, ecosystem change, and resilience.** *Front Ecol Environ* 2003, **1:**488–494.
38. Fraser EDG, Mabee W, Figge F: **A framework for assessing the vulnerability of food systems to future shocks.** *Futures* 2005, **37:**465–479.
39. Holling CS: **Understanding the complexity of economic, ecological, and social systems.** *Ecosystems* 2001, **4:**390–405.
40. McIntosh RP: **Concept and terminology of homogeneity and heterogeneity in ecology.** In *Ecological Heterogeneity.* Edited by Kolasa J, Pickett STA. New York: Springer-Verlag; 1991:24–46.
41. Baumgärtner S: **The insurance value of biodiversity in the provision of ecosystem services.** *Nat Resour Model* 2007, **20:**87–127.
42. Philippi T, Seger J: **Hedging one's evolutionary bets, revisited.** *Trends Ecol Evol* 1989, **4:**41–44.

43. Sasaki AS, Ellner S: **The evolutionarily stable phenotype distribution in a random environment.** *Evolution* 1995, **49**:337–350.

44. Tuljapurkar S: *Population Dynamics in Variable Environments.* New York: Springer-Verlag; 1990.

45. Tansey G, Worsley T: *The Food System: a Guide.* London: Earthscan; 1995.

46. O'Brien KL, Leichenko RM: **Double exposure: assessing the impacts of climate change within the context of economic globalization.** *Global Environ Chang* 2000, **10**:221–232.

47. FAO: *How to Feed the World in 2050.* Rome: Food and Agriculture Organization of the United Nations; 2009.

48. Department for Environment, Food and Rural Affairs: *England Farm Accounts.* London: DEFRA; 2007.

49. Markowitz HM: *Portfolio Selection: Efficient Diversification of Investments.* New York: Wiley & Sons; 1959.

50. Sharpe WF: *Portfolio Theory and Capital Markets.* New York: McGraw-Hill; 1970.

51. Figge F: **Bio-folio: applying portfolio theory to biodiversity.** *Biodivers Conserv* 2004, **13**:827–849.

52. McKillop DG: **The return-risk structure of lowland agriculture in Northern Ireland.** *Eur Rev Agric Econ* 1989, **16**:217–228.

53. Di Falco S, Chavas JP: **Crop genetic diversity, farm productivity and the management of environmental risk in rainfed agriculture.** *Eur Rev Agr Econ* 2006, **33**:289–314.

54. Di Falco S, Perrings C: **Crop genetic diversity, productivity and stability of agroecosystems.** *A theoretical and empirical investigation. Scot J Polit Econ* 2003, **50**:207–216.

55. Nix J, Hill P: *The John Nix Farm Management Pocketbook, 33rd edition.* 40th edition. Ashford: Wye College Press; 2010.

56. Centre for Ecology and Hydrology: *Land Cover Map:* http://www.ceh.ac.uk/data/lcm/LCM2000/shtm.

57. Department for Environment, Food and Rural Affairs: *Definitions of Terms Used in Farm Business Management.* London: DEFRA; 2010.

58. ESRI: *ArcGIS Desktop: Release 10.* Redlands, CA: Environmental Systems Research Institute; 2011.

59. Natural England: *Calculation of Stocking Rates and Recording of Grazing Livestock.:* Natural England; 2011.

60. Department for Environment, Food and Rural Affairs: *June 2009 Survey of Agriculture and Horticulture UK Final Results.* London: DEFRA; 2009.

61. University of Massachusetts: *FRAGSTATS: spatial pattern analysis program for categorical maps:* http://www.umass.edu/landeco/research/fragstats/downloads/fragstats_downloads.html].

62. Shannon C: *The Mathematical Theory of Communication.* Urbana: The University Of Illinois Press; 1949.

63. Taylor LR: **Bates, Williams, Hutchinson - a variety of diversities.** In *Diversity of Insect Faunas.* Edited by Mound LA, Waloff N. Oxford: Blackwell; 1978:1–18.

64. UK Gross Domestic Product (GDP) deflators: http://www.hm-treasury.gov.uk/data_gdp_fig.htm.

65. Alcock D, Hegarty RS: **Effects of pasture improvement on productivity, gross margin and methane emissions of a grazing sheep enterprise.** *Int Congr Ser* 2006, **1293**:103–106.

66. Morse S, Bennett RM, Ismael Y: **Genetically modified insect resistance in cotton: some farm level economic impacts in India.** *Crop Prot* 2005, **24**:433–440.

67. Pacini C, Wossink A, Giesen G, Vazzana C, Huirne R: **Evaluation of sustainability of organic, integrated and conventional farming systems: a farm and field-scale analysis.** *Agric Ecosyst Environ* 2003, **95**:273–288.

68. Fezzi C, Bateman IJ: **Structural agricultural land use modeling for spatial agro-environmental policy analysis.** *Am J Agr Econ* 2011, **93**:1168–1188.

69. Wiggering H, Dalchow C, Glemnitz M, Helming K, Muller K, Schultz A, Stachow U, Zander P: **Indicators for multifunctional land use - linking socio-economic requirements with landscape potentials.** *Ecol Indic* 2006, **6**:238–249.

70. Anderies JM, Janssen MA, Ostrom E: **A framework to analyze the robustness of social-ecological systems from an institutional perspective.** *Ecol Soc* 2004, **9**:18.

71. Borron S: *Buildng Resilience for an Unpredictable Future: How Organic Agriculture Can Help Farmers Adapt to Climate Change.* Rome: Food and Agriculture Organization of the United Nations; 2006.

72. Eakin H, Luers AL: **Assessing the vulnerability of social-environmental systems.** *Annual Review of Environment and Resources* 2006, **31**:365–394.

73. Benton TG, Vickery JA, Wilson JD: **Farmland biodiversity: is habitat heterogeneity the key?** *Trends Ecol Evol* 2003, **18**:182–188.

74. Fahrig L, Baudry J, Brotons L, Burel FG, Crist TO, Fuller RJ, Sirami C, Siriwardena GM, Martin JL: **Functional landscape heterogeneity and animal biodiversity in agricultural landscapes.** *Ecol Lett* 2011, **14**:101–112.

75. Howley P: **Landscape aesthetics: Assessing the general publics' preferences towards rural landscapes.** *Ecol Econ* 2011, **72**:161–169.

76. van Berkel DB, Verburg PH: **Spatial quantification and valuation of cultural ecosystem services in an agricultural landscape.** *Ecol Indic* 2012, doi:10.1016/j.ecolind.2012.06.025.

77. Benton TG, Gallani B, Jones C, Lewis K, Tiffin R, Donohoe T: *Severe Weather and UK Food Resilience. Report for UK Food Research Partnership.* London: Government Office for Science; 2012.

78. European Commission: *Establishing Rules for Direct Payments to Farmers Under Support Schemes Within the Framework of the Common Agricultural Policy.* Brussels: EC; 2011:1–105.

79. Concepción ED, Díaz M, Kleijn D, Báldi A, Batáry P, Clough Y, Gabriel D, Herzog F, Holzschuh A, Knop E, *et al*: **Interactive effects of landscape context constrain the effectiveness of local agri-environmental management.** *J Appl Ecol* 2012, **49**:695–705.

80. de Groot RS, Alkemade R, Braat L, Hein L, Willemen L: **Challenges in integrating the concept of ecosystem services and values in landscape planning, management and decision making.** *Ecol Complex* 2010, **7**:260–272.

81. Prager K, Reed M, Scott A: **Encouraging collaboration for the provision of ecosystem services at a landscape scale-Rethinking agri-environmental payments.** *Land Use Policy* 2012, **29**:244–249.

82. Goldman RL, Thompson BH, Daily GC: **Institutional incentives for managing the landscape: Inducing cooperation for the production of ecosystem services.** *Ecol Econ* 2007, **64**:333–343.

83. Franks JR, Mc Gloin A: **Environmental co-operatives as instruments for delivering across-farm environmental and rural policy objectives: lessons for the UK.** *J Rural Stud* 2007, **23**:472–489.

14

The value of trust in biotech crop development

Obidimma C Ezezika[1,2,3*], Kathryn Barber[1], Abdallah S Daar[1,4,5]

Abstract

Background: Agricultural biotechnology public-private partnerships (PPPs) have been recognized as having great potential in improving agricultural productivity and increasing food production in sub-Saharan Africa. However, there is much public skepticism about the use of GM (genetically modified) crops and suspicion about private sector involvement in agbiotech projects. This case study sought to understand the role of trust in the *Bacillus thuringiensis* (Bt) cotton in Burkina Faso project by exploring practices and challenges associated with trust-building, and determining what makes these practices effective from the perspective of multiple stakeholders.

Methods: We conducted semi-structured, face-to-face interviews to obtain stakeholders' understanding of trust in general as well as in the context of agbiotech PPPs. Relevant documents and articles were analyzed to generate descriptions of how trust was operationalized in this evolving agbiotech PPP. Data was analyzed based on emergent themes to create a comprehensive narrative on how trust is understood and built among the partners and with the community.

Results: We derived four key lessons from our findings. First, strong collaboration between research, industry and farmers greatly contributes to both the success of, and fostering of trust in, the partnership. Second, this case study also revealed the important, though often unrecognized, role of researchers as players in the communication strategy of the project. Third, effective and comprehensive communication takes into account issues such as illiteracy and diversity. Fourth, follow-up at the field level and the need for a multifaceted communications strategy is important for helping push the project forward.

Conclusions: Burkina Faso's well-established and effective cotton selling system laid the foundation for the implementation of the Bt cotton project – particularly, the strong dialogue and the receptivity to collaboration. Interviewees reported that establishing and maintaining trust among partners, researchers and the community in Burkina Faso greatly contributed to the success of the PPP. By addressing challenges to building trust and engaging in trust-building practices early on, improvements in the effectiveness of agbiotech PPPs are likely.

Background

As the first West African nation to adopt and commercialize a transgenic crop—particularly, *Bacillus thuringiensis* (Bt) cotton—Burkina Faso is a trendsetter in its region. The use of Bt cotton has increased dramatically in recent years, boasting the second highest growth rate in the world. In 2010, Bt cotton hectarage increased by 126% from the level in 2009. Currently, Bt cotton crops are estimated to cover 260,000 hectares of land and have been adopted by 65% of farmers [1]. Critical to the development

and implementation of Bt cotton in Burkina Faso was the partners' effective collaboration.

In this study, we focus on Burkina Faso, Africa's largest producer of cotton [2], and the role trust played in the country's adoption of Bt cotton (see Additional file 1 for additional background information on the Bt cotton project in Burkina Faso). The 1980s and 1990s proved to be difficult years for Burkina Faso's cotton industry, as annual yields drastically decreased due primarily to destruction by pests [3]. In 1999, Monsanto approached the Burkina government with the Bollgard GM cotton crop to address pest resistance to pesticides and increase cotton yields [4].

* Correspondence: obidimma.ezezika@srcglobal.org
[1]Sandra Rotman Centre, University Health Network and University of Toronto, Toronto, Ontario, Canada

A generally low level of infrastructure and development in Burkina Faso presented a unique set of challenges to project implementation. These challenges were addressed primarily through the establishment of partnerships between public and private stakeholders involved in the Burkinabè cotton industry. Such partnerships presented additional challenges to building trust, from which we have distilled four key lessons on trust-building in agbiotech public-private partnerships (PPPs).

The success of agbiotech projects depends on the ability of partners to engage in long-term collaborations to complete complex tasks. The issue of trust is especially critical to agricultural biotechnology, since the introduction of transgenic crops and involvement of multinational companies can be contentious and breed mistrust [5][6]. This case was chosen as one of eight case studies in a larger study investigating the role of trust in the adoption of GM crops in sub-Saharan Africa, placing particular emphasis on the conception, management and development of trust in agbiotech projects. Selection criteria for the case studies included a) representation of a variety of products and technological innovations, b) ensuring appropriate regional representation, c) ease of entry and availability of participants for interviews/focus group meetings, and d) potential of transferable knowledge (both successes and set-backs). The three specific goals of this study are to: 1) describe trust-building practices in the development of agbiotech projects; 2) describe the challenges associated with trust-building in PPPs; and 3) determine what makes these practices effective or ineffective. By identifying barriers to trust and trust-enhancing practices, this study provides insight for potential funders, researchers, farmers and others involved in agbiotech initiatives.

Methods

We received Research Ethics Board (REB) approval for conducting the case study from the University Health Network (UHN), University of Toronto before proceeding with the study. Data collection consisted of interviews with key informants; review of historical documents and research articles; and observations.

Interviewees were identified first by making a list of key individuals associated with the project based on the stakeholders identified within the research protocol. This list was then populated further through snowball sampling. We spoke with stakeholder informants who were familiar with the Bt cotton project in Burkina Faso. Potential interviewees were sent an invitation, which included an explanation of the case study series, to participate in the interview. Those who consented to participate were informed that the interview would be recorded, transcribed and then analyzed. Interviewees included small-scale Bt cotton farmers, and stakeholders

from the following organizations: Institut de l'Environnement et de Recherche Agricoles (INERA), Agence Nationale de Valorisation des Résultats de la Recherche (ANVAR), Association professionelle des societes cotonnières du Burkina Faso (APROCOB), Monsanto, and the West African Network for Communication on Agricultural Biotechnology (RECOAB). (see Additional file 2 for a list of the partners' roles and responsibilities)

The interviews took place in Bobo-Dioulasso, Sapouy and Ouagadougou, cities in Burkina Faso. The interviews followed a semi-structured, face-to-face format and each lasted approximately one and a half hours. The interview guide included questions on the interviewees' background, their understanding of the project, and their interpretation of the word *trust*. The interview explored perceptions of trust within the partnership and the public, apparent challenges to trust, and observed trust-building practices. Interviewees were also asked for advice on how to improve agbiotech PPPs (see Additional file 3 for sample questions from the interview guide).

The interviews were transcribed. The analysis was performed by reading through the transcripts several times, identifying trends and organizing them into major themes. A literature review of academic articles, news articles and publicly available project documents were also used in the writing of the report.

Results and discussion

With the intention of exploring the varying perceptions of trust, a definition of trust was solicited from our interviewees by asking about their general understanding of the word *trust*.

The interviewees' responses identify the presence of trust as a criterion for a good relationship, in which there is assurance, honesty, support, transparency and truth–elements that lend confidence and stability to successful long-term interactions. Trust was also characterized as a type of contract demanding the fulfillment of partner roles that are clearly defined by written agreements and regulations that structure interactions. In addition, trust was understood as an outcome of participation in particular groups or well-executed processes, such as stakeholder groups or the research process.

Based on the results of this study, we have derived four key lessons – from which partners in other agbiotech PPPs can learn and use as a guide for building and fostering trust.

1. Participatory plant breeding sows success: a strong connection between research, industry and farmers promotes on-going dialogue within the project

Burkina Faso has built a strong cotton selling system that connects farmers' needs to researchers' abilities and leverages the expertise of cotton companies. The initial partnership consisted of a collaboration between the Burkina agricultural research institute, INERA, Monsanto and Syngenta. Additional partners have also contributed their expertise, funding and platforms at different points during the project's development cycle.

Negative perceptions of Monsanto and GM crops

It is reported that Monsanto first approached Burkina Faso with GM crop information in 2000 to address the burden of pesticide resistance on the nation's cotton industry [7]. Field trials began in 2003 with a research agreement signed by Monsanto, Syngenta, and INERA. By 2007, Syngenta discontinued their involvement in the project, leaving Monsanto's Bollgard GM cotton crop to dominate field trials [8].

Skepticism grew over the potential consequences of such extensive private sector influence on Burkina Faso's cotton industry. One interviewee noted that negative perceptions of Monsanto arrived in Burkina Faso long before the introduction of Bt cotton. This negative perception permeated not only anti-GM communities but also the core partners themselves, who became highly suspicious of the motives and rationale for Monsanto's involvement in an agricultural project in Burkina Faso.

Recognizing motives, abilities and risks

Clear articulation of motives and risks was mentioned as an important trust-building practice to alleviate concerns pertaining to the development of PPPs. One researcher noted that a make-or-break factor in the success of the partnership was the candid disclosure of institutional motives early on in the project's development. In order for these national researchers to engage with new GM crops, any potential for risk had to be admitted in order for the partnership to proceed. It took various meetings for partners to feel comfortable with disclosing their institutional motivations, including the admission of profit-making motives. This practice improved transparency by opening channels of communication among partners, which helped alleviate suspicions and elevate levels of trust.

Collaboration is key

The interviews revealed that research in Burkina Faso is primarily funded not through grants from international organizations but through the sale of cotton on the international market. One interviewee reported that for every kilogram of cotton sold by cotton companies on the international market, one Franc is given to national research institutes. Through this exchange a vital connection is made among the farmers, research institutes and the commercial cotton companies. Interviewees associated the strong connection between the cotton research agenda and cotton farmers' needs to the funding relationship between cotton companies and national research institutes. Interviewees noted that the Burkina government had played a role in developing this funding relationship.

Our interviews also revealed that the farmers union, the Union Nationale des Producteurs du Coton de Burkina (UNPCB), held large stakes in the three major cotton companies: Société des Fibres Textiles du Burkina Faso (SOFITEX), Société Cotonnière du Gourma (SOCOMA), and FASO Cotton. One private sector interviewee noted that this arrangement gave the farmers a high degree of power in the cotton companies' affairs.

The case of Bt cotton in Burkina Faso demonstrates that the presence of a strong, inter-connected and collaborative partnership between industry, research and farmers has been invaluable in the development, implementation, and completion of the project. Levels of trust can be enhanced when there is a clear understanding of each partner's respective role, motivations, and contributions to the project. Furthermore, it is important to capitalize and build on pre-existing relationships and institutional structures as a means to establish and maintain trust in agbiotech PPPs.

2. Research is more than inquiry: researchers must collaborate with peers, journalists and the general public in a mutual and respectful relationship

Collaborative initiatives are most effective if a level of transparency is maintained through the timely dissemination of accurate and reliable information, the failure of which raises a key hurdle to project implementation issues pertaining to the public responsibility of researchers to communicate their findings.

Researcher disconnect

Researchers, sensitive to the volatile nature of public opinion toward agricultural biotechnology, were hesitant to speak to journalists about their scientific research and often directed journalists up the bureaucratic ladder. This not only weakened the informational content of communication strategies but also created unnecessary tension between researchers and journalists. Such a dynamic between the research community and the media works to limit civil society's access to appropriate and reliable sources of information on Bt cotton. Access to such information enhances transparency within the project and is imperative to the building of trust among all partners.

An issue that emerged as a challenge to trust building pertained not to the critical discourse over GM products but to the information that was used to substantiate these positions. A cotton company representative observed that much of the public discourse surrounding GM products was based on incorrect information. This incorrect information included beliefs that GM products will cause allergies, cause sterility and kill animals, to name a few.

Additionally, interviewees identified "activists" and "intellectuals" as groups strongly opposed to the project mainly due to a lack of reliable, scientifically-backed information on Bt cotton. The explanation offered for this highlights the intellectuals' disconnect from the farmers' fields. A belief pervaded that these groups had not visited the farms to see the Bt cotton in context and, consequently, had limited understanding of the process. In this case, a lack of transparency and correct information reaching civil society groups and the general public resulted in reduced levels of trust.

Further limiting the progress of the project was a noted lack of public confidence in Burkinabè researchers at project inception. One journalist interviewed made a comment reflecting the prominent view in Burkina Faso that a poor country like Burkina cannot produce high technology. This perception exacerbated public doubts about the future success of the project. The widespread view of Burkinabè researchers as "incapable" also pervaded Burkinabè researchers' interactions with their international research peers. One Burkinabè researcher noted from his experience that trust was often limited when scientists from developed countries behaved as if their African counterparts were incompetent. When the project initially began, a Burkinabè government researcher noted that this dismissive view of Burkinabè scientists had to be addressed before the project could continue.

An additional challenge to trust building in this project was posed by external forces. In the case of Burkina Faso, external influences are comprised of both France and the United States providing direction on GM crops. Not only do France and the United States have differing views on the introduction of GM crops in Burkina Faso—the former being opposed and the latter being in favor—but the Burkinabè government is caught in the middle having to deal with the views of its own research institutes. The contradicting direction provided by such external influences not only left the Burkinabè government ambivalent about what course of action to take on GM crops in its country but also made it more reluctant to trust the scientists at its own research institutes. A researcher from INERA commented: *this [disagreement characterizing the external influences] created some confusion within the government and the research institutes, which were caught in the middle of these conflicting opinions.*

Each of these factors—the lack of collaboration between researchers and media, negative public perceptions, limited confidence in researcher capabilities, and external country influences—presented significant challenges to project implementation and highlight the many public roles researchers must play to successfully navigate PPPs.

"Seeing-is-Believing" seminars

In an attempt to disseminate research information and dispel popular myths and misconceptions about biotechnology, the Bt cotton in Burkina Faso project launched a communications campaign, of which the "Seeing-is-Believing" seminars were a component. In these workshops,

members from civil society were invited to the test fields. Attendees heard lectures on different topics related to biotechnology and were subsequently invited to visit the testing sites [9]. The "Seeing-is-Believing" seminars allowed all members of the general public to visit the Bt cotton trial sites and witness the growth of the cotton as well as engage directly with individuals in discussions about the cotton. According to farmers interviewed, it was an effective trust-building practice.

3. Tell them often, in many ways: communicators must recognize the reality of illiteracy and diversity by developing dynamic, multi-lingual communication strategies

The accurate dissemination of research findings and information on Bt crops is rendered meaningless without the implementation of effective and comprehensive communications strategies.

The importance of effective communication

One interviewee expressed a strong opinion that communications is the first step in building trust. A strong effort was consistently made to develop an effective communication strategy in Burkina Faso. This statement rings true in countries like Burkina Faso where diverse ethnic, linguistic, and educational backgrounds exist. Three different national languages in addition to French are spoken in Burkina Faso. However, Burkina Faso has an adult literacy rate of about 29% [10], rendering written information useless to a large segment of the population. This limits the various media outlets that can be employed and thus presents tremendous barriers to the effective communication of information on biotechnology to Burkina Faso's diverse population. It is therefore unsurprising that print media is viewed as having limited effectiveness. Likewise, interviewees also highlighted the often limited access most people have to a media outlet such as the newspaper and the difficulties of translating new and complicated concepts, such as Bt, into local language.

"Seeing-is-Believing": media, language and literacy

While French is the primary language of communication in Burkina Faso, many individuals speak ethnic languages such as Mooré, Jula, and Gulmacema. Despite these language barriers, members of the Bt cotton project were able to inform the majority of farmers about their project through the use of an innovative multi-lingual, multimedia approach. Interviewees noted that, for those who are literate in languages other than French, the government translated the GM law into native languages as well. Similarly, information about biotechnology was made available in different media forms including newspaper, radio advertisements, television promotions and films.

4. Follow-up at the field level: researchers and farmers must engage in open and honest dialogue to maintain trust

In addition to disclosing information to civil society groups and the general public, it is essential for farmers and researchers to maintain an on-going dialogue in which questions and concerns about the product can be expressed. Adequate quality-assurance measures and customer service practices from seed providers are imperative to acquiring farmers' trust and ensuring their compliance to the best farming practices according to the needs of the new technology.

Maintaining cotton seed quality

A significant challenge to the project's success was posed by a problem regarding the seeds' physical quality. A technical issue of smaller seed and poor germination emerged and affected trust between the partners and the farmers. Some farmers rejected the Bt cotton seed solely because of this issue. The farmer stressed that this lack of trust did not come from a lack of trust in the GM crop but rather in the seeds themselves. This demonstrates a need for ongoing communication between farmers and other partners, particularly researchers and seed providers, as well as a need for enhanced customer-service provision. Farmers' concerns must be identified and addressed to ensure the maintenance of trust and the success of the project. Other farmers called for more follow-up from researchers and the seed providers at the field level in order to clarify questions they had. This practice, according to the farmers, was critical in building or undermining trust between them and the seed providers.

Potential for greater seed quality-assurance practices

A government researcher echoed the need for greater seed quality assurance practices. He stressed the importance of improving seed quality assurance processes to ensure farmers receive the best ones available. One farmer suggested importing the Indian approach to addressing issues pertaining to Bt cotton. He reported that India, faced with similar issues, had introduced an annual international conference convening Bt cotton researchers. Through this effort, they were able to create an appropriate venue in which the problems pertaining to their national Bt cotton production could be addressed and solved. The ability for national, commercial and research actors to recognize and remedy this issue will play a significant role in the continued adoption of Bt cotton in Burkina Faso.

Conclusions

Burkina Faso's progressive cotton selling system laid the foundation for the implementation of this project – particularly, the strong dialogue and the receptivity to collaboration. It further demonstrated the importance of capitalizing on existing institutional structures and relationships between industry, research and farmers. It is also important to recognize that researchers' roles are not limited to the lab. The challenge of establishing open, ongoing interactions between researchers and journalists must be addressed in order to disseminate accurate findings. Likewise, the generation of professional respect between Burkinabè and international researchers is essential for encouraging collaboration and information sharing, as well as assuring national and international confidence in Burkinabè research. Furthermore, the communication strategy of the Burkina effort attempted to reach as many people through as many means possible by including written, spoken and visual elements in several languages (including French and native Burkinabè languages), thereby broadening access to information as a means for building the foundations of trust. Although initial engagement with farmers and the public was noted as one of the strongest assets to the project, researchers' lack of follow-up on specific farmers' issues—seed germination problems, for example—led to broken trust with a number of farmers. Follow-up is therefore a critical element in seed adoption and the maintenance of trust among farmers, researchers and private companies. The lessons learned from this case study on Bt cotton in Burkina Faso can provide great insight to other agbiotech PPPs in sub-Saharan Africa. Interviewees reported that establishing and maintaining trust among partners, researchers and the community in Burkina Faso greatly contributed to the success of the PPP. By addressing challenges to building trust and engaging in trust-building practices early on, improvement in the effectiveness of agbiotech PPPs is likely.

Acknowledgements

The authors are grateful to each of the participants who contributed substantial time and effort to this study. The authors also thank Jocalyn Clark and Lauren Daley for comments on earlier drafts of the manuscript. This project was funded by the Bill & Melinda Gates Foundation and supported by the Sandra Rotman Centre, an academic centre at the University Health Network and University of Toronto. The findings and conclusions contained within are those of the authors and do not necessarily reflect official positions or policies of the foundation. This article has been published as part of *Agriculture & Food Security* Volume 1 Supplement 1, 2012: Fostering innovation through building trust: lessons from agricultural biotechnology partnerships in Africa. The full contents of the supplement are available online at http://www. agricultureandfoodsecurity.com/supplements/1/S1. Publication of this supplement was funded by the Sandra Rotman Centre at the University Health Network and the University of Toronto. The supplement was devised by the Sandra Rotman Centre.

Author details

[1]Sandra Rotman Centre, University Health Network and University of Toronto, Toronto, Ontario, Canada. [2]African Centre for Innovation and Leadership Development, Federal Capital Territory, Abuja, Nigeria. [3]Dalla Lana School of Public Health, University of Toronto, Toronto, Canada. [4]Grand Challenges Canada. [5]Dalla Lana School of Public Health and Department of Surgery, University of Toronto, Toronto, Canada.

Authors' contributions

Study conception and design: OCE. Data collection: OCE. Analysis and interpretation of data: OCE and KB. Draft of the manuscript: OCE and KB. Critical revision of the manuscript for important intellectual content: OCE and ASD. All authors read and approved the final manuscript.

Competing interests

The authors declare that they have no competing interests.

References

1. James C: **Global Status of Commercialized Biotech/GM Crops: 2010.** *ISAAA Brief* 2010, **Brief No. 42**:1-23.
2. U.S. Department of State: **Diplomacy in Action** [http://www.state.gov/r/pa/ei/bgn/2834.htm]. .
3. **Truth about Trade & Technology** [http://www.truthabouttrade.org/content/view/12259]. .
4. Birner R, Kone SA, Linacre N, Resnick D: **Biofortified Foods and Crops in West Africa: Mali and Burkina Faso.** *AgBioForum* 2007, **10(3)**:192-200.
5. Friedberg SE, Horowitz L: **Converging Networks and Clashing Stories: South Africa's Agricultural Biotechnology Debate.** *Africa Today* 2004, **51(1)**:3-25.
6. Stone GD: **Both Sides Now. Fallacies in the Genetic-Modification Wars, Implications for Developing Countries and Anthropological Perspectives.** *Current Anthropology* 2002, **43(4)**:611-630.
7. Birner R, Kone SA, Linacre N, Resnick D: **Biofortified Foods and Crops in West Africa: Mali and Burkina Faso.** *AgBioForum* 2007, **10(3)**:192.
8. Pollack A: **Monsanto buys Delta and Pine Land, Top Supplier of Cotton Seeds to the U.S.** *New York Times* 2006, Business.
9. James C: **Global Status of Commercialized Biotech/GM Crops: 2008.** *ISAAA Brief* 2008, **39**:1-243.
10. **International Human Development Indicators** [http://hdrstats.undp.org/en/indicators/101406.html]. .
11. Karembu M, Nguthi F, Ismail H: **Biotech Crops in Africa: The Final Frontier.** *ISAAA AfriCenter* 2009, 1-34.
12. Elbehri A, MacDonald S: **Estimating the Impact of Transgenic Bt Cotton on West and Central Africa: A General Equilibrium Approach.** *World Development* 2004, **32(12)**:2049-2064.
13. Cabanilla LS, Abdoulaye T, Sanders JH: **Economic cost of non-adoption of Bt-cotton in West Africa: with special reference to Mali.** *International Journal of Biotechnology* 2004, , **X**: 16.
14. Hema O, Somé HN, Traoré O, Greenplate J, Abdennadher M: **Efficacy of transgenic cotton plant containing the Cry1Ac and Cry 2Ab genes of Bacillus thuringiensis against Helicoverpa armigera and Syllepte derogata in cotton cultivation in Burkina Faso.** *Crop Protection* 2009, **28**:205.
15. **Convention on Biological Diversity** [http://www.cbd.int/biosafety/signinglist.shtml].
16. **SOFITEX** [http://www.diplomatie.gouv.fr/fr/IMG/pdf/Sofitex_Note_filiere_coton_Burkina.pdf].
17. Gupta A, Falkner R: **The Influence of the Cartagena Protocol in Biosafety: Comparing medico, China and South Africa.** *Global Environmental Politics* 2006, **6(4)**:23-55.
18. Paarlberg R: **GMO foods and crops: Africa's choice.** *New Biotechnology* **27(5)**:609-613.
19. Nubukpo K: **L'avenir des filieres cotonieres ouest africaines: quelles perspectives apres Cancun.** *Communication a la Commision Economique de la Francophonie, Paris* 2004, 1-18.
20. Navarro MJ: **Communicating Crop Biotechnology: Stories from Stakeholders.** *ISAAA Brief* 2009b, **40**:1-179.
21. Diallo L: **Analyse comparée des differentes politiques au Burkina Faso visant à differencier la qualité du coton pour mieux le valoriser sur le marché.** Montpellier, France; 2008.
22. Tao A: **Le Cotton bt a grande echelle en 2010.** *Le Faso* 2010.
23. **IFDC** [http://www.ifdc.org/Changing_Lives/Changing_Lives_Case_Studies/Biosafety_in_Burkina_Faso].
24. Devarakonda RK: **Cotton Dossier WIll Make or Break Doha Round.** *All Africa Press* 2010.

15

To label or not to label: balancing the risks, benefits and costs of mandatory labelling of GM food in Africa

Jessica Oh[1,3] and Obidimma C Ezezika[1,2,3,4*]

Abstract

There seems to be growing controversy among interest groups worldwide about whether genetically modified (GM) foods need to be labelled. There are also growing concerns, particularly among civil society groups, about the potential danger of GM foods, for which labels are being demanded. Particularly in Africa, the issue of labelling GM foods requires attention due to the rapid growth of agricultural biotechnology initiatives. Using Kenya as a case study, and based on interviews with key agricultural stakeholders and a review of the literature, we present five points to consider in discussions on how the need for mandatory GM labelling should be assessed. This framework encompasses, and is underpinned by, important considerations about ethics, consumer autonomy, costs, stigmatization, feasibility and food security as they pertain to agricultural biotechnology.

Keywords: agricultural biotechnology, genetically modified food, labelling, mandatory, Africa, Kenya

Introduction

Labelling of genetically modified (GM) foods has become an issue requiring attention in Africa due to the rapid growth of public-private partnerships operating agricultural biotechnology initiatives. Since partners must adhere to a given country's laws on biosafety and labelling, it is imperative that countries are also clear about the stipulations outlined in any GM labelling law as there is much concern among the private sector in Africa – be it warranted or not – that labels on GM products will raise fear and suspicion among the public and thereby stymie acceptance of agricultural biotechnology in the continent.

In our latest social audit[a] engagement with one such initiative, the Water Efficient Maize for Africa (WEMA) project, we asked key agricultural stakeholders[b] about any regulatory challenges facing the WEMA project. Among their responses, some stakeholders – particularly seed companies – voiced concern about the implications of the GM labelling regulations recently introduced in Kenya for

both the progress and future role of these initiatives. The WEMA project is a public-private partnership that involves funding partners such as the Bill & Melinda Gates and Howard G Buffett Foundations, as well as partner institutions such as the African Agricultural Technology Foundation, International Maize and Wheat Improvement Center, Monsanto, and national agricultural research systems in five African countries (Kenya, Mozambique, South Africa, Tanzania and Uganda). The goal of the partnership is to develop drought-tolerant and insect-protected maize and make these varieties available royalty-free to smallholder farmers in sub-Saharan Africa through seed companies.[c]

For the past decade, the introduction of GM ingredients in food products has been at the centre of highly controversial and polarizing debates in many countries. Because there is a wide gap between scientific judgment and public opinion on GM food, governments are confronted with the dilemma of how to regulate the marketing of GM food products. One option for regulating GM foods is to permit them but ensure segregation from their conventional counterparts, which implies creating two separate production tracks and introducing a labelling scheme to allow consumers to choose between GM and non-GM food products [1]. However, the choice of

* Correspondence: obidimma.ezezika@acild.org
[1]Sandra Rotman Centre, University Health Network and University of Toronto, Toronto, ON, Canada
[2]Dalla Lana School of Public Health, University of Toronto, Toronto, ON, Canada

the labelling scheme, mandatory or voluntary, is not uniform across countries. During the last decade, over 40 countries globally have required GM products to be labelled [2]. Some countries, such as the United States and Canada, have opted for a voluntary labelling scheme, while other countries such as European Union (EU) member states, Japan and Australia have opted for a mandatory labelling scheme, arguing that consumers have the right to know [3].

The issue of labelling GM foods has begun to gain salience in East Africa, specifically in Kenya, where GM crops are in the process of gaining approval for commercial production. We assess the cogency and practicality of the arguments for and against GM food labelling by using Kenya, which introduced mandatory GM labelling in May 2012, as a case study.

The debate on labelling of GM food products encompasses complex issues that are entangled in conflicting interests and beliefs about the production, consumption and implications of GM foods, as well as ethical concerns such as those over consumers' right to know what is in their food. We posit in this paper that the arguments underpinning the imperative to label GM foods need to be assessed within a framework in which ethics, consumer autonomy, costs, stigmatization, feasibility and food security are all considered. Based on the concerns about GM labelling raised by some stakeholders in our 2012 Social Audit, we seek to explore the logic behind these arguments to allow for an understanding of the implications of mandatory GM labelling in the context of Africa.

GM labelling regulations in Africa
Current status
To date, Burkina Faso, South Africa, Egypt and Sudan are the only African countries that have approved biotech crops for commercial production [4,5]. Nigeria, Kenya and Uganda currently have GM crops under confined field trials [4]. Of the aforementioned countries, only South Africa and Kenya have regulations for GM labelling in place.

In South Africa, the regulations governing the labelling of foods containing genetically modified organisms (GMOs) is outlined in Section 24 of the Consumer Protection Act, 2008 (No 68 of 2008) [6], which was signed into law on 24 April 2009 and came into effect on 31 March 2011 [7]. The Act sets out the minimum requirements to ensure adequate consumer protection and provides an overarching framework for all other laws that provide for consumer protection. Under the Act, Regulation 293 makes a provision for a 'may contain genetically modified ingredients' clause, which relieves companies from labelling foods if it is 'scientifically impractical or not feasible to test foods' for the presence of GMOs or GM ingredients [8]. On

9 October 2012, the Department of Trade and Industry published draft amendments to the Consumer Protection Act stipulating that all imported or locally produced food containing 5% or more GM components or ingredients must now be labelled as 'contains genetically modified ingredients or components' [9].

In 2009, Kenya passed the Biosafety Act, No 2 of 2009, which allows for commercialization of GM crops [10]. The Government of Kenya (GOK) first required the GM content of foods, feed or ingredients to be identified on product labels in 2010 when the Kenya Bureau of Standards published regulation KS 2225:2010 [11]. Most recently, the GOK, through Kenya Gazette Supplement No 48 of 2012,[d] introduced additional requirements on GM-containing foods – which are referred to as the 'Biosafety (Labeling) Regulations, 2012' [12]. In particular, the new regulations stipulate the following: 1) a reduction of the adventitious presence of unapproved events from 5 to 1%, 2) use of 'GMO-Free' labelling can only be used where the GM adventitious presence is below a 1% threshold, 3) labels should refer to the GM content in the same font size used for the other ingredients and trademarks, and 4) references to the CODEX Standard CAC/GL 76 2011 [11]. Violation of the newly gazetted regulations carries a fine of up to 20 million Kenyan shillings (approximately US$235,300) and/or imprisonment of up to ten years [12].

Reaction to the newly introduced GM labelling regulations in Kenya
Reactions to the new labelling regulations in Kenya are diverse, and have sparked intense public debate among pro- and anti-GM actors alike. We herein provide a general picture of what the public debate looks like and the various stakeholders who have voiced their views based on media reporting on the new labelling regulations in Kenya. Some in Kenya call the new regulations a 'victory' and 'consumer rights milestone' [13], as those in favour of the regulations usually assert that Kenyans have a 'right to know' what goes in their food and should be 'protect[ed]' [14].

On the other hand, millers are calling for an urgent review of the new regulation, which they claim is too prohibitive and likely to cause food shortages in the country. They argue that they will incur additional costs to meet the labelling requirements, which will have to be passed on to consumers, and that the new regulation will curtail free and fair trade in GM foods [15]. Kenyan scientists working in the field of biotechnology research have criticized the new regulation for lacking a scientific basis [16]. Experts and politicians have warned that such labelling may increase food prices, thereby hindering efforts to address food security needs in the country [15].

Balancing the interests at stake: five points to consider

Based on the reactions above, it is evident that the views of certain stakeholders are in conflict. Biased though these views may be, there will inevitably exist conflicting interests. For one, some consumers may continue to demand labels, and food companies – particularly millers – will continue to insist that labels will increase their costs. We outline the current debate and the conflicting – and often diametrically opposed – opinions and views on GM food labelling in the form of a number of points. We then explain how these competing interests can be woven into the discussion on GM food labelling regulations.

Right to know and consumer autonomy

Many have defined the rationale for GM labelling strictly as a 'right to know' issue – that consumers have a right to know what is in the foods they purchase. Embedded in the right-to-know argument is the notion of autonomy: GM labels allow consumers to exercise their freedom to choose (and avoid) which foods to purchase – be it GM, organic or conventional [17]. Many studies have unpacked GM labelling in terms of consumer autonomy [18-20]. Similarly, proponents of GM labels believe that labelling highlights the value of transparency by honouring all concerns about GM foods along a range of perspectives – from health or allergen to moral and religious [21]. Respecting consumers' autonomy of choice does not necessarily mean, however, that someone (e.g., producers, retailers or public authorities) is morally obliged to ensure that both GM and non-GM food products are available in the market [18,22].

On the other hand, this insistence on a right to know whether foods have been produced with GM ingredients seems to stem from the fear and uncertainty among the public about the unknown health and environmental effects of GM foods. According to proponents of mandatory GM labelling, labelling of new products – such as GM foods – that currently have no known risk of adverse effects to human health is a moral imperative from the viewpoint of public health ethics [21]. For example, anti-GM activists usually corroborate their claims with research that has shown that GM food is harmful to human health, such as the published study by French researchers showing evidence of cancer tumours in rats fed a diet of Monsanto's GM maize NK603 [23,24]. This study was immediately criticized by scientists, who mentioned a number of methodological problems – its small sample size, the chosen breed of rats (which are predisposed to developing tumours) and that the tumour rates did not increase in proportion to the dose of GMOs fed to the rats [25,26]. These findings also contradict those of other long-term studies in which GM foods were fed to a wide range of lab animals. Snell et al. [27] published a review of 24 similar long-term studies and concluded that those studies do not provide evidence of health hazards. In particular, when many in the scientific community and agencies such as the European Food Safety Authority have concluded that the study is of 'insufficient scientific quality to be considered as valid for risk assessment' [28], such studies should not serve as a valid guide to the safety of GM foods.

While consumer autonomy is the long-familiar rationale for GM labelling, some also argue that consumer choice and autonomy do not justify mandatory labelling of GM foods. Michael Reiss, for example, suggests that the principle of choice should be held both at the level of retailers and at the level of individual consumers: consumers who insist on knowing whether their food contains GMOs will seek out retailers that label and consumers who do not have a deep interest in knowing such information will not seek GM labels [29]. Similarly, Hansen argues that it is not obvious that autonomy by itself justifies mandatory labelling: 'Consumers have all the information they need if they simply assume that every product that is not specifically labelled "GM-Free" or "organic" may contain GM ingredients' [22].

Further, while GM labels have the potential to contribute positively to consumer education and awareness – as has been the case with nutrition labels – some argue that GM labels are only useful for the consumer who already reads the ingredients list. Gruere et al. state, 'Information concerning GMO content in a parenthesis in the ingredients list is not very likely to catch the eye of the average consumer' [3]. In South Africa, for example, in which a mandatory labelling law recently superseded the voluntary labelling scheme, and despite significant levels of GM food crop production in the country, the majority of South Africans are not aware of the existence of GM foods, nor are they aware that they are consuming GM foods [30]. Given studies such as this one, we suggest that, while GM labels can potentially increase consumer awareness about GMOs, improving consumers' knowledge and awareness of agricultural biotechnology can facilitate consumers' interpretation of GM labels, thereby maximizing the benefits – that is, choice and information – consumers obtain from labels.

Chris Macdonald deconstructs the right-to-know argument and points to the flip side by asking: are agri-food companies ethically obligated to provide labels? Macdonald, focusing on the Canadian context, argues that individual companies do not have an ethical obligation toward consumers to label their GM foods: 'although unilateral action in this regard might be admirable, an agri-food company has no ethical obligation to label its GM foods, given the current social, legal, scientific, and economic context' [31]. In other words, the lack of any scientific research showing that GM foods are harmful to human health makes moot the ethics argument: since

GM foods have thus far been proven to be safe, companies are not acting unethically if they fail to provide labels on their GM products.

Costs

A mandatory GM labelling law makes imperative a discussion about the costs of implementation. Labelling GM foods entails much more than the mere production of a sticker or label indicating the presence of GM ingredients; the different procedures that must be fulfilled at various stages of the entire food production chain potentially impose costs on farmers, traders, manufacturers, the government and, ultimately, consumers. These include the technicalities of grain production, handling, processing and storage. One study found that the bulk of costs incurred would be from measures of segregation and identity preservation to prevent or limit mixing within the non-GM supply chain [32]. In a study examining the economic effects of GMO regulations, Bullock and Desquilbet found that tolerance levels are a 'key element of costs of non-GMO segregation, and zero-tolerance levels may be impossible to obtain without major organizational and economic costs' [33]. Another study, however, shows that there is an insignificant difference in the number of products requiring GM labelling when comparing a threshold level of GM material of 1% and 5% [8]. The different results from these studies may be attributed to the difference in the countries examined.

Many, especially agri-food companies and food manufacturers, point out the significant costs involved in mandatorily labelling GM products. For example, a study conducted to capture the perspectives of major stakeholders in the Kenyan food industry concluded that, while most respondents deem important the traceability of GM products and ingredients, many do not support labelling of GM products because of the associated extra costs and the possibility of negative reactions from consumers [34]. While there have been studies done [5,35,36], though all *ex ante*, showing that mandatory labelling will create additional costs that will eventually be passed on to the consumer, actual increases in food prices due to mandatory GM labelling have not yet been reported. One study based on the current EU regulations in 2001 found that mandatory GM labelling results in an added per capita annual cost of approximately US$0.23 [37].

There is a need for more case-by-case assessments about the economic costs implied by mandatory GM labelling to obviate unsubstantiated claims verbalized by the food industry. Nevertheless, consumers should not accept, at face value, statements made by leaders in the food industry conveying that consumers will bear the costs of GM labels – which are often not impartial. The 'cost' argument seems to be overemphasized by the GM food industry to avoid the real fear of stigmatization – which we discuss in the next section. African governments can therefore benefit from credible assessments that allow them to determine the economic viability of regulating the mandate within their respective countries before introducing a labelling law, as well as to determine which stakeholders may actually be affected. More studies assessing the potential economic costs of mandatory labelling in Africa, however, are needed, since most of the cost experiments done pertaining to mandatory GM labelling are based on the experiences of countries outside of Africa.

Stigmatization

Opponents of mandatory GM labelling – which usually includes the agri-food industry – deem such labels as stigmatizing a product that does not deserve to be treated so. Moreover, our latest Social Audit of the WEMA project has revealed that the issue of labelling GM foods – particularly, the stigmatization of GM products and the potential impact on the cost of WEMA seeds – is a concern to stakeholders, particularly seed companies [38]. Regardless of whether or not these views are valid or authoritative, we included a discussion about stigmatization because it seems to underlie the biotech food industry's overall opposition to GM labels.

Stigma is defined as 'a mark placed on a person, place, technology, or product associated with a particular attribute that identifies it as different and deviant, flawed or undesirable', such that the perceived consequences of using a product or service 'exceed the direct physical harm to human beings and ecosystems to include more indirect effects on the economy, social institutions and well-being' [39]. Stigma is associated with negative attributions about the source of the mark, and there is indeed evidence that genetic modification is cognitively linked to negative objects and attributes – such as tampering and artificiality – and to concepts such as danger [40]. Baker and Burnham [41] suggest that mandatory labelling may raise concerns among consumers about GM foods and therefore stigmatize them, and raise the biotech food industry's fears about consumer rejection. This argument seems to suggest that ignorance of food ingredients is synonymous with acceptance of GM food products; however, it is not an unwarranted point given the widespread negative perceptions of GMOs.

As it stands, GM food labels have no relation to the safety of GMOs but are used to provide consumer choice. But given the current sociopolitical context in many African countries, labels on GM products at this time could give the impression of a possible danger, and, in turn, imply falsely that something is wrong with them. For example, some bring up the 'guilt by association' argument – that GM labels, particularly in the current climate, give the false impression that the food is less safe than conventional foods; conversely, a 'GMO-free' label may imply that such foods are safer or better than foods containing GMOs [42]. In such cases, the disclosing of GM ingredients through labelling may be more misleading than informative.

It has also been said that the stigmatizing of GM products that ensues from such impressions can significantly affect the efficiency of the market by reducing demand for GM products or inflating demand for their counterparts (conventional and organic) [39]. And there seems to be no significant difference in the reactions induced by a positive (e.g., 'contains GMOs') or negative (e.g., 'non-GMO') label – both have the potential to give rise to negative perceptions about GM foods [39]. One experimental study showed that consumers perceive GM labels as a negative signal, and that labels may therefore lower the expected market share of GM foods [43]. This is a problem because it implies that the labels on GM foods are not adequately and accurately informing consumers. Mandatorily labelling *all* food – to convey whether it is GM, organic or conventional (e.g., treated with pesticides) – may seem to be a logical solution to avoid stigmatizing only GM foods. Yet, in a context characterized by sensational reporting on GM foods and increasing anti-GM movements around the world, and in which it is increasingly recognized that 'organic' labels are used more for marketing purposes than to inform consumers, the foods labelled as 'GM' will be seen as least desirable; in this way, labels on GM foods can be considered stigmatizing.

While significant panels such as the World Health Organization, the International Council for Science and the Food and Agricultural Organization – to name a few – have reached the conclusion that GM foods do not pose risks to human health [44-46], media sensationalism and anti-GM activists continue to raise consumer fears by disregarding or downplaying the current scientific evidence on GM foods. Against this sort of backdrop, 'GM' labels can perpetuate and exacerbate fear of GM foods and engender more confusion. At the same time, when the biotech food industry amplifies the cost aspect of GM labels, they may be placing GMOs in a more undesirable light by adding to consumers' negative perceptions: not only may GM foods be unsafe for human consumption, but they may now be more expensive if they are labelled – the cost of which consumers will have to bear.

Feasibility

What is perhaps most germane in the African context is assessing the feasibility of labelling GM produce sold in informal markets such as open-air and roadside markets. These informal markets are crucial for household food security: they provide a direct source of income for the low-income vendors that run them; are an important outlet for small producers due to the less stringent requirements for product quality and packaging; and are an especially important outlet for poor consumers in both urban and rural areas because of familiarity, greater accessibility and ability to buy in smaller quantities [47]. It is unclear how the new regulations for GM labelling in Kenya will be enforced for the unpackaged produce sold in informal markets – though, there is mention that for products that are not pre-packaged, 'the words "genetically modified organisms" or "genetically modified (name of organism)" shall appear on, or in connection with, the display of the product' [12]. However, Kenya's new labelling regulations do not apply to foods sold by restaurants and food vendors. In Kenya's Biosafety (Labelling) Regulations, 2012, section 5d states that these regulations shall not apply to 'food intended for consumption prepared and sold from food premises and vendors' [12]. These regulations also state that the objectives thereof are a) to ensure that consumers are made aware that food, feed or a product is genetically modified so that they can make informed choices; and b) to facilitate the traceability of genetically modified organism products to assist in the implementation of appropriate risk management measures where necessary' [12]. It is unclear whether the exemption will apply to food vendors at open-air markets and roadside kiosks. If it does not, the implication is slightly concerning: the majority of the rural and urban poor in Africa rely on informal retail markets to obtain staple products and fresh produce, which means many people would not have access to GM labels and are thus left uninformed about whether their food is genetically modified.

Impact on food security and innovation

Studies on whether GM labels implemented in Kenya will increase the cost of food have yet to be done. Should GM labels increase costs, however, access to food will be negatively affected. Like the EU member states, many African governments have chosen to follow the precautionary approach toward regulating GM foods and crops, for reasons ranging from cultural ties and agricultural trade relations to the amount of bilateral foreign assistance Africa receives from Europe [48]. But the circumstances of Africa are very different than those of Europe. In Africa, the percentage of the population that might benefit directly from GM crops is much higher than in Europe, because 60% or more of all Africans are still farmers who depend directly on agriculture for income and subsistence, while farmers in Europe are highly productive even without using GM crops [48].

Innovations such as crop biotechnology offer one potential solution to combating hunger and malnutrition in developing regions. Without innovation in agriculture, the Green Revolution, for example, would never have achieved the remarkable success it did in Asia in contributing to a substantial increase in food production and reducing poverty. The private sector, through public-private partnerships, is already investing heavily in the research and development of biotech crops that will improve agricultural productivity in Africa. For example, the WEMA project has been firmly established in Kenya, South Africa and Uganda – among five participating countries – and is close to achieving its goal of providing small-scale farmers with drought-tolerant maize varieties. Maize is a staple food for over 300 million people in sub-Saharan Africa, many of whom are themselves growers of maize [48]. In Kenya, maize is important for both food security and income generation for almost 90% of the rural population [49], and drought-tolerant varieties are one response to population growth, unreliable rainfall and climate change – factors expected to increase drought risks to maize growers in Africa in the future [48].

Discussion

The points we outline in the paper are not to be understood as comprising an argument against mandatory labelling of GM foods or in support of GM crop development. Instead, our aim is to highlight, in light of the concerns raised by some stakeholders in our latest Social Audit of the WEMA project, the important task facing African governments of carefully weighing all the issues at stake in the debate over labelling GM foods. In particular, governments need to find a way to resolve the perceived conflict between their introduction of mandatory labelling laws – which seemingly hinder GM crop production – and their ambitious food security agendas that stress the need to invest in agricultural biotechnology.

In North America, widespread use and consumption of GM foods occurred before any debate about GM labelling even began. Consumers in African countries such as Kenya, however, have a chance to take part in, and shape, the debate on GM labelling as the introduction of GM foods is a relatively new phenomenon in Africa. That African governments seem inclined to take the precautionary approach to introducing GM foods and mandating labels is a rational move amidst the uncertainty surrounding the public health impact of GM foods and for safeguarding their agricultural exports to European markets. However, agricultural stakeholders for the most part interpret labelling regulations as implicit opposition by some key decision-makers within government to GM food production [50], which contributes to misunderstandings, accusations and perceived stigmatization. While the GOK's introduction of GM labelling regulations is commendable, it is equally important that the government consults various agricultural stakeholders so that the latter know whether, and how, they will be affected.

Many of the arguments put forth in favour of and against mandatory GM labelling appear to be *ad hominem*. The pro-labelling/anti-GMO group overemphasizes the 'right to know' aspect of GM labelling, perhaps to avoid supplying hard evidence that supports a 'real need' for labelling. The biotech industry, on the other hand, overemphasizes the cost aspect of GM labels, perhaps to avoid the real fear of stigmatization and, in turn, consumer rejection of GM food products. We believe it is beneficial for governments to consider the five points introduced in this paper to obviate perceptions of discriminating against or benefiting a particular group of stakeholders, which, we hope, will contribute to minimizing the spread of misinformation concerning agricultural biotechnology.

Endnotes

[a]Social auditing is a process whereby an audit team collects, analyses and interprets descriptive, quantitative and qualitative information from stakeholders to produce an account of a project's ethical, social, cultural and commercialization performance and impact.

[b]'Key agricultural stakeholders' here refer to stakeholders of the WEMA project whom we interviewed to produce the annual Social Audit reports from 2009 to 2012. For each report, the viewpoints of 100 people from across the five WEMA countries (Kenya, South Africa, Mozambique, Tanzania and Uganda) were collected using a quantitative questionnaire and a semi-structured interview guide. The people interviewed fell under the following stakeholder groups: technical resource and consultant; academics and scientists; legal; agricultural extension services; agricultural commercial enterprises; farmers' associations; technology funders; science and technology government departments; national agricultural research systems; regional organizations working with small-scale farmers; public/NGOs for public concerns; media; regional national authorities; project regional personnel and seed companies.

[c]To learn more about the WEMA project, see [51].

[d]Full title of the gazette: Kenya Gazette Supplement No 48 of 2012, Legal Notice No 40 (dated 25 May 2012).

Abbreviations

GM: genetically modified; GMO: genetically modified organism; GOK: Government of Kenya; WEMA: Water Efficient Maize for Africa.

Competing interests

The Sandra Rotman Centre, through the Ethical, Social, Cultural, and Commercialization (ESC²) Program, received funding from the Bill & Melinda Gates Foundation to evaluate (i.e., conduct a social audit on) the WEMA project mentioned in the paper.

Authors' contribution

OCE was responsible for the conception and design of the study, and for data collection. JO was responsible for the literature review and writing up of the manuscript. JO and OCE participated in analysis and interpretation of results. Both authors read and approved the final manuscript.

Acknowledgements

This study was funded by the Bill & Melinda Gates Foundation and supported by the Sandra Rotman Centre, an academic centre at the

University Health Network and University of Toronto. The findings and conclusions contained within are those of the authors and do not necessarily reflect official positions or policies of the foundation. We would like to thank Jerome Singh for very helpful comments and suggestions on earlier drafts of the manuscript. We are also grateful to the two anonymous reviewers whose comments and suggestions helped improve and clarify this manuscript.

Author details
[1]Sandra Rotman Centre, University Health Network and University of Toronto, Toronto, ON, Canada. [2]Dalla Lana School of Public Health, University of Toronto, Toronto, ON, Canada. [3]African Center for Innovation and Leadership Development, Federal Capital Territory, Abuja, Nigeria. [4]National Biotechnology Development Agency (Federal Ministry of Science and Technology), Umar Musa Yar' Adua Way/Airport Road, Lugbe Area, Abuja, Nigeria.

References

1. Dannenberg A, Scatasta S, Sturm B: **Mandatory versus voluntary labelling of genetically modified food: evidence from an economic experiment.** *Agric Econ* 2011, **42**:373–386.

2. Gruere G, Rao S: **A review of international labeling policies of genetically modified food to evaluate India's proposed rule.** *AgBioforum* 2007, **10**(1):51–64.

3. Gruere G, Carter C, Farzin Y: **What labelling policy for consumer choice? The case of genetically modified food in Canada and Europe.** *Can J Econ* 2008, **41**(4):1472–1497.

4. James C: *Global Status of Commercialized Biotech/GM Crops: 2011.* Ithaca, NY: ISAAA; 2011. Brief 43.

5. James C: *Global Status of Commercialized Biotech/GM Crops: 2012.* Ithaca, NY: ISAAA; 2011. Brief 44.

6. Republic of South Africa: **No. 68 of 2008: Consumer Protection Act, 2008.** *Govern Gazette* 2009, **526**(32186):1–186. Date published: 9 April 2009.

7. The South African Institute of Chartered Accountants: *Consumer Protection Act, No 68 of 2008.* Last Updated 2 September 2013.

8. Viljoen CD, Marx GM: **The implications for mandatory GM labelling under the Consumer Protection Act in South Africa.** *Food Control* 2013, **31**:387–391.

9. Republic of South Africa: **Notice 824 of 2012: Draft amendment regulations on Consumer Protection Act Regulations, 2011.** *Government Gazette* 2012, **568**(No. 35776):1–8. Date published: 9 October 2012.

10. Republic of Kenya: *The Biosafety Act, 2009*, Kenya Gazette Supplement. Nairobi, Kenya: Government Printer; 2009.

11. USDA Foreign Agricultural Service: *Kenya Agricultural Biotechnology Report*, USDA FAS GAIN. Washington, DC: USDA FAS; 2012. Gain Report.

12. Republic of Kenya: **Kenya Gazette Supplement No 48 of 2012.** *Govern Gazette* 2012, Legal Notice No. 40.

13. Mandatory GM labelling victory in Kenya: **Mandatory GM labelling victory in Kenya.** *Consum Int* 2012, 2 July 2012.

14. Were E: **Label GMO foods activists demand.** *Star* 2012, 17 October 2012.

15. Gichana A: **GMO labels will raise food prices, experts say.** *Star* 2012, 21 August 2012.

16. Achia G: **Kenya scientists criticise regulation on GMO labeling.** *Afr STI News* 2012, 2 July 2012.

17. Rubel A, Streiffer R: **Respecting the autonomy of European and American Consumers: defending positive labels on GM foods.** *J Agric Environ Ethics* 2005, **18**:75–84.

18. Siipi H, Uusitalo S: **Consumer autonomy and availability of genetically modified food.** *J Agric Environ Ethics* 2011, **24**:147–163.

19. Carter C, Gruère G: **Mandatory labeling of genetically modified foods: does it really provide consumer choice?** *AgBioforum* 2003, **6**(1–2):68–70.

20. Klintman M: **The genetically modified (GM) food labelling controversy: ideological and epistemic crossovers.** *Soc Stud Sci* 2002, **32**(1):71–91.

21. Lappe M: **Labeling should be mandatory.** *Nat Biotechnol* 2002, **20**:1081–1082.

22. Hansen K: **Does autonomy count in favor of labeling genetically modified food?** *J Agric Environ Ethics* 2004, **17**(1):67–76.

23. Maina A: **Kenyans have a right to know what they eat.** *Daily Nation* 2012, 14 October 2012.

24. Séralini G, Clair E, Mesnage R, Gress S, Defarge N, Malatesta M, Hennequin D, de Vendômois J: **Long term toxicity of a Roundup herbicide and a Roundup-tolerant genetically modified maize.** *Food Chem Toxicol* 2012, **50**(11):4221–4231.

25. Mestel R: *Study points to health problems with genetically modified foods.* Los Angeles Times; 2012. 20 September 2012.

26. Pollack A: *Foes of modified corn find support in a study.* New York Times; 2012. 19 September 2012.

27. Snell C, Bernheim A, Bergé J, Kuntz M, Pascal G, Paris A, Ricroch AE: **Assessment of the health impact of GM plant diets in long-term and multigenerational animal feeding trials: a literature review.** *Food Chem Toxicol* 2012, **50**(3–4):1134–1148.

28. European Food Safety Authority (EFSA). http://www.efsa.europa.eu/en/press/news/121004.htm.

29. Reiss M: **Commentary: labeling GM foods – the ethical way forward.** *Nat Biotechnol* 2002, **20**:868.

30. Botha G, Viljoen C: **South Africa: a case study for voluntary GM labelling.** *Food Chem* 2009, **112**:1060–1064.

31. MacDonald C, Whellams M: **Corporate decisions about labelling genetically modified foods.** *J Bus Ethics* 2007, **75**:181–189.

32. Carter C, Gruere G, McLaughlin P, MacLachlan M: **California's proposition 37: effects of mandatory labeling of GM food.** *Agric Resour Econ* 2012, **15**(6):3–8.

33. Bullock D, Desquilbet M: **The economics of non-GMO segregation and identity preservation.** *Food Policy* 2002, **27**:81–99.

34. Bett C, Ouma J, De Groote H: **Perspectives of gatekeepers in the Kenyan food industry towards genetically modified food.** *Food Policy* 2010, **35**:332–340.

35. De Leon A, Manalo A, Guilatco FC: **The cost implications of GM food labeling in the Philippines.** *Crop Biotech Brief* 2004, **IV**(No. 2).

36. Jaeger WL: *Economics Issues and Oregon Ballot Measure 27: Labeling of Genetically Modified Foods (EM 8817).* Corvallis, OR: Oregon State University; 2002.

37. National Economic Research Associates: *Economic Appraisal of Option for Extension of Legislation on GM Labeling, A Final Report for the Foods Standards Agency of the United Kingdom.* London, UK: NERA; 2001.

38. Social Ethical Cultural and Commercialization Program: *WEMA 2011 Social Audit Report: Ethical, Social, Cultural, and Commercialization (ESC2) Audit Report for the Water Efficient Maize for Africa (WEMA) Project 2011.* Toronto, Canada: Sandra Rotman Centre; 2012.

39. Ellen P, Bone P: **Strained by the label? Stigma and the case of genetically modified foods.** *Am Market Assoc* 2008, **27**(1):69–82.

40. Verdurme A, Viaene J: **Exploring and modeling consumer attitudes towards genetically-modified food.** *Qual Mark Res: Int J* 2003, **6**(2):95–110.

41. Baker G, Burnham T: **The market for genetically modified foods: consumer characteristics and policy implications.** *Int Food Agribus Manag Rev* 2002, **4**(4):351–360.

42. Eldred JS: **Analysis & perspective: labeling GMO-derived food ingredients: a recipe for misinformation.** *World Food Regul Rev* 1999, **14**:16.

43. Tegene A, Huffman W, Rousu M, Shogren J: *The Effects of Information on Consumer Demand for Biotech Foods: Evidence from Experimental Auctions.* Washington, DC: United States Department of Agriculture Economic Research Service; 2003. Technical Bulletin No. 1903.

44. Food and Agriculture Organization of the United Nations: *The State of Food and Agriculture 2003–2004: Agricultural Biotechnology Meeting the needs of the poor?.* Rome, Italy: FAO; 2004.

45. Mantell K: **WHO urges Africa to accept GM food aid.** *Sci Dev Net* 2002, 30 August 2002.

46. Persley G: *New Genetics, Food and Agriculture: Scientific Discoveries – Societal Dilemmas.* Paris, France: International Council for Science (ICSU); 2003:1–56.

47. McCullough EB, Pingali PL, Stamoulis KG: *The Transformation of Agri-Food Systems: Globalization.* Supply Chains and Smallholder Farmers: Routledge; 2008.

48. Paarlberg R: **GMO foods and crops: Africa's choice.** *New Biotechnol* 2010, **27**(5):609–613.

16

Building trust in biotechnology crops in light of the Arab Spring

Obidimma C Ezezika[1,2,3*], Abdallah S Daar[1,4,5]

Abstract

Background: The case of *Bacillus thuringiensis* (Bt) maize in Egypt presents a unique perspective on the role of trust in agricultural biotechnology (agbiotech) public-private partnerships (PPPs). This is especially relevant given the recent pro-democracy uprisings that spread throughout the Arab world that have significantly impacted the current political climate and status of both the public and private sector, and especially public-private collaborative initiatives. This case study aims to shed light on various trust-building practices adopted, and trust-related challenges faced, in the Bt maize project in Egypt.

Methods: We reviewed published materials on Bt maize in Egypt and collected data through direct observations and semi-structured, face-to-face interviews with stakeholders of the Bt maize project in Egypt. Data from the interviews were analyzed based on emergent themes to create a comprehensive narrative on how trust is understood and built among the partners and with the community.

Results: We have distilled five key lessons from this case study. First, it is important to have transparent interactions and clearly defined project priorities, roles and responsibilities among core partners. Second, partners need to engage farmers by using proven-effective, hands-on approaches as a means for farmers to build trust in the technology. Third, positive interactions with the technology are important; increased yields and secure income attributable to the seed will facilitate trust. Fourth, there is a need for improved communication strategies and appropriate media response to obviate unwarranted public perceptions of the project. Finally, the political context cannot be ignored; there is a need to establish trust in both the public and private sector as a means to secure the future of agbiotech PPPs in Egypt.

Conclusions: Most important to the case of Egypt is the effect of the current political climate on project success. There is reason to believe that the current political situation will dictate the ability of public institutions and private corporations to engage in trusting partnerships.

Background

Maize in Egypt

Considered as one of the principal crops in Egypt, maize is planted on approximately 728, 000 hectares of land, 75, 000 hectares of which is devoted to yellow maize while the remainder is designated to white maize [1]. Each year, 6.1 million tonnes of maize is produced domestically in Egypt. Moreover, 4.1 million tonnes of yellow maize is imported annually, valued at $US 1.3 billion [1].

* Correspondence: obidimma.ezezika@srcglobal.org
[1]Sandra Rotman Centre, University Health Network and University of Toronto, Toronto, Ontario, Canada
Full list of author information is available at the end of the article

Ajeeb-YG, the *Bacillus thuringiensis* (Bt) maize variety currently found in Egypt, was developed as a cross between MON810, a variety of genetically modified (GM) maize developed by Monsanto Company, and Ajeeb, a local Egyptian maize variety, by scientists working for the multinational agricultural biotechnology company Monsanto in South Africa [2,3]. Ajeeb-YG has been tested in Egypt since 2002 [4]. It is resistant to the three maize borers that pose a significant threat to conventional Egyptian maize seed varieties and has been shown to increase yield by up to 30% over conventional varieties when tested in field trials [5,6]. Ajeeb-YG and MON 810 thus seem able to provide significant benefits to farmers, consumers and the environment. In addition to a higher yield, use of the Bt technology also lowers use of insecticides; reduces potential exposure to insecticide; improves stalk lodging resistance; improves grain quality; lowers levels of mycotoxins; increases numbers of beneficial insects relative to insecticide-treated fields; and provides for greater flexibility in planting time. For Egyptian farmers, such benefits seem to render the switch to use Bt maize an appealing option [4].

In 1999, Monsanto initiated a joint project with the private Egyptian company, Fine Seeds International, for the development, commercialization and distribution of Bt maize in Egypt (see Additional file 1 for a list of additional project partners). From 2005 to 2008, the National Biosafety Committee (NBC) led the risk-assessment and testing process of Ajeeb-YG [1,7]. In 2008, Ajeeb-YG was approved for commercial use, making Egypt the first country in the Arab world to commercialize a biotech crop [2]. As of 2010, Egypt planted 2,000 hectares of Bt maize– an increase from 700 hectares in 2008 [8]. (see Additional file 2 for more details on the development of the Bt maize project).

The issue of trust
However, the development of the Bt maize variety in Egypt presented a number of challenges, primary among them being related to trust. Trust is central to agricultural biotechnology projects driven by public-private partnerships (PPPs) but is often difficult to earn due to the controversy over transgenic crops and public skepticism of multinational companies involved in such initiatives.

This case study focuses on the issue of trust in the conception, management, and development of Bt maize in Egypt. We believe that exploring trust in the context of agricultural development projects fueled by public-private collaboration is a valuable pursuit because it can provide insight to funders, researchers, farmers, and other stakeholders on how trust building can help contribute to success in future agbiotech endeavors.

Methods
Data was collected by conducting a literature review of academic articles, news articles and publicly available project documents of the Bt maize project in Egypt; using direct observation; and conducting semi-structured, face-to-face interviews with 13 stakeholders associated with the Bt maize Egypt project. These stakeholders represented participants from Fine Seeds International; Egypt Biotechnology Information Center; Monsanto; Central Administration for Seeds Application; Agriculture Genetic Engineering Research Institute (AGERI); Agricultural Research Center, Egypt; Agrifoods; and the Faculty of Agriculture (Saba Basha) at Alexandria University, Alexandria. The interviewees also included Bt maize farmers.

Interviewees were identified first by making a list of key individuals associated with the project based on the stakeholders identified within our case study research protocol.

This list was then populated further through snowball sampling by engaging with partners involved in the project and stakeholder informants who were familiar with the Bt maize project in Egypt through the Sandra Rotman Centre's Social Audit Project [9]. Potential interviewees were sent an invitation, which included an explanation of the case study series, to participate in the interview. Those who consented to participate were informed that the interview would be recorded, transcribed verbatim, and then analyzed.

The interview guide included questions on the interviewees' background, their understanding of the project, and their interpretation of the word *trust*. The interview explored perceptions of trust within the partnership and the public, apparent challenges to trust, and observed trust-building practices among the project partners and between the Bt maize project and the public. Finally, interviewees were asked to provide suggestions on how to improve agbiotech PPPs (see Additional file 3 for a list of sample interview questions).

The data from the interviews were analyzed by reading through the transcripts several times, identifying emerging trends and organizing them into major themes in order to create a comprehensive narrative on how trust is understood and built within the project and between the project and the community that it aims to serve.

We received Research Ethics Board (REB) approval for conducting the case study from the University Health Network (UHN), University of Toronto before proceeding with the study. Signed consent was obtained from each participant after providing information on the purpose of the study and stating that the interviews would be recorded.

Results and discussion
Stakeholders' understanding of trust
With the intention of soliciting authentic and holistic perceptions of trust from our interviewees, participants were led into discussing the issue of trust in the context of agbiotech PPPs after an initial question exploring their general understanding of the word *trust*. Overall, trust was predominantly described by participants as essential to the development of meaningful partnerships, a cornerstone of project success, and something that is built over time. Core elements of trust, as explained by interviewees, include transparency; open and honest communication; and respectful interactions among all project partners. Other key components of trust include building a common understanding regarding project goals and intended outcomes, as well as maintaining accountability, of which a central component is the establishment of clearly defined roles and responsibilities.

Based on the results of this study, we have derived five key lessons, from which partners in other agbiotech

PPPs can learn and use as a guide for building and fostering trust.

1. Transparent processes are central to effective project partnerships

The need for transparent interactions to secure trust among members of the partnership was the most dominant theme, particularly with regards to the partnership between Monsanto and Fine Seeds International, the two core private partners in the Bt maize project in Egypt. All participants acknowledged the importance of transparency in some capacity, and recognized that organizations and partners must be open and honest in their interactions if they are to create an environment conducive to building trust. An interviewee involved in raising public awareness about the application of biotechnology stated: *Transparency is most important for this technology, because it's a lot of debates about this technology. So if you're not transparent you will affect others.*

Most participants understood transparency to entail the disclosure and discussion of both the positive and negative aspects of the project. Furthermore, central to transparent processes is the establishment of clearly defined roles, responsibilities and priorities that are recognized and understood by all key players. When each actor has a solid understanding of their roles and responsibilities on an individual level, but especially on an institutional level, it works to enhance transparency and facilitate accountability among partners, both of which were described by participants as key elements of trust.

The particular partnership between Monsanto and Fine Seeds demonstrates the importance of transparency and having clearly defined priorities that are understood by all involved in the initiative. Fine Seeds initiated the interaction with Monsanto by approaching the multinational company to discuss herbicides and the potential for applying Roundup Ready, a top-selling herbicide made by Monsanto, to crops in Egypt. Monsanto initially refused the request, while placing pressure on Fine Seeds to engage in a biotechnology projector the development, commercialization and distribution of Bt maize instead. Despite initial conflict in the ideas and vision for Monsanto's role in Egyptian agriculture, Fine Seeds eventually agreed to take on the Bt maize project in Egypt in collaboration with Monsanto.

A unique feature of this particular case study is that only two core partners were responsible for bringing Bt maize to the commercial stages in Egypt. This contrasts with the other cases in our case study series [on trust in agbiotech PPPs] in which many key players contributed to the development and commercialization stages of those projects. While Monsanto, the owner and developer of the technology, was the larger of the two companies with substantial decision-making power regarding the future of biotechnology in Egypt, Fine Seeds was almost exclusively responsible

for mobilizing the project on Egyptian soil. It was Fine Seeds' duty to complete all regulatory processes and procedures leading to the commercialization and subsequent distribution of Bt maize in Egypt. All the same, Monsanto was largely perceived to be the driving force behind Bt maize in Egypt, despite Fine Seeds being the vehicle that mobilized the project from development to distribution. This was cause for some tension within the partnership. A representative from Fine Seeds described the relationship between Fine Seeds and Monsanto as being negatively affected by power imbalances and a lack of shared decision-making processes: *They [Monsanto] are our masters. That is talking in the old, before the revolution mentality. There is the big dictator and the small dictator sitting and meets all of us here. We [Fine Seeds] are the servants; we are the slaves in the private sector or whatever. And when we need anything to move, they need to put their say. So we go and beg and then they come down to our level and they try to listen to us and sometimes they say 'yes.'*

Additional challenges faced in the partnership between Fine Seeds and Monsanto are tied to the reality that decision-making processes in large multinational companies take time. Fine Seeds has expressed their frustration with the timely and costly process through which Monsanto reached an agreement as to how they should proceed with regard to biotechnology in Egypt. It seems that there is a lack of transparency and open and honest communication between the Cairo-based company Fine Seeds International and the multinational company Monsanto. At the time of our study, a participant suggested that Fine Seeds had been relatively uninformed as to where Monsanto's priorities lie regarding the future of Bt maize in Egypt: *In a huge multinational, decision making is very slow, very complicated... Very infrequently do they reach a beneficial decision after a lot of money has been spent on talking. I think that their problem is that they haven't really made up their mind in a business manner on do they or do they not want to be in Egypt. Are they or are they not interested in Egypt? Do they or do they not trust Egypt?*

Our study suggested that there is a clear gap between the visions of each of the core partners driving the Bt maize project in Egypt. There is also a need for respective project goals and priorities to be clearly expressed and shared among partners.

2. Farmer engagement is central to project success: the importance of raising awareness and opening the lines of communication

Closely related to the issue of transparency is the need for improving farmer engagement and raising awareness of the new biotechnology in question and the project in general. Genuine awareness can only be achieved if the project proceeds in a transparent manner, especially within and among higher-level institutions, such as public organizations and private corporations.

Engaging farmers in the research and testing process was also regarded by study participants as an essential tool, in line with the "Seeing-is-Believing" approach to building trust, which involves taking farmers and other community members to demonstration fields to show the increased yield and insect resistance properties of Bt maize over the conventional varieties.

A frequently acknowledged challenge to building trust within the Bt maize project in Egypt was the existing public fear of the new technology; this fear, however, was predominantly due to a general lack of awareness [3,10]. Several interviewees agreed that, when addressing the challenge of limited public awareness and farmer engagement, it is important to first validate, and not simply dismiss, farmers' concerns. Interviewees indicated that acknowledging the concerns of individuals who are skeptical of Bt maize in Egypt, and then alleviating these concerns using evidence-based practices, will increase the potential to build trust between the project and communities, thus leading to broader acceptance of the new technology. This can only be done if both the advantages and disadvantages of the new technology are discussed openly and honestly among all partners involved, especially with the farmers implementing Bt maize into their current farming practice. On this issue, an interviewee commented: *Because you are talking about biotech crops, then a big big part of it is how to manage it. You have to tell them exactly how to use it. You have to show them how to use it carefully so that you don't lose the benefits and that sort of thing. So here, the communication part is very important to building trust.*

We believe that allowing the farmers to gain first-hand experience with the technology is an ideal way to gain the farmers' trust since they are able to see the benefits and risks of the seeds for themselves. For example, field days and extension services, which are intended to engage and inform farmers, were recognized by interviewees as effective tools for building trust in the Bt maize project in Egypt. One representative from the Agricultural Research Centre commented: *So they see with their eyes. As they say, seeing is believing, so this is really where you convince the people. We had a hard time when we started the cotton project. Now the breeders have it... they have been exposed to the potential of the technology. They realize it now so they trust that it is working.*

3. Positive experience with the technology leads to trust: if the seed works, money comes, and trust will be built

It became apparent in our analysis of the data that the quality of the seed, considered in terms of its capacity to improve yields, is a fundamental basis for establishing trust between the farmers and the project. If the seed works to provide financial gain for the farmer, then little else is of concern. One Egyptian farmer who has implemented Bt maize into his farming practice described the adoption of Bt maize as being purely an effective, results-oriented business decision for him: *I know that it [the seed] is good, so that's what I use. I am a business man and I work for Fine Seeds and it gives me profit. If I saw any other company other than Fine Seeds that makes more profit, I would go to them.* Similar to the experience of other farmers, his trust in the new product has very little to do with the values and motives of the seed companies involved and everything to do with the ability of the seed technology to deliver on its promised benefits to farmers. In other words, good outcomes attributed to the product itself works to build trust between the community, namely farmers, and the technology. This supports the notion that positive first-hand experience with the seeds creates the foundation on which project partners and farmers can establish trust in the technology and with each other. Therefore, even more important than the motives and ideologies of corporations or organizations involved in the development of the seeds is the *ability* of the seed to deliver in terms of technological capacity.

4. Tainted public perception limits the project's progress: a need for improved communication and appropriate media response

In line with the need to improve public awareness and farmer engagement is a call for effective means of disseminating information that is accessible to a diverse population. Central to this issue is the media and the various tools used to the engage the public. One interviewee from the Egyptian Biotechnology Information Center (EBIC) explained the different methods of communication and forms of media used to engage the public and build awareness of the Bt maize project in Egypt: *We have developed many Arabic materials starting with newsletters– written newsletters with very small messages to the farmers. It can be read by the farmers and all kinds of public. We can also make brochures. I have also developed two books in Arabic. This is the kind of materials distributed.*

Despite these efforts to communicate information related to the Bt maize project in appropriate and creative ways, a common perceived challenge to increasing public awareness of biotechnology in Egypt was the media's delivery of inaccurate information to the public. While the dissemination of information through the media was acknowledged by participants as a predominant way of raising public awareness of the project, media coverage of Bt maize was also described as a potential way to erode trust if sources of information are unreliable and not backed by scientific evidence. This problem, combined with the fact that the media has overwhelming influence over public perception, presents an important challenge to stakeholders in the Bt maize project in Egypt.

In response to this challenge, participants discussed the importance of ensuring that the media has access to appropriate information and remains fully informed of the scientific evidence supporting GM crops. Engaging the media in various "Seeing-is-Believing" initiatives, such as field days and workshops similar to those used to engage farmers, helped to ensure that the dissemination of accurate information regarding agricultural biotechnology reaches communities. On this, a scientist with one of the research institutes in Egypt commented: *We have told them [the media] exactly what we do in order to assess the acceptability and risks. And it was very good. They [the media] have embraced the technology itself.*

Participants conveyed that effective engagement of the media will allow reliable information regarding agbiotech projects like Bt maize in Egypt to reach the public, thereby heightening awareness, improving acceptance, and establishing trust in the project. A scientist in particular stated: *I think the most difficult one is the media. And the media play the important role in launching the biotechnology in any country, because the people in media are not specialists in biotechnology. They will hear from you and they will also hear from others. So you have to deal with them honestly and inform them step by step by any development and by knowledge. When they are informed, they can write well and they can discover right and wrong.*

Furthermore, some interviewees recognized that changing public perception involves communicating information in a way that is easily understood. When sharing knowledge, one must give heed to the pre-existing knowledge and level of education of a given audience. One interviewee expressed the importance of framing the concept of genetically modified organisms (GMOs) and risk assessment of biotechnology initiatives in a more positive light by using terms that do not carry negative connotations. An example of this would be using the term "safety assessment" rather than "risk assessment" to describe certain regulatory procedures. An interviewee describes the importance of this approach in building trust: *There are a lot of things done for safety. I would rather talk about safety assessment because when you talk about risk, people think there is a risk and that sort of thing. That these things go through a rigorous procedure before they do come out [will] build this trust, as you say, to the public.*

Such choice of words (i.e., using *safety* over *risk*) decreases susceptibility to further skepticism, which may build trust and, in turn, facilitate public acceptance of the technology. Moreover, establishing partnerships can help uplift the reputation of one company through their association with another. One interviewee, for example, reported that Fine Seeds' ability to maintain a solid reputation in Egypt has improved the public's perception of Monsanto and overall approval of the partnership and project. The public—more specifically, the farmers—have placed their trust in Fine Seeds and are therefore able to

remain confident in Fine Seeds' partnership with Monsanto, despite initial hesitation to accept Monsanto as a trustworthy corporation.

5. The political climate cannot be ignored: where do public-private partnerships stand in the context of the new Egypt?

Perhaps the most important issue, unique to the Egyptian case and highly relevant to trust building in PPPs, is the recent uprisings that spread throughout the Arab world–the revolutionary wave of which is referred to as the "Arab Spring"– just after which this study was conducted (in the summer of 2011). Given the events that have transpired in Egypt, people are still adjusting to the new political climate, especially since the government plays an important role in facilitating partnerships [11].

A major challenge to the project, as indicated by participants, was the public image of the multinational seed company Monsanto and its largely contentious reputation preceding its engagement in Egyptian agriculture. The public's tainted perception of Monsanto is similar to their view of many private companies and businesses. In Egypt, the private sector is viewed by the public as corrupt, primarily due to the inner workings of the former Mubarak regime. One interviewee commented that the former Egyptian government would favor a few select businessmen, who later became business tycoons by way of government favoritism: *Those people who accepted such givings cannot be trusted by the people. And so even us little businessmen who cannot be discovered as corrupt or may think that we are honest, we also suffer. And this is why we don't even say the word 'business.' We don't like to be called 'businessmen' because it has a bad connotation.*

Evidently, the level of corruption in the government and a lack of transparency in Egypt's old regime lead to a lack of trust in the private sector. In line with the old regime, the private sector continues to view the public sector as corrupt; likewise, the public sector maintains a tainted image of private Egyptian companies. A representative from Fine Seeds acknowledged that more trust is needed between the two sectors, and that the trust-building process will take time: *So again, it is hard for us when we go to the public sector. We know that they are corrupt but in turn they think that we are also very corrupt. More trust [is needed]. It will take some time.*

According to the farmers' account of the results of the Bt maize field trials in Egypt, there is no doubt that the seed functions as expected. Yet, one interviewee verified that the issue is not a question of the technology and its viability; instead, it is a question of the political climate and whether or not it will permit the project to progress through the commercialization process. One interviewee expressed his concern about this issue, particularly emphasizing that the public sector should respond with increased transparency, accountability, and faith in their private partners in the wake of the old regime: *I don't think the question is a technical question, I think it is a political question. I think that the people in public organizations will need—and this will happen diagonally—will need a change of attitude. And will need to agree on a common pathway, a common direction for the good of the people of this country. And from my point of view, they need to be more transparent; they need to be accountable; and they need to believe more in the credibility of the private [sector].*

In line with the need to solve tensions between public and private actors is the need to establish common ground among the partners so they can engage in a more unified relationship. This means core project partners—like Monsanto and Fine Seeds, in the case of Egypt—must work together to maintain a consensus on project goals and intended outcomes. It is important for both sectors to converge on a common goal as a means for establishing trust, as described by one interviewee: *As I said, you [public and private sector] have to first talk together and understand each other and feel that you are working towards one goal that you have in mind. And once you have decided that you trust each other, then you go to the community to convince them. So here you have to first get together and understand each other and focus on what you want in a project or whatever product.*

In the same vein, it is important for the public institutions and private corporations involved in the project to agree upon and communicate the same message to the public regarding the agricultural biotechnology in question. If the partnership sends out mixed messages to the public, it may delegitimize, and therefore impede trust in, the project, as described by one interviewee: *It goes for both the public and the private sector. They have to work together when they are addressing the public or the community or whatever. One by themselves will not be enough. They have to both agree on how. If they are going to the media, at least you are talking the same language or saying the same thing, not saying something else that is a contradiction. Contradictions will make people not believe and then there is a problem there with trust.*

There is a clear need for a better alignment of goals within the public and private sector. Findings suggest that each contributing party cannot simply work in parallel on the same project with minimal interaction. There must be enhanced integration and engagement between public and private partners. This particular recommendation will perhaps be the most difficult to address in light of the recent uprisings and current political situation in Egypt. The obvious challenge rests in building an effective partnership between two distinct sectors that have historically not trusted each other. However, Egyptian scientists hope that the new regime will have favorable implications for the future of biotechnology development in the country [12].

Conclusions

The predominant theme drawn from the interviewees' responses is the issue of transparency – particularly, the need for public institutions and private corporations to establish a clear understanding of each partner's respective roles, responsibilities and priorities. Working transparently and engaging in open and honest communication with each other and with the public were found to be key elements of project success. Central to this, and specific to the partnership between Fine Seeds and Monsanto in Egypt, is the need for stakeholder priorities and intentions to be clearly communicated to all partners. This issue is intricately related to the need for improved farmer engagement and elevated levels of awareness on all fronts surrounding agricultural biotechnology, which depends primarily on the use of effective communications strategies. It also includes engagement of the media to ensure that they report to the general public information that is backed by scientific evidence. But most important to the case of Egypt is the effect of the current political climate on the success of a project that engages both the public and private sector. The revolutions sweeping the Arab world, although exciting for Egypt, render the establishment and maintenance of trust in any context challenging. This is especially difficult for two different sectors trying to engage in collaborative initiatives. Furthermore, although the commercialization of Bt maize in Egypt appears to have been successful, it did not meet a number of project goals. First, the Bt maize was only commercialized for animal feed and not human consumption. Second, the project was only allowed to introduce the gene to yellow maize—which, in comparison to white maize, is unpopular and not widely grown. Third, the adoption rate of Bt maize in Egypt has been slow and it is still beset by challenges related to trust. There is reason to believe that the current political situation will dictate the ability of public institutions and private corporations to engage in trusting partnerships. The question remains: how can a nation that once had limited trust in governmental institutions regain trust in the current state of affairs, and what implications does this have on collaborative initiatives such as the Bt maize project in Egypt that engage both the public and private sector?

Acknowledgements

The authors are grateful to each of the participants who contributed substantial time and effort to this study. Special thanks to Jennifer Deadman for data collection and Lauren Daley for data analyses. The authors also thank Jocalyn Clark and Jessica Oh for comments on earlier drafts of the manuscript.

This project was funded by the Bill & Melinda Gates Foundation and supported by the Sandra Rotman Centre, an academic center at the University Health Network and University of Toronto. The findings and conclusions contained within are those of the authors and do not necessarily reflect official positions or policies of the foundation.

This article has been published as part of *Agriculture & Food Security* Volume 1 Supplement 1, 2012: Fostering innovation through building trust: lessons from agricultural biotechnology partnerships in Africa. The full contents of the supplement are available online at http://www.agricultureandfoodsecurity.com/supplements/1/S1. Publication of this supplement was funded by the Sandra Rotman Centre at the University Health Network and the University of Toronto. The supplement was devised by the Sandra Rotman Centre.

Author details
[1]Sandra Rotman Centre, University Health Network and University of Toronto, Toronto, Ontario, Canada. [2]African Centre for Innovation and Leadership Development, Federal Capital Territory, Abuja, Nigeria. [3]Dalla Lana School of Public Health, University of Toronto, Toronto, Canada. [4]Grand Challenges Canada. [5]Dalla Lana School of Public Health and Department of Surgery, University of Toronto, Toronto, Canada.

Authors' contributions
Study conception and design: OCE. Data collection: OCE. Analysis and interpretation of data: OCE. Draft of the manuscript: OCE and ASD. Critical revision of the manuscript for important intellectual content: OCE and ASD. All authors read and approved the final manuscript.

Competing interests
The authors declare that they have no competing interests.

References
1. Karembu M, Nguthi F, Ismail H: **Biotech Crops in Africa: The Final Frontier.** *ISAAA AfriCenter* 2009, 1-34.
2. Chalony C, Moisseron J: **Research Governance in Egypt.** *Science, Technology & Society* 2010, **15(2)**:371-397.
3. Sawahel W: **First Egyptian approval of genetically modified corn raises questions.** *Checkbiotech* 2008.
4. Massoud M: **Effect of Bt corn on infestations of corn borers in Egypt.** In *From Promises to Practice: Applications of Science and Technology in Food, Healthcare, Energy and Environment. Volume Ch. 17.* Bibliotheca Alexandrina; Serageldin J, Masood E 2008:203-211, BioVision Alexandria, 12-16 April 2008, c2010.
5. James C: **Global status of Commercialized biotech/GM Crops: 2008.** *ISAAA Brief No. 39* 2008.
6. Massoud M: **The influence of encoding Bt corn hybrids (MON 810 event) on the infestation of the corn borers in Egypt.** *The 3rd International Conference of Plant Protection Research Institute* 2005, **83(2)**:469-496.
7. Adenle AA: **Adoption of commercial biotech crops in Africa.** In *IPCBEE. Volume 7.* IACSIT Press, Singapore; 2011.
8. James C: **Global Status of Commercialized Biotech/GM Crops: 2010.** *Brief No. 42* 2010.
9. Ezezika OC, Thomas F, Lavery JV, Daar AS, Singer PA: **A Social Audit Model for Agro-biotechnology Initiatives in Developing Countries: Accounting for Ethical, Social, Cultural and Commercialization Issues.** *Journal of Technology Management and Innovation* 2009, **4(3)**:24-33.
10. Abdallah N: **GM crops in Africa: challenges in Egypt.** *GM Crops & Food* 2010, **1(3)**:116-119.
11. Ayele S, Wield D: **Science and technology capacity building and partnership in African agriculture: perspectives on Mali and Egypt.** *Journal of International Development* 2005, **17(5)**:631-646.
12. Abdallah NA: **Biotech crops and the Egyptian revolution: Where we stand.** *GM Crops & Food* 2011, **2(2)**:83-84.

17

Rice production constraints and 'new' challenges for South Asian smallholders: insights into *de facto* research priorities

Adam John[1,2*] and Matthew Fielding[2,3]

Abstract

Background: The international community and national agricultural research systems (NARS) recognize the importance of supporting smallholders in order to reduce poverty and promote the food security status of some of the most vulnerable groups in the world. South Asia has the largest food-insecure population in the world, and in several farming systems in the region, rice is the most important staple crop. This study examined the extent to which agricultural research has prioritized the greatest factors that constrain smallholder productivity in those farming systems. It also explored the degree to which research has connected production constraints and environmental challenges faced by rice smallholders.

Results: Estimated congruency ratios suggested that peer-reviewed research has been heavily skewed towards abiotic production constraints. Meanwhile, socio-economic production constraints had received relatively little attention from the research community, even though socio-economic constraints account for more than 22% of rice yield losses in the South Asian farming systems examined. Furthermore, although research publications have tended to concentrate on the most important rice production constraints and linked those constraints to challenges identified by environmental disciplines, there are many medium and small production constraints which have received little research attention. This is despite the fact that the sum of these less severe constraints represents the largest contribution to total rice yield losses.

Conclusions: While national and international research bodies are well aware of the challenges smallholders face, there seems to be a lack of coordination in setting research priorities, since there are many areas, particularly in the social sciences field, which are not receiving the research attention that they warrant, when compared to the opportunity improvements in this sector could provide—as demonstrated in this study. This suggests that steps need to be taken in providing the research community with incentives and support in understanding these 'needs' to increase the impact of their research. Increasing the level of accountability of research institutions to smallholders' and rural populations' needs and promoting participatory farmer-focused research may help in improving research coordination and improving livelihoods by reducing poverty.

Keywords: Production constraints, Rice, Agricultural research, Food security, Smallholders, South Asia

* Correspondence: adam.dr.john@hotmail.com
[1]Institute of Agricultural and Food Policy Studies, Universiti Putra Malaysia, Putra Infoport, Jalan Kajang-Puchong, Serdang, Selangor 43400, Malaysia
[2]Stockholm Environment Institute, Linnégatan 87 D, Stockholm 115 23, Sweden
Full list of author information is available at the end of the article

Background

Smallholders are a vital part of the international agricultural community, but they have historically been neglected in most national and international forums. Supporting smallholders has been a strategic priority for the International Fund for Agricultural Development (IFAD) since it was set up in the early 1970s in response to a food crisis, and now a broader range of international organizations say they are putting smallholders at the top of their agenda [1]. An example of this increased attention is the UN's establishment of 2014 as the international year of family farming. The Consultative Group on International Agricultural Research (CGIAR) reevaluated its strategy in 2008 to focus its research more towards poor farmers' needs [2]. There has even been a growth in smallholder research interest from non-traditional private donors, including the Syngenta Foundation for Sustainable Agriculture and the Yara Foundation [3]. In the political arena, promoting sustainable smallholder agriculture was on the agenda of the G20 meeting in 2012, which emphasized the need for greater market participation to complement productivity gains [4]. As a high-level panel noted in 2013 [1], there has been a vicious circle of poor research and extension for low-income farmers, and it needs to be broken.

The study examines where research relevant to production constraints is focused, and how that fits with where yield losses are known to take place. The study looks specifically at research on rice in South Asia, since the region is home to one of the world's largest food-insecure populations, and rice is the most important crop there. The study also aims to identify whether research has made connections between different areas of production constraints and environmental concerns which are recognized as new challenges for smallholders.

This article begins by providing an overview of linkages between smallholders, food security, and agricultural research support for smallholders. We then describe the methodology used to quantify research and how network analysis was used to show interactions within different research fields. We then present and discuss the results and conclude with some implications of the study and its relevance to agricultural research policy.

Supporting smallholders promotes global food security

Defining what constitutes a smallholder is not merely concerned with setting a threshold in terms of hectares. The High Level Panel of Experts highlights that a smallholding is 'small' in the sense that it has scarce resources which are barely enough for a smallholder to satisfy their basic needs [1]. For the majority of smallholders, especially those in South Asia, the threshold is far below 2 ha. There are an estimated 500 million smallholder farms around the world, which employ about 2.5 billion people either on a full-time or part-time basis [5]. In South Asia, 75% of farmers are smallholders [6]. In India, the number of smallholdings rose from 70 million in 1970–1971 to 121 million in 2000–2001, which has put downward pressure on farm size, shrinking the average smallholding from 2.3 to 1.32 ha in the same period [7]. Women commonly take responsibility for producing food crops, particularly where smallholder farms produce both food for the household and cash crops [8]. Smallholders produce the bulk of food in developing countries, and their contribution is growing [9]. For instance, smallholders produce 70% of the food supply in Africa [10], and 80% of the food consumed in sub-Saharan Africa and Asia is produced by smallholders [11].

Paradoxically, although smallholders play a significant role in global food production, they are net food buyers [12] and make up the majority of the world's undernourished and poor population [1,13-16]. This is why supporting smallholder agriculture is recognized as playing a major role in reducing hunger and malnutrition. Lipton [17] found no examples of agricultural development that alleviated poverty without improving smallholder productivity. A review by the UN Food and Agricultural Organization (FAO) and the World Bank [18] concluded that international development goals such as halving hunger and poverty would not be achieved without policies that prioritize improving the productivity of small farms. The Sustainable Development Network concluded that millennium development goal 1, to eradicate extreme poverty and hunger, would not be achieved without focusing on poor farmers in rural areas [14].

It is important to acknowledge that increasing smallholder productivity not only improves smallholders' food security but also global food security because they produce such a large share of developing countries' food supplies. Its importance for global food security is expected to rise because of a growing world population [1]. One study [19] estimated that more than 50% of the food needed to feed the projected nine billion world population in 2050 will be produced by smallholders. In terms of poverty reduction, supporting smallholder agriculture is expected to have a greater impact in sub-Saharan Africa and South Asia, as developing non-agricultural sectors are seen as more important for poverty reduction in East Asia and Latin America [20].

Agricultural research for smallholders

The massive agricultural research and technology transfer effort of the 1960s and 1970s, often referred to as the 'green revolution', led to dramatic increases in agricultural productivity. Smallholders in developing countries, especially in Asia and Latin America, benefited substantially from these advances in agricultural research as well as from strong extension services [21]. However, many of these production gains resulted in environmental degradation [5]. The current food security challenge is as great as 40 years ago but now with the need to address sustainability and deal with climate change. Not only does food production need to increase substantially in order to meet future demand, but with climate change, there is considerable concern that we may even struggle to sustain current food production levels. For instance, studies have outlined how agricultural production is predicted to decline in developing countries [22,23]; one estimated that the decline could be as much as 20% [14]. Therefore, agricultural productivity needs to rise both to meet new demand and to offset expected climate-related yield losses in some regions [14].

International bodies have outlined that research to benefit smallholders and promote food security needs to tackle both traditional and new challenges. Traditional areas of interest include strengthening land tenure rights [5,24] and extension services [5,14], improving infrastructure [24], making smallholder agriculture more market-oriented [3,5,24], and improving smallholders' access to inputs [2,5,14,24]. Newer areas of focus include adaptation to climate change [3,5,25], biodiversity and natural resource management [5,25], crop diversification and nutrition security [1,2,25,26], multifunctional agriculture [5], sustainable intensification [5], and promoting food crops [1,2,24,26,27].

Although smallholders share common constraints and challenges, such as those mentioned above, it is important to recognize that they are not a homogeneous group nor are they equally affected by common challenges. For instance, in East Africa, smallholders in hotter, low-lying areas are predicted to see their yields decline due to climate change, whereas smallholders located at higher elevations and lower average temperatures may even see their yields rise in the future because of climate change [28]. In addition, a study focusing on the Sudanian and Sahelian savannahs of West Africa showed that the impact of temperature and rainfall due to climate change is expected to have very different consequences for the regions of West African countries located in the two savannahs [29]. For instance, their results suggested that increases in temperature would have a more adverse effect on millet and sorghum grown in the southern part of Senegal than in the north of the country. Such examples support the need to consider the heterogeneity of smallholders when making policy decisions, which was stressed in a recent FAO report [4].

Agricultural research in South Asia

National agricultural research systems (NARS) such as India's were some of the first countries in the developing world to develop strong relationships with international agricultural research centers (IARC), such as the International Rice Research Institute (IRRI) [30]. This may be partly due to the fact that South Asian NARS were established earlier than in most other developing countries. For example, the Indian Centre of Agricultural Research (ICAR) was formed in 1929, [31]. Agricultural research in South Asia has traditionally largely focused on productivity-enhancing technologies for staple grains such as rice and wheat, which has been successful at meeting South Asian countries' food production goals [32]. For instance, India has managed to increase its production of food grains fourfold since the 1960s [7]. This was achieved through partnerships between the likes of IRRI and ICAR in India whereby the former provided new high-yielding gene lines for the South Asian NARS to develop into end products for their domestic farmers [30].

It has been argued, however, that the focus on rice and wheat in irrigated and rainfed favorable regions may have come at the expense of other important food crops as well as less favorable, resource-poor regions [32]. This has been followed by questions being raised about whether different groups of South Asians benefited disproportionately from the agricultural developments of the green revolution. A CGIAR science council assessment [32] found that while overall, agricultural research had been very effective at reducing poverty, sometimes richer South Asian households benefited more than poorer ones, widening income gaps. The latter study also raised the issue of 'hidden hunger'—micronutrient deficiencies which were not satisfactorily addressed by the green revolution breakthroughs.

Other concerns have been raised as well about the impact of agricultural research in South Asia. A recent report by the Asia-Pacific Association of Agricultural Research Institutions (APAARI) and the International Food Policy Research Institute (IFPRI) concluded that, with the exception of India, agricultural research for development in South Asia has generally been neglected, with only a slight improvement since the 2008 food crisis [6]. The report also highlights that public sector investments in agricultural research in Bangladesh and Nepal in 1996–2009 were proportionally lower than in India and other developing countries, such as Brazil. What is more, the rate of growth in agricultural research fell between 1996 and 2009 in Bangladesh, Nepal, and India, though it has risen again in India since 2006. Agricultural growth in South Asia has also been weaker in recent decades. The APAARI and IFPRI assessment [6] attributes the South Asian NARS' limited impact on weak institutional capacities and irregular funding.

The CGIAR assessment [32] finds that food grain productivity should continue to be prioritized but argues that new efforts must also encompass natural resource management and sustainability. It also notes that unlike the green revolution technologies, which were best-suited to areas with favorable cultivating conditions, research now needs to focus on increasing productivity in least favorable areas (LFAs) as well. The APAARI and IFPRI report [6] makes similar points and suggests that the public sector will play the dominant role in supporting LFAs, because the private sector shows little interest in these regions.

The APAARI and IFPRI assessment [6] also finds that agricultural research for development will have to triple or quadruple in the coming years for South Asia to achieve its food production goals. It says the participation of stakeholders in agricultural research is fundamental to achieving agricultural development goals, so farmer-participatory focused research should be a top priority for South Asia. Expanding on research priorities to cover areas such as natural resource management is also seen as important by South Asian NARs. ICAR's 'Vision 2030' [7] sees halting land degradation and rehabilitating degraded land and water resources in India as a top research priority. Promoting agricultural diversification and involving social sciences more in agricultural research challenges are other priorities raised by ICAR. Ultimately, supporting smallholders must remain a key component of agricultural research in South Asia [6,7].

Methods
Farming systems approach
Farming systems analysis is a conceptual framework which considers the heterogeneity of smallholders and is useful for designing appropriate agricultural development strategies. It defines the interactions between a household, its activities, and the resource base as a farm system, where the biophysical, socio-economic, and human elements are interdependent [18,33]. The FAO and World Bank [18] have further defined a population of individual farm systems with broadly similar characteristics and constraints as a farming system. They identified eight broad farming system types covering the developing world and categorized these into 63 farming systems covering Latin America, sub-Saharan Africa, South Asia, and East Asia. Despite the amount of detail in their farming systems analysis, they acknowledge that there is some degree of heterogeneity within each of the specified farming systems, and the boundaries between farming systems across geographic areas can be quite loose. Nevertheless, a farming systems approach can provide an insightful take on the idiosyncratic nature of smallholder agriculture.

A 2008 study [33] identified the most food-insecure and drought-prone among the 63 farming systems in order to demonstrate where agricultural research should be prioritized. The authors identified 15 farming systems with more than 2.5 million stunted children each, and

they detected the 13 major food crops that these vulnerable farming systems largely relied on. The two farming systems with the highest amount of stunted children were the South Asian rice-wheat (28.3 million) and rainfed mixed (24.5 million) farming systems. In fact, five of the top ten farming systems with the highest number of stunted children were in South Asia.

The International Maize and Wheat Improvement Center (CIMMYT) [34] interviewed panelists with regional expertise for 12 of the 15 food-insecure and drought-prone farming systems highlighted by [33] to identify the top ten production constraints for maize cultivation in each of the farming systems. While droughts were shown to be a common constraint for all of the farming systems in the [33] study, due to the heterogeneity that exists between farming systems, CIMMYT wanted to exploit the tacit knowledge of maize research and extension experts. The panel of experts assessed the relative importance of several groups of production constraints, which could be categorized into four groups: biotic, abiotic, crop management, and socio-economic constraints. A 2010 study [35] used the same methodology but expanded the analysis by including six other food crops important to the 13 of the 15 food-insecure farming systems identified by [33]: wheat, rice, sorghum, cowpea, chickpea, and cassava. In the case of the five South Asian farming systems, the authors identified rice as the most important staple crop in four of the five systems, with the exception being the dry rainfed farming system, where sorghum is the most important staple grown in terms of total area harvested.

Like the CIMMYT study, [35] was able to identify the major production constraints for the six food crops in 13 farming systems. Furthermore, the authors indicated the cost of each constraint in terms of its contribution to the yield gap between the highest achieved crop yield and the average crop yield for each food crop analyzed in each of the food-insecure farming systems. What is striking about the two studies [34,35] is the diversity of constraints and their relative importance for each crop in each farming system. A shortcoming of these studies, which [35] acknowledges, is that while studies tend to assume that constraints are additive, in practice, many constraints interact and should be seen as multiplicative. For instance, improving soil fertility may also improve water availability (due to less runoff), so the impact on increased yields may be amplified. Nevertheless, the studies illustrate the extent of heterogeneity in smallholder agriculture and help confirm why a farming systems approach is useful for setting priorities in smallholder agricultural research and policy.

This study focuses on the production constraints of four of the food-insecure South Asian farming systems described in [35] where rice is the dominant crop: rice-wheat, rainfed mixed, rice, and highland mixed. The [35] analysis also identified the top ten constraints to rice production, which account for only about half of the total rice yield losses in the South Asian farming systems. This study extends on [35] by including the top 24 rice production constraints that were identified by the authors in [35], which they did not publish but made available for this study. The advantage of including the top 24 production constraints is that together, they cover 80%–90% of total rice yield losses, providing a more holistic perspective.

As noted in the introduction, agricultural research today needs to focus not only on the traditional production constraints, but also on 'new' constraints, most of which may be seen as environmental challenges, such as climate change, natural resource management, ecosystem services, biodiversity, and sustainability. This study considers those constraints, which are being given more attention by global development agencies, as well as two other key factors also prioritized by those agencies: diversification and food security.

Quantifying agricultural research

Measuring scientific output has its own field of study, called scientometrics, which provides various types of quantitative methods based on bibliometric and patent indicators [36]. This study uses publication analysis, which can be used as an indicator of scientific activity [37]. We perform a straight publication count of peer-reviewed journal articles cited by Scopus which include the keyword criteria captured by queries performed in the advanced search options of Scopus. There are several abstract and citation databases of peer-reviewed literature to choose from, but Scopus is the largest, and it covers a broad spectrum of academic disciplines in the physical and social sciences [38] which is why it was used in this study. A weakness of Scopus is that its coverage of publications only starts in 1995, but the period covered by this study begins in 1997, the year after the Rome Declaration for World Food Security.

We counted the journal articles published between 1997 and 15 February 2014 that had focused on rice production constraints for South Asian farming systems as outlined in [35] as well as the 'new' challenges for smallholders as specified in many international development reports. Keywords for the search were derived from the description of production constraints in [35]. The keywords used to represent each production constraint, along with the rice production constraints descriptions, can be found in Additional file 1: Table S4. To be included in the count, the publication had to have the keywords in its title or in the publication keywords. Since we are only interested in publications focusing on rice, the word 'rice' was also specified in the Scopus queries as being present in the publications' title or as a keyword.

It is important to recognize that this is not a definitive publication search. While Scopus is the most extensive citation-based database, there will naturally be other quality publications which were not captured in the search. What is more, the specification of the keywords puts further restrictions on the publication selection process. There will surely be publications which are perhaps relevant to South Asian rice smallholders but which were not captured in the publication search because they used different keywords. However, the inclusion of too many keywords would have diluted the relevance of the captured publications, so a trade-off had to be made.

The final criterion of the selection of the publications was the location. South Asian farming systems as defined in [18] cover seven countries: India, Pakistan, Bangladesh, Sri Lanka, Nepal, Bhutan, and Afghanistan, therefore the Scopus queries were confined to these countries. The term 'smallholder' was not included in the queries; this is because it was found that the majority of journal articles focusing on rice production constraints in South Asia did not use the term, even though it was evident from the abstracts that the articles focused on smallholders. This may be explained by the fact that the majority of farmers in South Asia are smallholders, so it may seem redundant to use the term when discussing South Asian farmers. An example of one of the queries used in Scopus is given in the top cell of Additional file 1: Table S4. The entire list of journal articles retrieved from the Scopus search is stored in the spreadsheet in Additional file 2.

To clearly represent the results, congruency ratios were estimated for each constraint. A congruency ratio is a common way of assessing the efficiency of research resource allocation [39]. In this study, the measure is the ratio between the percentage of journal articles which focused on the constraint and the role of the constraint as a percentage of total yield losses. A ratio score of more than 1 may suggest that there is a surplus of articles which have focused on the particular constraint, whereas a ratio of less than 1 may suggest the contrary.

Network analysis

Network analysis or social network analysis was originally developed in the sociology field in the 1970s [40]. Other disciplines in the social sciences as well as the physical sciences such as biology have since incorporated network analysis as an empirical tool. Network analysis is primarily interested in representing relationships between actors using graphs. Interactions between pairs of actors or nodes are represented in the graph with a line, called an edge. The characteristics of a network graph can be described using global graph metrics or individual actor properties [41]. The former is used to describe the network as a whole, while the latter provides insights into the influence or connectedness of the actors. For the purposes of this study, the network analysis is restricted to describing the actor properties of the network which represent rice production constraints.

Centrality measures are a common way of capturing the status of actor properties. There are several centrality measures whose application depends on the purpose of the study; however, their binding characteristic is that they all quantify how central each actor is within a network. For instance, closeness centrality measures how long it takes an actor to spread information to the other actors within the network whereas 'betweenness' centrality measures how many times an actor connects two other actors along the shortest path in the network [42]. Degree centrality is perhaps the most straightforward centrality measure which is an actor's sum of interactions with other actors within the network. Degree centrality suffices for our study, as we are simply interested in identifying the diversity of interactions that agricultural research has found between rice production constraints.

Network graphs are also used as a visual tool. Since networks can become quite complex due to the number of actors and interactions which exist, the graphs are arranged using the Fruchterman-Reingold force-directed algorithm. An advantage of the Fruchterman-Reingold force-directed graph over geometric ones for representing a network is that the positioning of the actors in the graph depends on the interactions which exist, whereby interacting actors are located close to one another. The graph may therefore be seen as giving a crude indication of the centrality of each actor. Actors which interact with many other actors may be seen as clustering together whereas actors with few interactions should not be within close proximity to other actors.

Results
Preliminary observations of time series data
Figure 1 shows the number of journal articles cited by Scopus and published between 1997 and 2013 which were found to focus on at least one rice production constraint in South Asia. Figure 1 also shows the number of journal articles which focused on 'new' challenges, as described by [1]. It is clear that there has been a general increase in the number of publications per year during the observed period. However, the third time series in Figure 1, which represents the total number of agricultural and social science publications cited by Scopus related to South Asia, shows that there has been an increase in publications in general. Therefore, it could be argued that the increase in publications focused on production constraints is simply due to the general increase in publications. It is therefore difficult to determine from such a crude observation whether the food

crisis in 2008 created a significant response in agricultural research in dealing with the food insecurity situation.

Figure 2 breaks down the number of journal articles focusing on production constraints into abiotic, biotic, management-related, and socio-economic constraints. One can observe that all four categories have been following an increasing trend, albeit at different magnitudes. Publications focusing on socio-economic constraints are lagging behind the other three categories while the category with the highest number of publications for each year between 1997 and 2013 was abiotic production constraints.

De facto rice research priorities in South Asia
The main motivation for this study was to attempt to reveal whether the focus of research relevant to South Asian agriculture, as identified by peer-reviewed journal articles, has been geared towards the most important production and non-production constraints of the most food-insecure South Asian smallholders. Table 1 displays the yield loss contribution of each production constraint, which is a weighted average of the four South Asian farming systems where rice makes up the largest area of land harvested. The yield losses for each of the four South Asian farming systems can be found in Additional file 1: Table S1. The weighted average takes into account the absolute number of stunted children residing in each farming system, which was taken from [33].

The table also indicates the number and percentage of journal articles which were found to primarily focus on each production constraint. An article was recorded as focusing on a particular production constraint if the production constraint keyword, as identified in Table 1, was included in the publication's title or as an indexed keyword. The last column gives the congruency ratios, which is the ratio between the percentage of relevant articles found and the weighted average yield loss. A tentative interpretation of a congruency ratio of more than 1 is that there is a surplus of research focusing on that particular production constraint whereas a ratio of less than 1 may suggest a deficit of research focusing on the particular production constraint.

Table 1 classifies the production constraints into four groups: socio-economic, abiotic, biotic, and management-related where each individual production constraint is ordered in descending order by its contribution to the total yield loss. The congruency ratios suggest that research related to socio-economic production constraints is severely lacking. The seven socio-economic constraints combined were the largest cause of South Asian rice yield losses, yet less than 8% of the research articles in the study's sample focused on at least one of the socio-economic constraints. The congruency ratios for irrigation and information access were 0, since no articles focusing on these constraints

Rice production constraints and 'new' challenges for South Asian smallholders: insights into de facto...

207

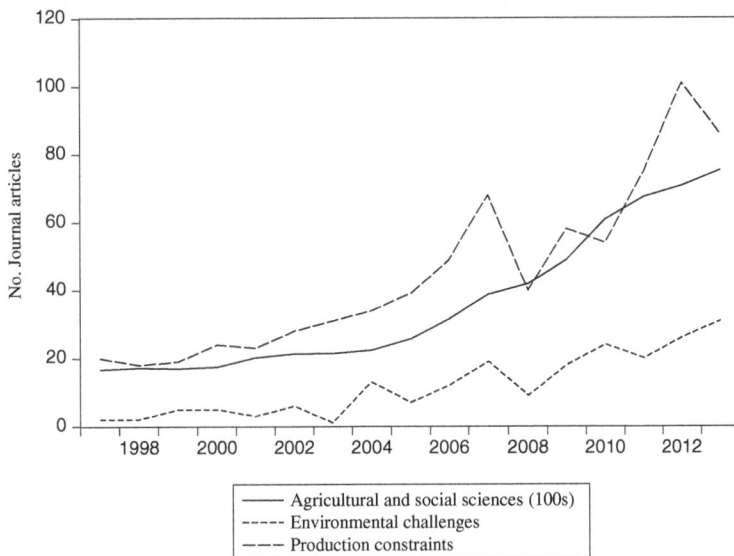

Figure 1 Journal articles cited by Scopus, 1997–2013.

were found in the Scopus search. The five remaining socio-economic constraints all had congruency ratios which scored well below 1. Therefore, one may argue that research is lacking in all areas of socio-economic constraints affecting South Asian rice farmers.

On the opposite side of the scale, abiotic constraints were prioritized in 45% of the Scopus cited journal articles, even though this production constraints category represented less than 22% of total yield losses in food-insecure South Asian farming systems. Abiotic constraints received by far the most attention in the literature. The three largest abiotic constraints—drought, soil fertility, and nitrogen, which make up a combined 15.7%

of the yield gap—were prioritized by 7.1%, 7.9%, and 22.6% of the journal articles, respectively. Although abiotic constraints were the most numerous (11), five of them represent less than 1% of total rice yield losses. However, despite their relative unimportance to rice yield losses for the most food-insecure South Asian farming systems, some have arguably received a surplus of research attention. For instance, all five of these abiotic constraints had a congruency ratio of greater than 1, while three of them scored higher than 5. In fact, 10 of the 11 abiotic constraints had a congruency ratio greater than 1, suggesting that there may be an over-emphasis on abiotic constraints research.

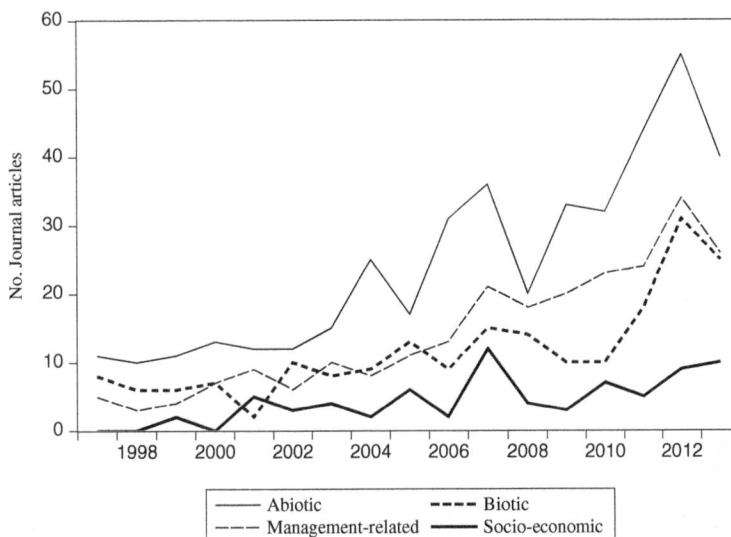

Figure 2 Journal articles cited by Scopus categorized into production constraints, 1997–2013.

Table 1 Scopus cited journal articles prioritizing most important rice production constraints (1997–2014)

Major rice production constraints in South Asia	Weighted average yield loss %	Relevant articles		Congruency ratios
		No.	%	
Socio economic	22.2	73	7.8	0.35
Difficult access to sufficient irrigation water	4.5	0	0	0
Unavailability of quality seed	3.8	15	1.6	0.42
High cost of irrigation	3.4	20	2.1	0.63
Nitrogen fertilizer expensive/in short supply	3.3	16	1.7	0.52
Inadequate farmer knowledge/training	2.8	15	1.6	0.56
Insufficient access to agricultural information	2.6	0	0	0
High price of inputs other than nitrogen	2.1	10	1.1	0.50
Abiotic	21.9	421	45.0	2.05
Drought or intermittent water stress	6.9	66	7.1	1.03
Soil fertility depletion	5.7	74	7.9	1.38
Nitrogen deficiency	3.1	212	22.6	7.19
Flooding of low lying fields	1.5	28	3.0	2.02
Deficiency or toxicity of micronutrients	1.3	11	1.2	0.90
Phosphorus unavailability	1.1	32	3.4	2.98
Cyclone/typhoon damage	0.5	5	0.5	1.06
Soil physical/structural degradation	0.5	27	2.9	5.76
High temperature stress	0.4	20	2.1	6.07
Potassium deficiency	0.4	10	1.1	3.03
Low temperature (cold) stress	0.1	8	0.9	11.58
Biotic	18.2	200	21.4	1.18
Weed competition	6.6	85	9.1	1.38
Leaf and stem pests	4.5	33	3.5	0.78
Leaf, stem, and panicle diseases	3.4	75	8.0	2.35
Rodent damage	1.8	6	0.6	0.35
Storage pests	1.0	1	0.1	0.11
Root and soil diseases	0.9	0	0	0
Soil insects	0.5	2	0.2	0.44
Management-related	21.5	244	26.1	1.21
Inadequate water management	6.0	119	12.7	2.12
Inappropriate/poor nutrient/fertilizer use	4.1	75	8.0	1.93
Late planting of crop	2.7	20	2.1	0.80
Use of low yielding or old variety	2.6	10	1.1	0.41
Poor crop rotations and sequences	1.7	10	1.1	0.62
Inappropriate/poor insect/disease management	1.6	14	1.5	0.92
Field crop establishment difficulties	1.2	2	0.2	0.18
Inappropriate/poor weed management	1.1	29	3.1	2.78
Poor seedling nursery management	0.4	0	0	0
Total	83.8	777	83.8	

Note that the category totals are not necessarily the sum of the individual production constraints in each category. This is because some articles have focused on more than one production constraint.

Biotic production constraints combined make up the smallest contribution to total yield losses (18.2%); however, some of the individual biotic constraints are substantial sources of yield losses. For instance, weeds are one of the largest single production constraints for South Asian rice cultivation, representing 6.6% of the yield gap. Researchers obviously recognize the importance of this constraint, as it has a congruency ratio of 1.38. The same could be said for leaf, stem, and panicle diseases such as blast and sheath blight, as this constraint scored a congruency ratio of 2.35. However, the second greatest biotic constraint, leaf and stem pests, seems to have been overlooked. It represents 4.5% of total rice yield losses, yet it has a congruency ratio of 0.78. Rodent damage is perhaps another area which demands further research attention, as it represents 1.8% of total rice yield losses, yet its congruency ratio was 0.35. Furthermore, contrary to less severe abiotic constraints, less severe biotic constraints such as storage pests and soil diseases and insects received congruency ratios closer to 0 than 1.

Management-related production constraints were prioritized by 26.1% of the journal articles in the sample and had a congruency ratio of 1.21. It is important to note, however, that the bulk of research has focused on the two largest management constraints: water and fertilizer use. Seed varieties, planting, and crop rotation are three medium-sized management-related constraints which scored congruency ratios of less than 1. The only medium-to-small management-related constraint to receive a surplus congruency ratio score was weed management. The others all had deficit scores.

According to the congruency ratios, several of the major rice production constraints affecting yields for the most vulnerable South Asian smallholders have been addressed by research. Two major exceptions to this are irrigation access and leaf and stem pests. However, [40] showed that smallholders are faced with a broad set of constraints which need to be considered collectively in order to increase yields. This argument is just as valid in the case of South Asian rice smallholders. What Table 1 shows is that there are a lot of small and medium socio-economic, biotic, and management-related production constraints which, when combined, are at least as important as the sum of the most severe constraints. Despite this balance in yield loss contribution, the congruency ratios show there has been a significant imbalance in research focus.

Linking production constraints in the research

Degree centrality was calculated for the entire 1997–2014 period and the pre- and post-2008 periods. The links or interactions found between production constraints are illustrated in Figure 3, where each node represents a rice production constraint. The color of the nodes represents which type of production constraint it is, e.g., biotic is green. Moreover, the size of the node indicates how much the production constraint it represents contributes to total yield losses. In other words, the larger the node, the more significant is its contribution to yield losses. The width of the lines or edges connecting the nodes indicates how

many research articles focused on both of those two particular production constraints.

Table 2 gives the degree of centrality for each of the production constraints and ranks the constraints. The constraints without a degree of centrality either had no journal articles focused on them, as already shown in Table 1, or have not been linked to any of the other rice production constraints by research articles.

Another objective of this study was to provide an insight into where research has linked different production constraints faced by South Asian rice farmers, to evaluate whether research is being conducted in isolation, or progress has been made in breaking down research silos. Degree centralities are used for this purpose. Over the entire 1997–2014 period, 28 of the production constraints were linked to at least one other production constraint by a research article. The most connected constraint was nitrogen; the least connected was rodents and soil insects.

As already seen in Table 1, three of the seven socio-economic production constraints had not been considered by the Scopus cited literature. Of the four remaining socio-economic constraints which had been considered in the literature, all had been connected to other production constraints. Nitrogen supply had the highest degree of centrality, ranking sixth overall for the entire period. Cost of irrigation and farmer knowledge were jointly ranked ninth, while seed quality ranked 15th for the entire period. However, comparing the centrality results for the pre- and post-2008 periods shows that these three socio-economic constraints have moved up in the overall ranking since 2008.

Five of the abiotic constraints were ranked among the ten most connected constraints. These included all four abiotic constraints which contributed to more than 1% of total rice yield losses. However, drought, which along with weeds, contributes the most to total yield losses overall, ranked lowest out of these five constraints. Interestingly, even some of the abiotic constraints, such as potassium and heat stress, which contributed less than 1% to total yield losses, were relatively well connected with other constraints, jointly ranking ninth for the entire period.

Weeds were the only biotic constraint which was relatively well connected in research articles. The other four biotic constraints, which had been linked in the literature to other constraints, ranked at the bottom, even though three of these constraints, namely (1) leaf and stem pests; (2) leaf, stem, and panicle diseases; and (3) rodents, collectively contribute roughly 10% of total rice yield losses for South Asian smallholders.

The two largest management-related production constraints, water management and fertilizer use, ranked in the top three most connected constraints for the entire

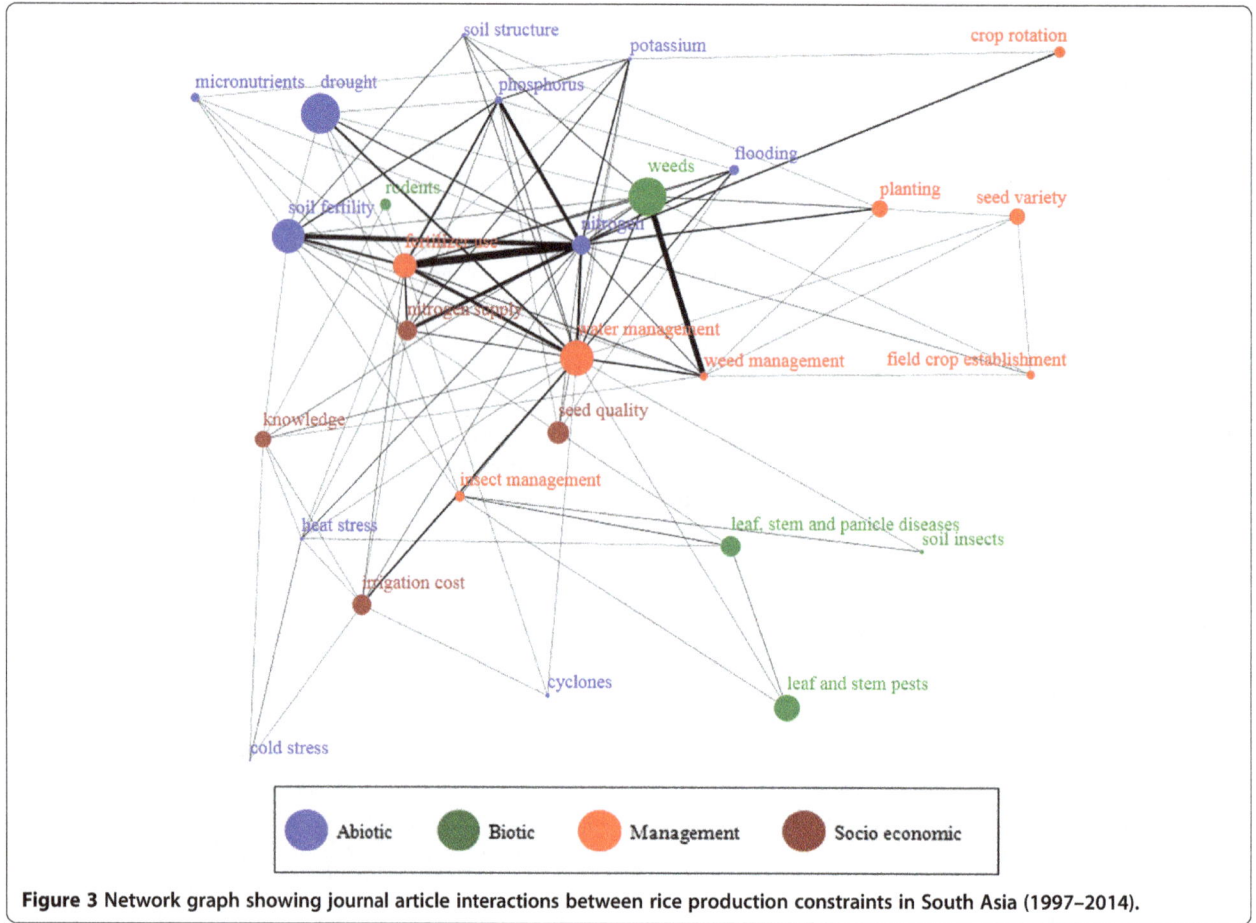

Figure 3 Network graph showing journal article interactions between rice production constraints in South Asia (1997–2014).

period. However, the next four most important management constraints (planting, seed varieties, crop rotation, and insect and disease management) did not rank very high, with three of them ranking near the bottom of the list despite the fact that the four combined make up a similar contribution to total yield losses as aggregated losses due to water management and fertilizer use.

Of the seven production constraints which contribute to 4% or more of the total yield losses, four ranked high in the degree of centrality, suggesting that research in a wide field of disciplines is considering these constraints. However, there have not been extensive links made for the major biotic constraint: leaf and stem pests. The medium-sized socio-economic and abiotic constraints, representing 2%–4% of yield losses, are positioned in the middle of the ranking. In contrast, the medium-sized biotic and management-related constraints ranked low, suggesting the disciplines that study them are relatively isolated from those focused on other major sources of rice production constraints for the most vulnerable South Asian smallholders.

Linking production constraints with new challenges

A second set of networks was specified which included the 'new' challenges discussed earlier. The rankings are not given since it does not seem sensible to compare production constraints with these non-production constraints, which should instead be seen as complementary to the former. Additional file 3: Figure S3 shows that for the entire period, the five environmental challenges—biodiversity, climate change, ecosystem services, natural resource management, and sustainability—as well as food security were relatively centrally located in the network. Comparing Additional file 3: Figure S4 and Additional file 3: Figure S5, which represent the pre- and post-2008 periods, respectively, shows how research has made significant progress in linking these 'new' challenges when considering rice production constraints in South Asia. One can observe how the nodes representing the five environmental challenges and food security have moved in substantially to the center of the network since 2008. This increase in linking these non-production challenges to other disciplines in the literature can also be seen by the degree of centrality results in Table 3. The degree

Table 2 Degree centrality results for rice production constraints

Rice production constraints in South Asia	Yield losses (weighted average)	Degree centrality (ranking in brackets)		
		1997–2014	1997–2007	2008–2014
Socio economic	22.2			
Difficult access to sufficient irrigation water	4.5	-	-	-
Unavailability of quality seed	3.8	6 (15th)	3 (14th)	5 (13th)
High cost of irrigation	3.4	8 (9th)	3 (14th)	8 (8th)
Nitrogen fertilizer expensive/in short supply	3.3	11 (6th)	7 (6th)	8 (8th)
Inadequate farmer knowledge/training	2.8	8 (9th)	1 (22nd)	8 (8th)
Insufficient access to agricultural information	2.6	-	-	-
High price of inputs other than N	2.1	-	-	-
Abiotic	21.9			
Drought or intermittent water stress	6.9	7 (13th)	3 (14th)	6 (11th)
Soil fertility depletion	5.7	12 (4th)	7 (6th)	10 (3rd)
Nitrogen deficiency	3.1	18 (2nd)	17 (1st)	16 (1st)
Flooding of low lying fields	1.5	7 (13th)	4 (9th)	6 (11th)
Deficiency or toxicity of micronutrients	1.3	5 (17th)	4 (9th)	1 (25th)
Phosphorus unavailability	1.1	9 (7th)	5 (8th)	9 (6th)
Cyclone/typhoon damage	0.5	3 (23rd)	-	3 (17th)
Soil physical/structural degradation	0.5	5 (17th)	3 (14th)	3 (17th)
High temperature stress	0.4	8 (9th)	4 (9th)	5 (13th)
Potassium deficiency	0.4	8 (9th)	8 (4th)	4 (16th)
Low temperature (cold) stress	0.1	3 (23rd)	-	3 (17th)
Biotic	18.2			
Weed competition	6.6	12 (4th)	8 (4th)	10 (3rd)
Leaf and stem pests	4.5	3 (23rd)	2 (19th)	2 (23rd)
Leaf, stem, and panicle diseases	3.4	4 (19th)	4 (9th)	3 (25th)
Rodent damage	1.8	2 (24th)	-	2 (23rd)
Storage pests	1.0	-	-	-
Root and soil diseases	0.9	-	-	-
Soil insects	0.5	2 (24th)	-	2 (23rd)
Management-related	21.5			
Inadequate water management	6.0	19 (1st)	16 (2nd)	15 (2nd)
Inappropriate/poor nutrient/fertilizer use	4.1	13 (3rd)	12 (3rd)	10 (3rd)
Late planting of crop	2.7	4 (19th)	3 (14th)	3 (17th)
Use of low yielding or old variety	2.6	4 (19th)	1 (22nd)	3 (17th)
Poor crop rotations and sequences	1.7	2 (24th)	2 (19th)	1 (25th)
Inappropriate/poor insect/disease management	1.6	6 (15th)	2 (19th)	5 (13th)
Field crop establishment difficulties	1.2	4 (19th)	1 (22nd)	3 (17th)
Inappropriate/poor weed management	1.1	9 (7th)	4 (9th)	9 (6th)
Poor seedling nursery management	0.4	-	-	-

A dash (−) indicates that no interactions were made for that particular rice production constraint by the Scopus cited journal articles captured in the analysis. Rice production constraints with the same degree of centrality are given the same rank.

of centrality increased for all of the factors apart from diversification after 2008, and in fact, more than doubled in some instances.

Most frequent connections made by research

A feature of the network diagram in Figure 3 which has not been addressed yet is the width or thickness of the

Table 3 Degree centrality results for new challenges

New challenges	Degree centrality results		
	1997–2014	1997–2007	2008–2014
Environmental challenges			
Biodiversity	13	2	13
Climate change	10		10
Ecosystems	24	15	19
Natural resource management	14	5	14
Sustainability	15	5	14
Crop diversification	8	4	5
Food security	19	5	18

lines connecting the nodes. So far, we have discussed the degree or number of interactions that each constraint has. However, the frequency of interactions, that is to say, how many journal articles made a link between two particular constraints, also provides an insight into *de facto* research priorities. Figure 4 enlists the most common interactions identified by research among production constraints as well as between production and non-production constraints for each of the three time periods.

The constraints in Figure 4 are color-coordinated by their constraint type. The columns on the left display the most frequent interactions found between rice production constraints. The most common link made in the research was between nitrogen and fertilizer use, with 40 articles linking these two constraints between 1997 and 2014. It is clear that red and blue, that is, management-

related and abiotic production constraints, rank as the most frequent interactions found in the literature. This is not surprising, since the results in Table 1 indicated that it was these two production constraint types which were shown to have been prioritized by journal articles. Another interesting point is that it tends to be the production constraints which ranked highest in terms of degree centrality that also have the highest frequency of journal articles linking them to other constraints. This pattern can also be seen in Figure 3, where the thickest lines tend to be connecting the production constraints which are the most centrally positioned in the network.

The interactions between production and non-production constraints, as shown on the right side of Figure 4, also tend to include these same abiotic and management-related production constraints. What this shows is that rice research relevant to South Asian smallholders tends to prioritize a relatively small group of abiotic and management-related production constraints, which can be seen not only in the diversity of links made in the research, as captured by the degree of centrality results, but also in the frequency of these interactions, as show in Figure 4.

Discussion

Smallholder farming has been at the top of the international community's agenda since the food crisis of 2007–2008, yet the resulting progress, as predicted by [18], has only been small. The need had been clearly defined, and funding is available, so why are there certain areas of production constraints that receive proportionally

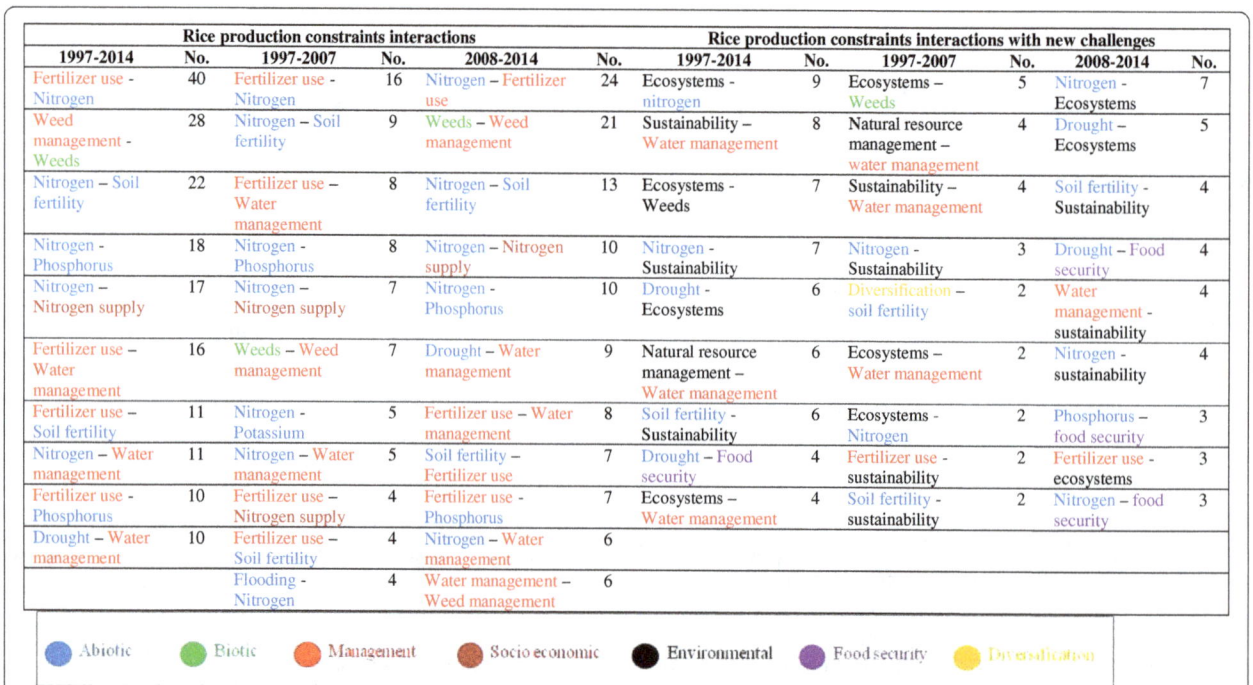

Rice production constraints interactions						Rice production constraints interactions with new challenges					
1997-2014	No.	1997-2007	No.	2008-2014	No.	1997-2014	No.	1997-2007	No.	2008-2014	No.
Fertilizer use - Nitrogen	40	Fertilizer use - Nitrogen	16	Nitrogen – Fertilizer use	24	Ecosystems - nitrogen	9	Ecosystems – Weeds	5	Nitrogen - Ecosystems	7
Weed management - Weeds	28	Nitrogen – Soil fertility	9	Weeds – Weed management	21	Sustainability – Water management	8	Natural resource management – water management	4	Drought – Ecosystems	5
Nitrogen – Soil fertility	22	Fertilizer use – Water management	8	Nitrogen – Soil fertility	13	Ecosystems - Weeds	7	Sustainability – Water management	4	Soil fertility - Sustainability	4
Nitrogen - Phosphorus	18	Nitrogen - Phosphorus	8	Nitrogen – Nitrogen supply	10	Nitrogen - Sustainability	7	Nitrogen - Sustainability	3	Drought – Food security	4
Nitrogen – Nitrogen supply	17	Nitrogen - Nitrogen supply	7	Nitrogen - Phosphorus	10	Drought - Ecosystems	6	Diversification – soil fertility	2	Water management - sustainability	4
Fertilizer use – Water management	16	Weeds – Weed management	7	Drought – Water management	9	Natural resource management – Water management	6	Ecosystems – Water management	2	Nitrogen - sustainability	4
Fertilizer use – Soil fertility	11	Nitrogen - Potassium	5	Fertilizer use – Water management	8	Soil fertility - Sustainability	6	Ecosystems - Nitrogen	2	Phosphorus – food security	3
Nitrogen – Water management	11	Nitrogen – Water management	5	Soil fertility – Fertilizer use	7	Drought – Food security	4	Fertilizer use – sustainability	2	Fertilizer use - ecosystems	3
Fertilizer use - Phosphorus	10	Fertilizer use – Nitrogen supply	4	Fertilizer use - Phosphorus	7	Ecosystems – Water management	4	Soil fertility - sustainability	2	Nitrogen – food security	3
Drought – Water management	10	Fertilizer use – Soil fertility	4	Nitrogen – Water management	6						
		Flooding - Nitrogen	4	Water management – Weed management	6						

● Abiotic ● Biotic ● Management ● Socio economic ● Environmental ● Food security ● Diversification

Figure 4 Interactions made by Scopus cited journal articles between rice production constraints and new challenges.

little research attention? GAFSP and CGIAR have been primed to strategically invest in key areas of agricultural research, and have shown that the progress made against poverty and malnutrition with even small improvements in agricultural efficiency are worthwhile and significant. For instance, a 1% increase in agricultural production is expected to lead to a five times larger increase in poverty reduction than a 1% increase in GDP [43]. Since agricultural improvements for smallholders are seen as so effective in meeting international poverty and food security goals, why are there so many areas of potential yield gains which are not being addressed by peer-reviewed research? For instance, in the case of South Asian rice productivity, socio-economic production constraints, which make up over 22% of the entire yield losses, could be the source of new increases in yields if research priorities are re-evaluated to consider this neglected research area.

Obsession with yields

Commitments made at the World Food Summit in 1996 and in MDG 1.C have not been fully realized. The proportion of undernourished people has declined globally, but this average ignores substantial variation across regions. Agricultural research in South Asia has largely focused on yield-enhancing technologies for rice and wheat that have been successful at meeting countries' food production goals [32]. Yet these yield increases must be seen in context. This area is one of the most drought-prone and food-insecure areas in the world. Within the dominant rice-wheat and rainfed mixed farming system areas, there are 28.3 and 24.5 million stunted children, respectively. In fact, five of the top ten farming systems with the most recorded stunting were in South Asia. Clearly, an overt focus on yield has gotten results, but whom do these results benefit? How much do they contribute to reducing stunting and poor nutrition within the region?

Of course, yield is in itself a broad term, and there are several constraints that contribute to yield losses. However, this study has highlighted the fact that research tends to prioritize certain types of production constraints, overlooking other constraints which collectively are substantial contributors to total yield losses, such as socio-economic, biotic, and management-related constraints.

This study has shown that while there is relatively 'low-hanging fruit' awaiting research—7% of articles reviewed focusing on socio-economic constraints which account for 22% of the yield gap—there is no system in place to direct high-quality academic research into areas where the results would have maximum impact in terms of addressing the food security concerns of some of the world's largest malnourished population. So while individual factors such as nitrogen deficiency have a significant impact on production (and are especially relevant within intensive, commercial farming), the gains to be made from further study

are dwarfed by those to be made through research into other lesser-known constraints that have a proportionately larger impact.

Communications

The central finding of this study is that there is a disconnect between the types of research that are prioritized and funded in peer-reviewed academic research at many universities and research institutes, and the types of information needed to address 'real-world' problems. Of course, these are not mutually exclusive, but the findings of this study suggest that there are known areas of research that receive very little attention but which can have a proportionally larger impact. This disconnect can be seen between the stated commitments made at events such as the World Food Summit in 1996, and within the MDGs, and the *de facto* prioritization of research within academia.

The disconnect manifests itself through a lack of communication of information needs from the implementation side and a lack of communication and knowledge-brokering between research-generating institutions and research users. Newer sources of funding such as the Bill & Melinda Gates Foundation and the Yara Foundation have led to a shortcutting of the peer-review system, directly taking research knowledge to implementation. However, it remains to be seen whether these institutions have gone too far in ignoring the benefits of contributing to journal submissions for the peer-reviewed system.

It should be in the interest of research institutions in South Asia, particularly the NARS, to address this disconnect. For instance, ICAR explicitly states that supporting smallholder agriculture and sustainably managing natural resources goes hand in hand. However, the network analysis conducted in this study suggests that research has not yet begun to build bridges between many of the production constraints and these new environmental challenges.

Conclusions

This study was motivated by the growing commitments in international and national agendas to reduce poverty and increase the status of the food security situation of the most vulnerable groups in the world through supporting smallholder agriculture. Its aim was to determine whether academic research was focusing on the greatest causes of yield losses in the most food-insecure farming systems in South Asia. It also sought to determine whether research was considering environmental constraints that the international community (IPCC, FAO, IFAD) identifies as new challenges for smallholders (such as climate change). The study searched for relevant journal articles cited by Scopus and used yield loss estimates from the literature to calculate congruency ratios. These ratios were then interpreted to determine where there had been a surplus or deficit in agricultural research for South Asian rice farmers who make up a sizable section of the world's most food-insecure. Degree centralities were also estimated to see which different types of rice constraints had been linked together by journal articles.

The results suggest that while research had focused on several of the most severe production constraints, many medium-sized and large constraints have not received the warranted research attention. The most striking finding was the shortage of research focused on socio-economic production constraints, despite that being the most important category of production constraints to South Asian rice farmers in terms of contributing to yield losses. An implication of the study is that if international and national development organizations want research to better support smallholder agriculture, research on overlooked and underfunded production constraints will need to be addressed through better coordination between research organizations and the institutions that fund the research. The paper's findings highlight specific areas which perhaps deserve more attention from agricultural research in light of their relative importance to current rice yield losses for South Asian smallholders.

It is important to highlight the methodological limitations of the study. While there are weaknesses in the way the journal articles were quantified, there are also issues with the quality of research in terms of its usefulness to smallholder farmers. The international community and researchers are beginning to recognize that the way they conduct their research significantly affects technology adoption by farmers. A high-level panel [1] argued that the way research is conducted needs to be participatory and empowering for smallholders. Moreover, to achieve this, research systems need to be held accountable to smallholders, whereby their funding depends on the impact of their research. This study has not considered the quality of peer-reviewed research in this respect, but such an analysis would no doubt improve our understanding of how effective research has been in dealing with the production and environmental constraints that smallholder face.

Such a change in how research is conducted is also recognized by NARS such as ICAR, which has set the promotion of farmer-participatory research as a priority in its mission statement. In fact, in the case of India, the problems raised in this study, such as a lack of research from the social sciences and understanding the relevance of natural resource management to agricultural production seem to be well understood since those topics are clearly identified among South Asian agricultural research priorities.

What is also clear from the results of this study, however, is that there is a large gap between official and *de facto* research priorities. In defense of the research community, it is understandable that there is a path dependency, where research tends to be pursued in areas which complement South Asia's research resource base—for instance, pursing abiotic rather than socio-economic research. This tradition derives from a legacy of viewing agricultural production constraints as being within the realm of the natural sciences which explains why agricultural departments in universities and research institutions are set up to deal with abiotic constraints in great detail. In contrast, when it comes to the socio-economic production constraints, as these draw heavily on the social sciences, there are simply not the personnel within these departments to tackle these issues in the same depth. Therefore, if the under-researched priority areas are to be addressed in a meaningful way, institutional transformations in South Asia's agricultural research institutions are needed. Furthermore, funding and donor institutions and governments have a critical role in supporting the research organizations to ensure that these research gaps—as have been shown in this paper—are addressed.

One crucial role for organizations which fund South Asian agricultural research lies within the assessment of the impact of their research investments, and this is already happening to a large extent. Hazell [32] argues that the returns from South Asian agricultural research investments in terms of productivity are widely studied in the literature. However, he goes on to point out that the same cannot be said for impact assessment in terms of poverty reduction, which is surprising since poverty reduction tends to be one of the key drivers behind justifying funds for agricultural research. Donors therefore also have their role to play in monitoring funded research in order to increase the level of accountability of agricultural research organizations for their research output and impact on poverty and food security.

Ending on a positive note, with the emergence of cross-cutting topics such as food security, it has become easier to justify collaborations across traditional academic divisions, especially, as we have shown in this paper, given the fact that the different academic fields tend to have an equal standing in importance in terms of reducing rice production losses.

Additional files

Additional file 1: Table S1. Operationalizing production constraints for Scopus queries and **Table S2.** yield losses for South Asian farming systems.

Additional file 2: Dataset used in the analysis taken from the Scopus queries.

Additional file 3: Additional network diagrams. Figure S1. Network graph showing journal article interactions for rice production constraints in South Asia, 1997–2007. **Figure S2.** Network graph showing journal article interactions for rice production constraints in South Asia, 2008–2014. **Figure S3.** Network graph showing journal article interactions for rice production constraints and new challenges in South Asia, 1997–2014. **Figure S4.** Network graph showing journal article interactions for rice

production constraints and new challenges in South Asia, 1997–2007.
Figure S5. Network graph showing journal article interactions for rice production constraints and new challenges in South Asia, 2008–2014.

Abbreviations

APAARI: Asia-Pacific Association of Agricultural Research Institutions; CGIAR: Consultative Group on International Agricultural Research; CIMMYT: International Maize and Wheat Improvement Center; FAO: Food and Agricultural Organization; GAFSP: Global Agriculture and Food Security Program; GDP: gross domestic product; IARC: international agricultural research centers; ICAR: Indian Centre of Agricultural Research; IFAD: International Fund for Agricultural Development; IFPRI: International Food Policy Research Institute; IPCC: Intergovernmental Panel on Climate Change; IRRI: International Rice Research Institute; LFA: Least Favorable Areas; MDG: Millennium development goals; NARS: National Agricultural Research Systems; UN: United Nations.

Competing interests

The authors declare that they have no competing interests.

Authors' contributions

Both authors were involved in the conception of the idea and design of the study. AJ collected and analyzed the data. Both authors interpreted the data and drafted, read, and approved the final manuscript.

Acknowledgements

We would like to thank Dr. Louise Karlberg for providing useful suggestions during the conception of this article, Drs. Stephen Waddington and Li Xiaoyun for the data they shared with us, and Marion Davis for reviewing an earlier draft. We would also like to thank the anonymous reviewer and editors of the journal for their input. SIANI (the Swedish International Agricultural Network Initiative) provided the publication fees for the article which we are sincerely grateful for. Naturally, the authors take full responsibility for the content of the article.

Author details

[1]Institute of Agricultural and Food Policy Studies, Universiti Putra Malaysia, Putra Infoport, Jalan Kajang-Puchong, Serdang, Selangor 43400, Malaysia. [2]Stockholm Environment Institute, Linnégatan 87 D, Stockholm 115 23, Sweden. [3]Swedish International Agricultural Network Initiative (SIANI), Linnégatan 87 D, Stockholm 115 23, Sweden.

References

1. HLPE: **Investing in smallholder agriculture for food security.** In *A report by the High Level Panel of Experts on Food Security and Nutrition of the Committee on World Food Security, Rome*; 2013.
2. CGIAR: *Changing Agricultural Research in a Changing World. A Strategy and Results Framework for the Reformed CGIAR.* Montpellier: Consultative Group on International Agricultural Research; 2013.
3. IFAD: *IFAD Strategic Framework 2011–2015.* Rome: International Fund for Agricultural Development; 2011.
4. FAO: *Biotechnologies at Work for Smallholders: Case Studies from Developing Countries in Crops, livestock and Fish.* Rome: Food and Agricultural Organization; 2013.
5. IFAD: *Smallholders, Food Security and the Environment.* Rome: International Fund for Agricultural Development; 2013.
6. APAARI and IFPRI: *Priorities for Agricultural Research for Development in South Asia*; 2013.
7. ICAR: *Vision 2030.* New Delhi: Indian Council of Agricultural Research; 2011.
8. World Bank, FAO, and IFAD: *Gender in Agriculture Sourcebook.* Washington, DC: World Bank, Food and Agricultural Organization, and International Fund for Agricultural Development; 2009.
9. Koohafkan P: *Globally Important Agricultural Heritage Systems.* Beijing: Presentation at the International Forum on Globally Important Agricultural Heritage Systems (GIAHS); 2011:9–11.
10. IAASTD: *Agriculture at a Crossroads: Sub-Saharan Africa (SSA) Report (Vol. V).* Washington, DC: International Assessment of Agricultural Knowledge, Science and Technology for Development, Island Press; 2009.
11. IFAD: *Viewpoint: Smallholders Can Feed the World.* Rome: International Fund for Agricultural Development; 2011.
12. UN: *The Millennium Development Goals Report, United Nations New York*; 2013.
13. Millennium Project UN: *Halving hunger: It can be done. Task Force on Hunger.* London, Sterling, VA, USA: Earthscan; 2005.
14. GAFSP: *Framework Document for a Global Agriculture and Food Security Program*; 2009. Sustainable Development Network; 2009. The World Bank: Washington, D. C.
15. IFAD: *Rural Groups and the Commercialization of Smallholder Farming: Targeting and Development Strategies (draft). (Issues and Perspectives from a Review of IOE Evaluation Reports and Recent IFAD country Strategies and project Designs).* Rome: International Fund for Agricultural Development; 2011.
16. FAO: *The State of Food and Agriculture, Investing in Agriculture.* Rome: Food and Agricultural Organization; 2012.
17. Lipton M: *The Family Farm in a Globalizing World: the Role of Crop Science in Alleviating Poverty.* Washington, D.C: International Food Policy Research Institute (IFPRI); 2005.
18. Dixon J, Gulliver A, Gibbon D: *Farming Systems and Poverty. Improving Farmers' Livelihoods in a Changing World.* Rome and Washington DC: Food and Agricultural Organization and World Bank; 2001.
19. Crowley E: *Ending Poverty: Learning from Good Practices of Small and Marginal Farmers.* Rome: Self Employed Women's Association Exposure and Dialogue and the Food and Agricultural Organization; 2013.
20. Hasan R, Quibria MG: **Industry matters for poverty: a critique of agricultural fundamentalism.** *Kyklos* 2004, **57:**253–264.
21. Ellis F: *Small farms, Livelihood Diversification, and Rural–Urban Transitions: Strategic Issues in Sub-Saharan Africa.* Washington: International Food Policy Research Institute; 2005.
22. Cline WR: *Global Warming and Agriculture: Impact Estimates by Country.* Washington, D.C: Peterson Institute; 2007.
23. Gornall J, Betts R, Burke E, Clark R, Camp J, Willett K, Wiltshire A: **Implications of climate change for agricultural productivity in the early twenty-first century.** *Philos Trans R Soc Lond B Biol Sci* 2010, **365:**2973–2989.
24. HLTF: *High Level Task Force on the Global Food Security Crisis. Outcomes and Action for Global Food Security.* New York, USA: Excerpts from the Comprehensive Framework for Action; 2008.
25. HLPE: *Food Security and Climate Change.* Rome: A report by the High Level Panel of Experts on Food Security and Nutrition of the Committee of World Food Security; 2012.
26. FAO: *The State of Food and Agriculture (SOFA): Food Systems for Better Nutrition.* Rome: Food and Agricultural Organization; 2013.
27. FAO: *Agriculture, Food and Nutrition for Africa: A Resource Book for Teachers of Agriculture.* Rome: Food and Agricultural Organization; 1997.
28. Thornton PK, Jones PG, Alagarswamy G, Anderson J: **Spatial variation of crop yield response to climate change in East Africa.** *Glob Environ Change* 2009, **19:**54–65.
29. Sultan B, Roudier P, Quirion P, Alhassane A, Muller B, Dingkuhn M, Ciais P, Guimberteau M, Traore S, Baron C: **Assessing climate change impacts on sorghum and millet yields in the Sudanian and Sahelian savannahs of West Africa.** *Environ Res Lett* 2013, **8**(1).
30. Evenson R, Gollin D: **Contributions of national agricultural research systems to crop productivity.** In *Handbook of Agricultural Economics.* 3rd edition. Edited by Evenson R, Pingali P; 2007:2420–2447.
31. Garg KC, Kumar S, Lal K: **Scientometric profile of Indian agricultural research as seen through science citation index expanded.** *Scientometrics* 2006, **68**(1):151–166.
32. Hazell PBR: *An Assessment of the Impact of Agricultural Research in South Asia since the Green Revolution.* Rome, Italy: Science Council Secretariat; 2008.
33. Hyman G, Fujisaka S, Jones P, Wood S, De Vicente MC, Dixon J: **Strategic approaches to targeting technology generation: Assessing the coincidence of poverty and drought-prone crop production.** *Agric Syst* 2008, **98**(1):50–61.
34. Gibbon D, Dixon J, Flores D: *Beyond Drought Tolerant Maize: Study of Additional Priorities in Maize. Report to Generation Challenge Program.* Mexico DF, Mexico: CIMMYT; 2007:42.
35. Waddington SR, Li X, Dixon J, Hyman G, De Vicente MC: **Getting the focus right: production constraints for six major food crops in Asian and African farming systems.** *Food Secur* 2010, **2**(1):27–48.

36. Pouris A, Pouris A: **The state of science and technology in Africa (2000–2004):**
     a scientometric assessment. *Scientometrics* 2009, **79**(2):297–309.
37. Hassan S, Haddaway P, Kuinkel P, Degelsegger A, Blasy C: **A bibliometric**
     study of research activity in ASEAN related to the EU in FP7 priority
     areas. *Scientometrics* 2012, **91**:1035–1051.
38. Kosecki S, Shoemaker R, Baer CK: **Scope, characteristics, and use of the U.**
     S. Department of Agriculture's intramural research. *Scientometrics* 2011,
     88:707–728.
39. Stads G, Roozitalab MH, Beintema NM, Aghajani M: **Agricultural research in**
     Iran. Policy, investments, and institutional reform. In *ASTI country report*.
     Washington, D. C: International Food Policy Research Institute (IFPRI) and
     Agricultural Extension, Education and Research Organization (AEERO); 2008.
40. Newman MEJ: *Networks: An Introduction.* Oxford, UK: Oxford University Press; 2010.
41. Zuo M, Hua X, Wen X: **Who is the best connected researcher? An analysis**
     of co-authorship networks of knowledge management from 2000 to
     2010. In *The 19th International Conference on Industrial Engineering and*
     Engineering Management. Edited by Ershi Qi, Jian Sheng, Runliang Dou;
     2013:761–770.
42. Newman MEJ: **A measure of betweenness centrality based on random**
     walks. *Soc Netw* 2005, **27**:39–54.
43. Christiaensen L, Demery L, Kuhl J: **The (evolving) role of agriculture in**
     poverty reduction – an empirical perspective. *J Dev Econ* 2011, **96**:239–254.

PERMISSIONS

LIST OF CONTRIBUTORS

Chikelu Mba and Kakoli Ghosh
Plant Genetic Resources and Seeds Team, Plant Production and Protection Division, Food and Agriculture Organization of the United Nations (FAO), Rome, Italy

Elcio P Guimaraes
International Centre for Tropical Agriculture (CIAT), Cali, Colombia

Kerri L Steenwerth
Crops Pathology and Genetics Research Unit, Agricultural Research Service, United States Department of Agriculture (ARS/USDA), c/o Department of Viticulture and Enology, RMI North, Rm. 1151, 595 Hilgard Lane, Davis, CA 95616, USA

Amanda K Hodson
Department of Land, Air and Water Resources, University of California at Davis, One Shields Avenue, Davis, CA 95616, USA

Arnold J Bloom
Department of Plant Sciences, University of California at Davis, One Shields Avenue, Davis, CA 95616, USA

Michael R Carter
Department of Agricultural and Resource Economics, University of California at Davis, One Shields Avenue, Davis, CA 95616, USA

Pablo Tittonell
Plant Sciences, Wageningen University, 6700AN Wageningen, the Netherlands

Stephen M Wheeler
Department of Landscape Architecture, University of California at Davis, One Shields Avenue, Davis, CA 95616, USA

Obidimma C Ezezika
Sandra Rotman Centre, University Health Network and University of Toronto, Toronto, Ontario, Canada
African Centre for Innovation and Leadership Development, Federal Capital Territory, Abuja, Nigeria

Dalla Lana School of Public Health, University of Toronto, Toronto, Canada

Abdallah S Daar
Sandra Rotman Centre, University Health Network and University of Toronto, Toronto, Ontario, Canada
Grand Challenges Canada
Dalla Lana School of Public Health and Department of Surgery, University of Toronto, Toronto, Canada

Meenakshi Santra, Shawna B Matthews and Henry J Thompson
Cancer Prevention Laboratory, Colorado State University, 1173 Campus Delivery, Fort Collins, CO 80523, USA

Ola T Westengen
Centre for Development and the Environment, University of Oslo, Blindern, NO0317 Oslo, Norway
Centre for Ecological and Evolutionary Synthesis (CEES), Department of Biosciences, University of Oslo, Blindern, NO-0316 Oslo, Norway

Anne K Brysting
Centre for Ecological and Evolutionary Synthesis (CEES), Department of Biosciences, University of Oslo, Blindern, NO-0316 Oslo, Norway

Satyendra Nath Mishra
Institute of Rural Management, Anand 388001 Gujarat, India

Sudip Mitra and Pooja Singh
School of Environmental Sciences, Jawaharlal Nehru University, New Delhi 110067, India

Latha Rangan and Subashisha Dutta
Indian Institute of Technology-Guwahati, Assam 781039, India

Mekuanent Tebkew
Department of Natural Resource Management, University of Gondar, Gondar, Ethiopia

Zebene Asfaw
School of Forestry, Wondo Genet College of Forestry and Natural Resources, Hawassa University, Shashemene, Hawassa, Ethiopia

Solomon Zewudie
Department of Biology, Addis Ababa Science and Technology University, Addis Ababa, Ethiopia

Joseph Gichuru Wang'ombe
Maastricht School of Management, Maastricht, Netherlands
African Population & Health Research Center, Nairobi 00100, Kenya

Meine Pieter van Dijk
Maastricht School of Management, Maastricht, Netherlands
UNESCO-IHE Institute for Water education, Delft, Netherlands

Felix Klaus, Catrin Westphal and Teja Tscharntke
Agroecology, Department of Crop Sciences, University of Göttingen, Grisebachstraße 6, D-37077 Göttingen, Germany

Björn Kristian Klatt
Agroecology, Department of Crop Sciences, University of Göttingen, Grisebachstraße 6, D-37077 Göttingen, Germany
Centre for Environmental and Climate Research, University of Lund, Sölvegatan 37, SE-22362 Lund, Sweden

Justin Mabeya
Sandra Rotman Centre, University Health Network and University of Toronto, Toronto, Ontario, Canada

Obidimma C Ezezika
Sandra Rotman Centre, University Health Network and University of Toronto, Toronto, Ontario, Canada
African Centre for Innovation and Leadership Development, Federal Capital Territory, Abuja, Nigeria
Dalla Lana School of Public Health, University of Toronto, Toronto, Canada

Tannis Thorlakson
Sustainability Science Program, Harvard University, Cambridge, MA, USA

Henry Neufeldt
World Agroforestry Centre (ICRAF), Nairobi, Kenya
CGIAR Research Program on Climate Change, Agriculture and Food Security (CCAFS)

Erwin Bergmeier
Department of Vegetation & Phytodiversity Analysis, Albrecht-von-Haller Institute of Plant Sciences, Georg-August University of Göttingen, D-37073 Göttingen, Germany

Nur Rochmah Kumalasari
Department of Vegetation & Phytodiversity Analysis, Albrecht-von-Haller Institute of Plant Sciences, Georg-August University of Göttingen, D-37073 Göttingen, Germany
Department of Nutrition and Feed Technology, Faculty of Animal Science, Bogor Agricultural University (IPB), Bogor, Indonesia

David J Abson
FuturES Research Center, Leuphana Universität, Lüneburg, Germany
Sustainability Research Institute, University of Leeds, Leeds, UK

Evan DG Fraser
Sustainability Research Institute, University of Leeds, Leeds, UK
Department of Geography, University of Guelph, Guelph, Ontario, Canada

Tim G Benton
Institute of Integrative and Comparative Biology, University of Leeds, Leeds, UK

Kathryn Barber
Sandra Rotman Centre, University Health Network and University of Toronto, Toronto, Ontario, Canada

Obidimma C Ezezika
Sandra Rotman Centre, University Health Network and University of Toronto, Toronto, ON, Canada
Dalla Lana School of Public Health, University of Toronto, Toronto, ON, Canada
African Center for Innovation and Leadership Development, Federal Capital Territory, Abuja, Nigeria

National Biotechnology Development Agency (Federal Ministry of Science and Technology), Umar Musa Yar' Adua Way/Airport Road, Lugbe Area, Abuja, Nigeria

Jessica Oh
Sandra Rotman Centre, University Health Network and University of Toronto, Toronto, ON, Canada
African Center for Innovation and Leadership Development, Federal Capital Territory, Abuja, Nigeria

Adam John
Institute of Agricultural and Food Policy Studies, Universiti Putra Malaysia, Putra Infoport, Jalan Kajang-Puchong, Serdang, Selangor 43400, Malaysia
Stockholm Environment Institute, Linnégatan 87 D, Stockholm 115 23, Sweden

Matthew Fielding
Stockholm Environment Institute, Linnégatan 87 D, Stockholm 115 23, Sweden
Swedish International Agricultural Network Initiative (SIANI), Linnégatan 87 D, Stockholm 115 23, Sweden

Index